Cooperative C Complex Network Systems with Dynamic Topologies

Cooperative Control of Complex Network Systems with Dynamic Topologies

Guanghui Wen

Wenwu Yu

Yuezu Lv

Peijun Wang

CRC Press
Taylor & Francis Group
Boca Raton London New York

CRC Press is an imprint of the
Taylor & Francis Group, an **informa** business

First edition published 2021
by CRC Press
6000 Broken Sound Parkway NW, Suite 300, Boca Raton, FL 33487-2742

and by CRC Press
2 Park Square, Milton Park, Abingdon, Oxon, OX14 4RN

ISBN: 978-1-032-01913-0 (hbk)
ISBN: 978-1-032-01917-8 (pbk)
ISBN: 978-1-003-18098-2 (ebk)

Typeset in Nimbus
by KnowledgeWorks Global Ltd.

Contents

Preface

Complex network systems (CNSs) roughly refer to networking systems composed of numerous interconnected individuals and could exhibit fascinating collective behaviors far beyond the individual's inherent properties. Many social and engineering systems could be abstracted and modeled as CNSs, where the tight interactions among neighboring entities are indispensable to achieve various global phenomena. Representative examples of CNSs include scientific collaboration networks, the Internet, power grids, multiple unmanned aerial systems, and transportation systems, to name just a few.

Extensive attention has been given, and many efforts have been devoted to the investigations of the coordination of CNSs. Such a research field would not only deepen our understanding about the emergence mechanism of macroscopical behaviors of CNSs, such as flocking, but also could prompt the applications of theoretical results in network science to solve practical engineering problems, such as the allocation in distributed sensor networks, task collaboration of multiple robots, distributed energy management in smart grid, etc. Noteworthily, different from the stability analysis of a single control plant, analyzing the collaborative behavior for CNSs is much more challenging and exciting. The collaborative behavior will not merely depend on the inherent dynamics of individual agents, but also on the network topological configuration that they interact with each other. However, the communication topologies for CNSs are not always fixed. They are dynamic and switching due to various practical factors, such as adding or deleting links in network evolutions, external interferences or attacks on communication channels, and limited sensing radius in some engineering networks. For instance, as a typical CNS, power grids are subject to transmission lines' or communication lines' switching during their operations. Although the profound theory of hybrid dynamical systems and switched systems have been established in prior works, most of these methodologies could not be directly applied to CNSs' scenarios, and the topology information within CNSs deserves dedicated analysis. Thus, it is imperative to investigate the cooperative behaviors of CNSs under such switching topologies.

Therefore, in this book, we thoroughly dissect the cooperative control of complex networks under dynamic communication topologies. We begin our research trip by introducing the concepts, state-of-the-art, and promising future directions of synchronization of complex networks and consensus of multiagent systems (MASs) with switching topologies in Chapter 1. Chapter 2 gives the mathematical notations we will apply throughout the book and preliminaries on matrix theory, algebraic graph, and switching systems, which would act as crucial tools and pave the way to deal with the subsequent analysis effectively. First, we focus on consensus for the general linear

CNSs with switching topologies. Chapter 3 studies both the leaderless and leader following consensus for linear CNSs with directed switching topologies. To deal with MIMO (multi-input multi-output) linear CNSs with unknown disturbances and directed switching topologies, Chapter 4 further investigates the consensus disturbance rejection technique with UIO (unknown input observer), fully distributed control protocols with dynamic coupling strengths are developed. Next, we move forward to the investigations of nonlinear CNSs. Chapter 5 delves into consensus tracking of CNSs with Lipschitz-type and Lorenze-type under directed switching topologies, respectively. Chapter 6 considers the consensus tracking problem of CNSs with higher-order dynamics and directed switching topologies, where we also explore the case with occasionally missing control inputs. The distributed $\mathcal{H}_\infty$ consensus problem of CNSs under switching directed topologies is further investigated in Chapter 7, where results for both higher-order linear dynamics with continuous communication and Lipschitz nonlinear dynamics with aperiodic sampled-data communications are established. To cope with the challenge when full state information of neighboring agents is unavailable, Chapter 8 addresses the consensus tracking of CNSs under directed switching topology by constructing effective distributed observers. We also consider a more general case where communication configurations of distributed observers and controllers could be independent and asynchronously switching. Chapter 9 focuses on the distributed cooperative consensus tracking of linear CNSs with a high-dimensional leader, and gives sufficient conditions to ensure a finite $\mathcal{L}_2$-gain performance when confronted with unknown disturbances. The neuro-adaptive consensus for CNSs under directed switching topologies and unknown dynamics is addressed in Chapter 10, where we elaborate on the practical tracking with a high-dimensional leader and containment problem with multiple leaders. Lastly, the resilient consensus of CNSs with switching topologies in the presence of input saturation or malicious attack is studied in Chapter 11.

We would like to gratefully acknowledge the effort and support to our research on cooperative control of complex network systems by our colleagues and students. Particularly, we are indebted to Professor Guanrong(Ron) Chen at City University of Hong Kong for his leadership in the area of complex network systems and for the inspiring discussions on cooperative control of complex network systems. We are also indebted to Professor Xinghuo Yu at RMIT University, Professor Zhisheng Duan at Peking University, Professor Wei Ren at University of California, Riverside, Professor Valeri Ougrinovski at The University of New South Wales, Canberra, and Professor Guoqiang Hu at Nanyang Technological University for many constructive discussions on research ideas. In addition, we would like to acknowledge IEEE, John Wiley & Sons, Elsevier, and Taylor & Francis for granting us the permission to reuse some materials from our publications copyrighted by these publishers in this book. Finally, we gratefully acknowledge the support of our research on cooperative control of complex network systems by the Natural Science Foundation of Jiangsu Province of China under Grant BK20170079, the National Natural Science Foundation of China under Grants 61722303, 61673104, and 62073079.

Guanghui Wen, Wenwu Yu, Yuezu Lv
Southeast University
Nanjing, China
October 2020

Peijun Wang
Anhui Normal University
Wuhu, China
October 2020

CHAPTER 1

Introduction

This chapter overviews some recent research progress in cooperative control of complex network systems over directed switching communication topologies. Distributed cooperation of complex network systems (CNSs), including synchronization of complex networks and consensus control of multiagent systems (MASs), has been a very active research topic in a wide variety of scientific communities, ranging from applied mathematics to physics, engineering to biology, even sociology. In Section 1.1, CNSs include MASs and complex networks are introduced. In Section 1.2, definitions of synchronization of complex networks and consensus of MASs are given, moreover, some differences between these two topics are briefly summarized. In Section 1.3, the research progress of synchronization of complex networks with switching topologies are presented. In Section 1.4, the research progress of consensus of MASs with switching topologies are presented. In Section 1.5, we conclude this chapter by presenting some future works from our own viewpoint.

1.1 COMPLEX NETWORK SYSTEMS

Far from being separate entities, many natural, social, and engineering systems can be considered as CNSs associated with tight interactions among neighboring entities within them [3, 10, 18, 29, 39, 50, 92, 120, 130, 141, 175, 194, 201, 211, 219]. Roughly speaking, a CNS refers to a networking system that consists of lots of interconnected agents, where each agent is an elementary element or a fundamental unit with detailed contents depending on the nature of the specific network under consideration [175]. For example, the Internet is a CNS of routers and computers connected by various physical or wireless links. The cell can be described by a CNS of chemicals connected by chemical interactions. The scientific citation network is a CNS of papers and books linked by citations among them. The WeChat social network is a CNS whose agents are users and whose edges represent the relationships among users, to name just a few.

With the aid of coordination with neighboring individuals, a CNS can exhibit fascinating cooperative behaviors far beyond the individuals' inherent properties. Prototypical cooperative behaviors include synchronization [38, 95, 101, 177], consensus [76, 118, 128], swarming [48, 115], flocking [117, 161]. In this book, we focus on the CNSs which include complex networks and MASs as special cases. A lot of new research challenges have been raised about understanding the emergence mechanisms responsible for various collective behaviors as well as global statistical properties of CNSs [3, 15, 114, 178]. Network science, as a strong interdisciplinary research field, has been established at the first several years of the 21st century [110]. It is increasingly recognized that a detailed study on cooperative dynamics of CNSs would not only help researchers understand the evolution mechanism for macroscopical cooperative behaviors, but also prompt the application of network science to solve various engineering problems, e.g., design of distributed sensor networks [135], formation control of multiple unmanned aerial vehicles [37], distributed localization [89], and load assignment of multiple energy storage units in modern power grid [191].

Among the various cooperative behaviors of CNSs, synchronization of complex networks and consensus of MASs are the most fundamental yet most important ones. Synchronization of complex networks exhibits the cooperative behavior that the states of all entities within these networks achieve an agreement on some quantities of interest. Compared with stability analysis of an isolated control plant, synchronization behavior analysis in CNSs are much more challenging as the synchronization process is determined by the evolution of network topology as well as the inherent dynamics of individual units within these network systems [96, 102, 121, 198, 199]. As a topic closely related to synchronization of complex networks, the consensus of MASs has recently gained much attention from various research fields, especially the system science, control theory, and electrical engineering communities [22, 65, 88, 116, 128]. In the remainder of this chapter, we will review some existing results on achieving synchronization of complex networks and consensus of MASs over dynamically changing communication topologies.

1.2 DEFINITIONS OF SYNCHRONIZATION AND CONSENSUS

Before moving forward, the definition of consensus of MASs is given. Moreover, the synchronization of complex networks can be defined similarly.

Consider an MAS which consists of N agents. Without loss of generality, we label the N agents as agents $1, \ldots, N$, respectively. The dynamics of agent i, $i = 1, \ldots, N$, are represented by

$$\dot{x}_i(t) = f(t, x_i(t), u_i(t)), \tag{1.1}$$

where $x_i(t) \in \mathbb{R}^n$ and $u_i(t) \in \mathbb{R}^m$ represent, respectively, the state and the control input, $f(\cdot, \cdot, \cdot) : [t_0, +\infty) \times \mathbb{R}^n \times \mathbb{R}^m \mapsto \mathbb{R}^n$ represents the nonlinear dynamics of agent i. A particular case is the general linear time-invariant MASs with the dynamics of agent i are described by

$$\dot{x}_i(t) = Ax_i(t) + Bu_i(t), \ i = 1, \ldots, N, \tag{1.2}$$

where $A \in \mathbb{R}^{n \times n}$ and $B \in \mathbb{R}^{n \times m}$ represent, respectively, the state matrix and control input matrix. For convenience, throughout this book, we call MAS (1.1) to represent the MAS whose dynamics are described by (1.1).

Definition 1.1 *Consensus of the MAS* (1.1) *is said to be achieved if for arbitrary initial conditions* $x_i(t_0)$, $i = 1, \ldots, N$,

$$\lim_{t \to \infty} \|x_i(t) - x_j(t)\| = 0, \quad i, j = 1, \ldots, N. \tag{1.3}$$

The definition of consensus for MAS (1.1) given by Eq. (1.3) does not concern about the final consensus states. However, it is sometimes important to make the states of all agents in the considered MASs to finally converge to some predesigned trajectory, especially from the viewpoint of controlling various complex engineering systems. To ensure the states of all agents in MAS (1.1) converge to some desired states, a target system (may be virtual) is introduced to the network (1.1) as

$$\dot{s}(t) = f(t, s(t)) \tag{1.4}$$

for some given initial value $s(t_0) \in \mathbb{R}^n$. Under this scenario, we call agent i whose dynamics are described by (1.1) the follower i, $i = 1, \ldots, N$, and call the agent whose dynamics are described by (1.4) the leader.

Definition 1.2 *Consensus tracking (or leader following consensus) of the MAS with the followers given by* (1.1) *and the leader given by* (1.4) *is said to be achieved if for some given initial conditions* $s(t_0)$ *and* $x_i(t_0)$, $i = 1, \ldots, N$,

$$\lim_{t \to \infty} \|x_i(t) - s(t)\| = 0. \tag{1.5}$$

The existence and uniqueness of the solutions of system (1.1) will be discussed in Chapter 2.

Remark 1.1 *The mathematical definitions for synchronization of complex networks and consensus of MASs are precisely similar. However, some differences between these two topics are briefly summarized as follows from our viewpoint.*

(1) *A complex network typically contains a great number of individual nodes (e.g., the Internet) while the scale of an MAS may be relatively quite small (e.g., a team of several robots).*

(2) *The objective of synchronization control is to make the states of a large-scale network achieve state agreement under some given inner linking matrices by selecting only the coupling strength, while the objective of consensus is to make the states of agents achieve state agreement by designing the gain matrices as well as the coupling strength.*

(3) *Significant attention has been paid to revealing the relationship between the statistical properties (e.g., the degree distribution, the average path length, and the symmetry) of network topology and the synchronizability of complex networks within the context of synchronization in complex networks, while in the context of consensus of MASs, much attention has been focused on addressing the relationship between the algebraic properties (e.g., the algebraic connectivity for undirected interaction topology and the general algebraic connectivity for directed interaction topology) of interaction topology and the consensusability.*

Without causing any confusion, synchronization of complex networks and consensus of MASs are referred to as consensus of CNSs in this book.

Practical applications of consensus of CNSs: Achieving consensus in CNSs is critical for controlling these CNSs and thus helpful in dealing with various distributed control problems for practical network systems. For instance, reaching consensus of velocities for all individual agents is a precondition in achieving flocking in various second-order CNSs [117]. In another instance, frequency synchronization of multiple generator units within a power system is one of the most important issues in power system stability control [221]. In addition, clock synchronization among sensors within wireless sensor networks is highly desirable in their applications [154].

1.3 SYNCHRONIZATION OF COMPLEX NETWORKS WITH SWITCHING TOPOLOGIES

In the field of complex networks' synchronization with switching topologies, a wide range of research has been recently focused on dealing with issues related to the switchings and their effects on synchronization.

There has been increasing recognition that each topology candidate's properties and the switching strategy for topologies play essential roles in achieving synchronization for complex networks with switching topologies. The analytical approaches for synchronization of continuous- and discrete-time complex networks with switching topologies are generally different. Mathematically, the continuous-time complex network with switching topologies is a special kind of those with time-varying topology. However, it is preliminarily assumed in some existing works on synchronization of continuous-time network systems with time-varying topology that the connection links evolve continuously over time with a known bound for the changing rate [103] or with a time-varying Laplacian matrix being simultaneously diagonalizable [11]. Thus, the techniques developed in these works to solve synchronization problem of complex networks with special time-varying topology are generally hard to apply to that with switching topologies, especially to the case with directed switching topologies.

Specifically, averaging-based approaches were developed to analyze synchronization of continuous-time complex networks with fast switching topologies [7, 140] while multiple Lyapunov functions (MLFs)-based approaches were developed to analyze synchronization of continuous-time complex networks with slowly switching topologies (especially for the case with directed switching topologies) [190]. Furthermore, MLFLs-based approaches were usually employed to analyze synchronization of continuous-time complex networks with switching topologies under delayed or sampled-data coupling [90, 187]. Common Lyapunov function (CLF)- and functional (CLFL)-based approaches are applicable only to some special continuous-time complex networks with switching topologies such as each possible topology candidate is undirected [222].

For discrete-time CNSs with switching topologies, global synchronization for non-autonomous linear complex networks with randomly switching topologies was studied in [200] by developing a kind of approaches from ergodicity theory for nonhomogeneous Markovian chains. A method based on the Hajnal diameter of infinite coupling matrices was proposed in [97] to analyze the local synchronizability of a class of discrete-time complex networks with directed switching topologies. Synchronization of discrete-time complex networks with undirected switching topologies and impulsive controller was studied in [73] by constructing MLFs. Globally almost sure synchronization for discrete-time complex networks with switching topologies was investigated in [51] by using the super-martingale convergence theorem. For more recent related works, one can refer to the survey.

1.4 CONSENSUS OF MASS WITH SWITCHING TOPOLOGIES

Since the pioneer works [65] in which heading consensus of the linearized Vicsek's model was analyzed, consensus of MASs with switching topologies has attracted increasing attention from a wide range of scientific interests.

Consensus of first-order MASs with switching topologies: In the year of 2004, consensus problem of continuous time first-order (integrator-type) MASs with directed switching and balanced topology was formulated and studied in [116]. Due to the balanced property of each possible topology candidate, a common Lyapunov function was constructed in [116] for analyzing the convergence behaviors of disagreement vector. Consensus of both continuous- and discrete time first-order MASs with directed switching topologies was further studied in [128] where each possible topology candidate is not required to be balanced. By using graphical approaches, some interesting issues on consensus of a class of first-order MASs with switching topologies were further addressed in [13]. By employing a CLFL based approach, it was proven in [83] that average consensus in continuous time first-order MASs with time delayed protocol can be achieved if each topology candidate is strongly connected and balanced, and some linear matrix inequalities hold. Note that most of the aforementioned results are concerned with consensus of first-order MASs with deterministically switching topologies. However, considering the underlying topology may randomly switch among a set of topology candidates in some practical applications, there have been a number of results focusing on consensus of first-order MASs with randomly switching topologies [54, 155, 156].

Consensus of second-order MASs with switching topologies: Based on the stability results for switched systems provided in [108], some dwell time (DT) based criteria for consensus of continuous time second-order MASs under directed switching topologies were established in [129] where it was revealed that consensus can be achieved if each topology candidate contains a directed spanning tree and the DT for switchings among different topology candidates is larger than a threshold value. When the graph describing the communication topology among followers is undirected, it was proven in [59] by constructing a CLF that leader-following consensus could be achieved if the topology jointly contains a directed spanning tree. Later, leader-following consensus problem of MASs with switching jointly reachable interconnection and transmission delays was solved in [234] by designing the switching laws among topology candidates, where the dynamics of the leader are described by first-order integrator. Note that the switching mode for topology evolution of the MASs studied in [234] is a kind of state-dependent switching. By constructing a CLFL, Lin and Jia [85] showed that leaderless consensus of MASs with time-delayed protocol could be achieved if the underlying topology is undirected and jointly connected. Leaderless consensus of MASs with time-delayed protocols under directed switching topologies was further studied in [124]. Note that there is no specific restriction for the value of the DT for switching signals in the consensus criteria provided in [59, 85, 124] as CLF- and CLFL-based approaches were respectively adopted. By constructing a CLF, Wen *et al.* [181] obtained some sufficient criteria for achieving consensus in MASs with intermittent communication. Note that the underlying communication topology of the closed-loop MASs with intermittent communication can be seen as a directed switching topologies with two topology candidates: A strong connected graph and the null graph. More recently, pulse-modulated intermittent control which unifies impulsive control and sampled control was proposed in [93] to solve the consensus problem of MASs under intermittent communications.

For discrete time second-order MASs, by using the convergence property of infinite products of stochastic matrices, it was shown in [84] that consensus can be guaranteed if the union of switching graphs frequently contains a directed spanning tree. By assuming that each possible topology is fully connected, some consensus criteria for consensus of MASs with heterogeneous sampling periods were provided from the approach of estimating the eigenvalues of stochastic matrices [23]. In [87], Lin *et al.* studied consensus of MASs with nonconvex velocity and control input constraints under a directed switching topologies. It was shown in [87] that consensus can be achieved if the joint graph of the switching communication graphs has a directed spanning tree among each time interval of certain bounded length.

Consensus of MASs with general linear dynamics and switching topologies: In [186], under the assumption that each possible topology candidate contains a directed spanning tree with leader as the root, a novel MLF was constructed by using the M-matrix theory. And it was shown in [186] that leader-following consensus can be ensured if the DT for switchings among different topology candidates is larger than a derived positive scalar. Under the assumption that the inherent linear dynamics of agents are stabilizable and each possible topology candidate is undirected and connected, it was shown in [166] by using the CLF-based approach that consensus of the single input linear MASs with an arbitrarily given switching signal for underlying topology can be achieved if the feedback gain matrix of the consensus protocol is suitably designed. Leaderless consensus of multiple-input linear MASs with directed switching topologies as well as its disturbance rejection issue were addressed in [189] by assuming that the possible strongly connected topology graphs share a common left eigenvector of the Laplacian matrices associated with zero eigenvalue.

Note that most of the aforementioned criteria for consensus of general linear MASs with (directed) switching topologies are derived based on the assumption that the switching frequency among different topology candidates is sufficiently slow, i.e., the DT for switchings among different topology candidates should be larger than a positive quantity depending on both the inherent dynamics of agents and the properties of topology candidates (see e.g., [186]). However, in some cases, it is possible to achieve consensus in general linear MASs with fast switching topologies [67]. By using averaging theory, it has been shown in [67] that leaderless consensus in the linear MASs with output-coupling can be guaranteed under sufficiently fast switching topologies if the consensus problem of the MASs with the corresponding fixed averaging network topology can be solved via designing output-coupling protocols.

Compared with consensus of continuous-time general linear MASs with switching topologies, consensus of discrete-time general linear MASs with switching topologies has received relatively less attention in the last years. In [145], with the assumption that the system matrix of the inherent dynamics of agents is neutrally stable, both leaderless and leader-following consensus problems of discrete time general linear MASs under switching topologies were studied based on a generalized version of Barbalat's lemma. By assuming that the inherent linear dynamics of each agent are controllable and observable, output consensus problem for a class of discrete-time heterogeneous linear MASs with directed switching topologies and time delays was studied in [206] by designing a kind of distributed predictor-based controller.

Most of the above-mentioned works are concerned with consensus of linear MASs with deterministically switching topologies. Note that consensus of linear MASs with randomly switching topologies has also been considered in the literature [49,213]. Specifically, consensus problems of continuous- and discrete-time linear MASs with Markovian switching topologies were studied in [213] by constructing a kind of stochastic MLFs. Then, robust consensus of continuous-time linear MASs with Markovian switching topologies subject to unknown jumping modes was investigated in [49].

1.5 EXTENSIONS AND APPLICATIONS OF CNSS WITH SWITCHING TOPOLOGIES

In the above sections, we have surveyed some recent developments in the analysis and synthesis of CNSs with switching topologies, mainly focusing on the synchronization and consensus behaviors and comparison to complex networks and MASs' scenarios. The above survey is by no means complete. However, it depicts the whole general framework of coordination control for CNSs with dynamic communication networks and lays the fundamental basis for other exciting and yet critical issues concerning CNSs with switching topologies. These extensions still deserve further study, although a variety of efficient tools have been successfully developed to solve various challenging problems in those active research fields. Next, we elaborate on several state-of-the-art extensions and applications of CNSs with dynamic topologies.

Resilience analysis and control of complex cyber-physical networks. Most of the units in various network infrastructures are cyber-physical systems in the Internet of Things era. One of the essential and significant features of the cyber-physical system is integrating and interacting with its physical and cyber layers. As a new generation of CNS, the complex cyber-physical network has received drastic attention in recent years. Specifically, the CNSs' paradigm provides an excellent way to model various large-scale crucial infrastructure systems, such as power grid systems, transportation systems, water supply networks, and many others [4]. These systems all capture the basic features that large numbers of interconnected individuals through wired or wireless communication links, and many essential functions of these large-scale infrastructure systems fall under the purview of coordination of CNSs. Disruption of these critical networked infrastructures could be a real-world effect across an entire country and even further, significantly impacting public health and safety and leading to massive economic losses. The alarming historical events urgently remind us to seek solutions for maintaining certain functionality of CNSs against malicious cyber-attacks (i.e., resilience or cybersecurity). It is critically essential to exploit security threats during the initial design and development phase.

Noteworthily, any successful cyber or physical attack mentioned above on complex cyber-physical networks may introduce undesired switching dynamics (e.g., loss of links due to DoS attacks or human-made physical damages) to the operation of these networks [194]. Inspired by the pioneering work [194], [168] further investigated the distributed observer-based cyber-security control of complex dynamical networks. This work considered the scenario that the communication channels for controllers and observers might both subject to malicious cyber attacks, which aim to block the information exchanges and result in disconnected topologies of the communication networks. New security control strategies are proposed, and an algorithm to properly select the feedback gain matrices and coupling strengths has been given. The asynchronous attacks in these two communication channels were explored in [169], where the attacks can be launched independently and may occur at different time intervals. Recently, [69] studied the distributed cooperative control for DC cyber-physical microgrids under communication delays and slow switching topologies would destruct the system's transient behaviors at the switching time instants. The average switching dwell-time-dependent control conditions were given to ensure the exponential stability of the considered cyber-physical systems. For the event-triggered communication scenario, [26] studied the distributed consensus for general linear MASs subjected to DoS attacks. By the switched and time-delay system approaches, one constraint was provided to illustrate the convergence rate of consensus errors and uniform lower bound of non-attacking intervals of DoS attacks.

On the other hand, switching communication topologies may be an effective and promising candidate for detecting and defending against various cyber attacks. This will inspire us to apply the related theory of CNSs under switching topology to deal with cyber-security problems. An attack detection strategy was proposed in [104] for detecting zero-dynamics attack (ZDA) in a networked control system, where the detection is constructed based on Luenberger observer and carefully crafted switching policy for communication graphs. Using detectability conditions, they proved that the strategic topology-switching algorithms could detect intelligent attackers. To effectively reconstruct the states in the networked system when confronted with multiple sensor attacks and also the disturbances, [233] designed the distributed adaptive observers by employing a resilient switching scheme, and it was shown that the control performances could be ensured with the switched-type observers and the associated control protocols.

Distributed optimization of CNSs with switching topologies. Distributed optimization problem of CNSs with fixed topology has been studied under various scenarios where only the information about the local objective function and relative state (or output) information between its own and the neighbors' are available to each individual. An interesting yet challenging problem is how to efficiently solve the distributed optimization problem over CNSs with switching topologies. Recent years have witnessed exciting advances in this research field. [111] studied distributed optimization problem with diminishing step-sizes and under directed time-varying networks, and it was shown that the convergence could be ensured if switching topologies are uniformly jointly strongly connected. Afterward, [112] considered the distributed optimization over the uniformly jointly strongly connected switching topologies and designed the Push-DIGing algorithm, which absorbs the push-sum protocol and the DIGing structure. For time-varying weight-balanced case and the unbalanced case in distributed optimization, one could refer to the results in [86] and [208], respectively, where they also consider the communication time delays. Specifically, [86] then studied the distributed constrained optimization problem with communication time delays and nonidentical constraint sets. It was shown that the convergence of the subgradient projection algorithm could be guaranteed when switching topologies are uniformly jointly strongly connected even with large time-delays. [208] addressed the distributed optimization with the push-sum strategy and showed that the convergence could be ensured if the directed switching topologies are uniformly jointly strongly connected even with large bounded time-delays. Recently, the distributed resource allocation problem with dynamic topologies was explored in [35], where the almost sure convergence was derived when the underline topologies were uniformly jointly connected. The distributed energy management for microgrids was considered by [41], and push-pull based algorithm were developed for fixed strongly connected directed graphs and dynamic topologies which are uniformly jointly strongly connected. It is noteworthy that this research field is still active and looking forward to more prosperous and more in-depth results, such as investigations for algorithms under various event-triggered communication schemes for saving communication resources.

Finite-time coordination control of CNSs with switching topologies. To date, many distributed protocols have been developed to solve asymptotical coordination problems (including consensus, synchronization, rendezvous, and flocking problems) of CNSs with switching topologies. However, in some practical applications, it is desirable to design distributed protocols for CNSs such that the coordination objective can be completed in finite time. For CNSs with fixed topology, various efficient protocols have been designed based on tools from sliding mode control theory to complete the goal of finite-time coordination. It is interesting but challenging to see how to design sliding mode controller-based protocols for CNSs with switching topologies such that the goal of finite-time coordination can be guaranteed. Recent results gave some answers to the above issues. [133] studied the almost-surely practical finite-time leader-following formation tracking problem, where the agents have Lipchitz nonlinear dynamics and under time-varying weighted topologies. The closed-loop system's signals remain bounded in probability with the back-stepping sliding-mode controllers. [36] addressed the adaptive sliding-mode control for multi-robot systems under external disturbances, where time-varying communication topologies were considered since the communication among robots changes continuously along with time. A polytopic model was formulated for the switching topologies, and the proposed adaptive sliding-mode control could mitigate the impacts of the disturbance and improve tracking performance even under such continuously switching topologies.

On the other hand, different from the existing studies focusing on the finite-time control, where the convergence time is closely related to the initial consensus errors, the fixed-time control may be more applicable for some applications since it has the settling time independent of the initial values and could be directly calculated or predesigned. Some results were proposed in recent years, focusing on the distributed fixed-time control of CNSs with switching topologies. The fixed-time consensus problem of MASs under directed and switching communication topology was considered in [235]. The finite settling time's explicit bounds were addressed, and the proposed protocol remains effective, provided that the sum of specific time intervals regarding strongly connected information flow is larger than an estimated bound. The fixed-time consensus of MAS with discontinuous nonlinear inherent dynamics was investigated in [113], where distributed protocols were developed to realize the fixed-time consensus over both fixed and switching topologies. The fixed-time cooperative control of switching CNSs still needs further explorations. Interesting topics include developing practical distributed fixed-time controllers for general nonlinear CNSs under dynamic topologies or under malicious cyber attacks.

Bridging the gap between consensus/synchronization under fast switching topologies and that under slowly switching topologies. Consensus or synchronization of CNSs under fast switching topologies has been studied from averaging theory, while that with slowly switching topologies has been generally studied from MLFs-based approaches. Lastly, it is worth mentioning that an interesting topic is to develop a unified framework to deal with consensus/synchronization problem of CNSs under fast switching topologies and slowly switching topologies. Another interesting problem is to study how to reduce the conservatism of the consensus/synchronization criteria or algorithm to select proper control gains derived by tools from averaging theory or MLFs based stability analysis. It is worth mentioning that our recent consensus tracking results [195] theoretically proved that the proposed Lyapunov inequality based criteria process much less conservatism than those derived from the M-matrix theory, and the results are also applicable for switching topologies that frequently contains a directed spanning tree. This would inspire us to construct less conservative sufficient conditions regarding the coordination control and extend the results to broader application scenarios.

CHAPTER 2

Preliminaries

This chapter presents some preliminaries used in this book. In Section 2.1, notations are presented. Section 2.2 begins by introducing the matrix theory that includes Schur complement lemma, Finsler's lemma, Gershgorin's disc theorem, and some other Lemmas. Then the Barbălat lemma and the $\mathcal{K}$ function are presented. In Section 2.3, algebraic graph theory is presented that includes directed (undirected) graph, connected graph, strongly connected graph, directed spanning tree, adjacency matrix, Laplacian matrix. Specifically, the nonsingular M-matrix theory is presented which will play a crucial role in constructing the MLFs. In Section 2.4, stability theory of switched systems is given. This section begins by introducing the Carathéodory's solution of switched systems. Then the MLFs-based methods are presented, both dwell time and average dwell time stability analysis methods of switched systems are given. Note that this chapter provides some necessary tools for understanding the subsequent chapters of this book, which are especially important for a fresh graduate.

2.1 NOTATIONS

$\mathrm{diag}\{a_1, a_2, \ldots, a_n\}$	a diagonal matrix with diagonal entries from a_1 to a_n
$\subseteq$	a subset of
$\in$	belongs to
cos	cosine function
$\emptyset$	empty set
Eq.	equation
$\|x\|$	Euclidean (or 2-) norm of a real vector x
$\backslash$	excludes
$\forall$	for all
$\mathrm{Im}(z)$	imaginary part of z
$\|A\|$	induced 2-norm of a real matrix A
$\|A\|_p$	induced p-norm of a real matrix A
∞	infinity
$\|A\|_\infty$	∞-norm of a real matrix A
$\cap$	intersection

$\otimes$	Kronecker product
$\prod$	left product
lim	limit
LMI	linear matrix inequality
$\mapsto$	maps to
max	maximum
min	minimum
MAS	multi agent systems
$\mathbf{1}_n$	n-dimensional column vector with each element being 1
$\mathbf{0}_n$	n-dimensional column vector with each element being 0
I_n	$n \times n$-dimensional identity matrix
$\partial f(x(t))/\partial x(t)$	partial derivative of f with respect to state variable $x(t)$
$\|x\|_p$	p-norm of a real vector x
$\mathrm{Re}(z)$	real part of z
$\mathbb{C}$	set of complex numbers
$\mathbb{C}^+$	set of complex numbers with positive real parts
$\mathbb{Z}$	set of integers
$\mathbb{R}^{m\times n}$	set of $m \times n$-dimensional real matrices
$\mathbb{N}$	set of natural numbers
$\mathbb{R}^n$	set of n-dimensional column real vectors
$\mathbb{R}^+$	set of positive real numbers
$\mathbb{R}$	set of real numbers
sgn	signum function
sin	sine function
$\sum$	summation
$t \nearrow t_k$	t tends to t_k from the left
$t \searrow t_k$	t tends to t_k from the right
$\rightarrow$	tends to
$\dot{f}(t)$	the derivative of f with respect to the variable t
$\|A\|_F$	the Frobenius norm of a real matrix A
$\lambda_i(A)$	the i-th eigenvalue of matrix A
$\lambda_{\max}(A)$	the largest eigenvalue of real symmetric matrix A
$A \geq B$	the matrix $A - B$ is nonnegative definite
$A > B$	the matrix $A - B$ is positive definite
$\lambda_{\min}(A)$	the smallest eigenvalue of real symmetric matrix A
$\mathbb{L}_\infty$	the space of functions with finite L_∞ norm
$\mathbb{L}_p$	the space of functions with finite L_p norm
$\exists$	there exists
x^T	transpose of a real vector x
$\bigcup$	union
$x > (\geq)y, \ x, y \in \mathbb{R}^n$	x is greater (not less) than y in element-wise comparision

2.2 MATRIX THEORY AND ORDINARY DIFFERENTIAL EQUATION

Without mentioning any fundamental matrix theory, some Lemmas which will be used in this book are included in this section.

Lemma 2.1 *[62] **(Gershgorin's disc theorem)** Let $B = [b_{ij}] \in \mathbb{R}^{N \times N}$ and $R'_i(B) = \sum_{j=1, j \neq i}^{N} |b_{ij}|$, $i = 1, \ldots, N$. Then all eigenvalues of B are located in the union of N discs*

$$\bigcup_{i=1}^{N} \{ z \in \mathbb{C} : |z - b_{ii}| \leq R'_i(B) \}.$$

Furthermore, if a union of k of these N discs forms a connected region that is disjoint from the remaining $N - k$ discs, then there are exactly k eigenvalues of B in this region.

Lemma 2.2 *[62] Suppose that matrix $B = [b_{ij}] \in \mathbb{R}^{N \times N}$ has $b_{ij} \leq 0$ for all $i \neq j$, $i, j = 1, \ldots, N$. Then, the following statements are equivalent:*

(1) *All eigenvalues of B have positive real parts;*

(2) *B^{-1} exists and B^{-1} is nonnegative;*

(3) *There exists a diagonal matrix $\Phi = \mathrm{diag}\{\phi_1, \ldots, \phi_N\}$ with $\phi_i > 0$, $i = 1, \ldots, N$, such that $B^T \Phi + \Phi B > 0$;*

(4) *B is a nonsingular M-matrix,*

where B^{-1} is said to be nonnegative if all its entries are nonnegative.

Lemma 2.3 *(Lemma 3.5 of [6]) If $A \in \mathbb{R}^{N \times N}$ is symmetric, then all the eigenvalues of A are real.*

Lemma 2.4 *(Lemma 3.9 of [6]) If $A \in \mathbb{R}^{N \times N}$ is semi-positive definite, then there exists a unique semi-positive definite matrix $B \in \mathbb{R}^{N \times N}$ such that $B^2 = A$. The matrix B is called the square root of A and is denoted by $A^{\frac{1}{2}}$.*

Lemma 2.5 *(Page 28 of [6]) If $A, B \in \mathbb{R}^{N \times N}$, then AB and BA have the same eigenvalues.*

Lemma 2.6 *(Theorem 4.2.2 of [62]) Let $A \in \mathbb{R}^{N \times N}$ be symmetric. Then $\lambda_{\min}(A) x^T x \leq x^T A x \leq \lambda_{\max}(A) x^T x$, for all $x \in \mathbb{R}^N$, $\lambda_{\min}(A) = \min_{x \neq \mathbf{0}_N} \frac{x^T A x}{x^T x} = \min_{x^T x = 1} \frac{x^T A x}{x^T x}$, and $\lambda_{\max}(A) = \max_{x \neq \mathbf{0}_N} \frac{x^T A x}{x^T x} = \max_{x^T x = 1} \frac{x^T A x}{x^T x}$.*

Lemma 2.7 *If $A \in \mathbb{R}^{N \times N}$ is positive definite and $B \in \mathbb{R}^{N \times N}$ is symmetric, then, for all $x \in \mathbb{R}^N$, the following inequality holds:*

$$\lambda_{\min}(A^{-1}B) x^T A x \leq x^T B x \leq \lambda_{\max}(A^{-1}B) x^T A x.$$

Proof 2.1 *As A is positive definite, it follows from Lemma 2.4 that $A^{\frac{1}{2}}$ is well defined and is positive definite. Let $y = A^{\frac{1}{2}}x$. Then $x^T A x = y^T y$ and $x^T B x = y^T A^{-\frac{1}{2}} B A^{-\frac{1}{2}} y$, where $A^{-\frac{1}{2}}$ is the inverse matrix of $A^{\frac{1}{2}}$. So it suffices to show*

$$\lambda_{\min}(A^{-1}B) = \min_{y \neq \mathbf{0}_N} \frac{y^T A^{-\frac{1}{2}} B A^{-\frac{1}{2}} y}{y^T y}, \tag{2.1}$$

and

$$\lambda_{\max}(A^{-1}B) = \max_{y \neq \mathbf{0}_N} \frac{y^T A^{-\frac{1}{2}} B A^{-\frac{1}{2}} y}{y^T y}. \tag{2.2}$$

Since $(A^{-\frac{1}{2}} B A^{-\frac{1}{2}})^T = A^{-\frac{1}{2}} B A^{-\frac{1}{2}}$, according to Lemma 2.5, it suffices to show $\lambda_{\min}(A^{-1}B) = \lambda_{\min}(A^{-\frac{1}{2}} B A^{-\frac{1}{2}})$ and $\lambda_{\max}(A^{-1}B) = \lambda_{\max}(A^{-\frac{1}{2}} B A^{-\frac{1}{2}})$. On the other hand, we learn from Lemma 2.3 that both $\lambda_{\min}(A^{-\frac{1}{2}} B A^{-\frac{1}{2}})$ and $\lambda_{\max}(A^{-\frac{1}{2}} B A^{-\frac{1}{2}})$ are real. By using Lemma 2.5, we can obtain that

$$\lambda_{\min}(A^{-\frac{1}{2}} B A^{-\frac{1}{2}}) = \lambda_{\min}((A^{-\frac{1}{2}} B) A^{-\frac{1}{2}})$$
$$= \lambda_{\min}(A^{-\frac{1}{2}}(A^{-\frac{1}{2}} B)) = \lambda_{\min}(A^{-1}B)$$

and

$$\lambda_{\max}(A^{-\frac{1}{2}} B A^{-\frac{1}{2}}) = \lambda_{\max}((A^{-\frac{1}{2}} B) A^{-\frac{1}{2}})$$
$$= \lambda_{\max}(A^{-\frac{1}{2}}(A^{-\frac{1}{2}} B)) = \lambda_{\max}(A^{-1}B).$$

Then the proof is completed.

Lemma 2.8 *[70] For matrices A, B, C, and D with appropriate dimensions, one has*

(1) $(A \otimes B)^T = A^T \otimes B^T$;

(2) $A \otimes (B + C) = A \otimes B + A \otimes C$;

(3) $(A \otimes B)(C \otimes D) = AC \otimes BD$;

(4) $(A \otimes B)^{-1} = A^{-1} \otimes B^{-1}$, *for any given invertible matrices A and B.*

Lemma 2.9 *[8] Suppose that $A \in \mathbb{R}^{N \times N}$ is a positive definite matrix and $B \in \mathbb{R}^{N \times N}$ is a symmetric matrix. Then, for any given semi-positive definite matrix $W \in \mathbb{R}^{r \times r}$ and vector $x \in \mathbb{R}^{Nr}$, the following inequality holds:*

$$x^T (B \otimes W) x \geq \lambda_{\min}(A^{-1}B) x^T (A \otimes W) x.$$

Lemma 2.10 *For any given $x, y \in \mathbb{R}^n$, $P > 0$, and matrices A, B of appropriate dimensions, one has*

$$2x^T A B y \leq x^T A P A^T x + y^T B^T P^{-1} B y.$$

Proof 2.2 *Since $P > 0$, it follows from Lemma 2.4 that $P^{\frac{1}{2}}$ is well defined and is positive definite. By using the fact that $x^T x \geq 0$ holds for $\forall\ x \in \mathbb{R}^n$, we get*

$$0 \leq (P^{\frac{1}{2}} A^T x - P^{-\frac{1}{2}} By)^T (P^{\frac{1}{2}} A^T x - P^{-\frac{1}{2}} By)$$
$$= x^T A P A^T x - x^T A B y - y^T B^T A^T x + y^T B^T P^{-1} By.$$

So the Lemma easily follows.

Lemma 2.11 *[12] (**Schur complement lemma**) Suppose $A = A^T \in \mathbb{R}^{n \times n}$, $B = B^T \in \mathbb{R}^{m \times m}$, and $C \in \mathbb{R}^{n \times m}$. The condition*

$$\begin{bmatrix} A & C \\ C^T & B \end{bmatrix} > 0$$

is equivalent to any one of the following conditions:

(1) *$B > 0$ and $A - CB^{-1}C^T > 0$;*

(2) *$A > 0$ and $B - C^T A^{-1} C > 0$.*

Lemma 2.12 *[32] (**Finsler's lemma**) Let $x \in \mathbb{R}^n$, $A = A^T \in \mathbb{R}^{n \times n}$, and $B \in \mathbb{R}^{m \times n}$ that satisfies $\text{rank}(B) = r < n$, then the following statements are equivalent:*

(1) *$\exists\ \mu \in \mathbb{R}$ such that $A - \mu B^T B < 0$;*

(2) *$\exists\ C \in \mathbb{R}^{n \times m}$ such that $A + CB + B^T C^T < 0$.*

Lemma 2.13 *[105] (**Barbălat lemma**) If $f, \dot{f} \in \mathbb{L}_\infty$ and $f(t) \in \mathbb{L}_p$ for some $p = [1, \infty)$, then $\lim_{t \to \infty} f(t) = 0$.*

Definition 2.1 *[153] A real-valued function $\alpha : [0, +\infty) \mapsto [0, +\infty)$ is said to be of class $\mathcal{K}$ if it is continuous, strictly increasing, and $\alpha(0) = 0$. If in addition, α is unbounded, then it is said to be of class $\mathcal{K}_\infty$. A real-valued function $\beta : [0, +\infty) \times [0, +\infty) \mapsto [0, +\infty)$ is said to be of class $\mathcal{KL}$ if $\beta(\cdot, t)$ is of class $\mathcal{K}$ for each fixed $t \geq 0$, and $\beta(r, t)$ is decreasing to zero as $t \to \infty$ for each fixed $r \geq 0$.*

We shall write $\alpha \in \mathcal{K}_\infty$ and $\beta \in \mathcal{KL}$ to indicate that α is a class $\mathcal{K}_\infty$ function and β is a class $\mathcal{KL}$ function, respectively.

2.3   ALGEBRAIC GRAPH THEORY

Suppose a CNS consists of N nodes (agents) which interact with each other through a communication or sensing network or a combination of both. It is natural to model the interactions among the N nodes (agents) by undirected or directed graphs. Without loss generality, the N nodes can be labeled as node $1, \ldots, N$. Let $\mathcal{V} = \{1, \cdots, N\}$ be the set of nodes. Then the directed graph is described by $(\mathcal{V}, \mathcal{E})$, where the set of edges $\mathcal{E} \subseteq \mathcal{V} \times \mathcal{V}$ represent the interactions among the N nodes. For notational simplicity, the graph $(\mathcal{V}, \mathcal{E})$ is denoted by $\mathcal{G}$. The edge $(j, i) \in \mathcal{E}$ if and only if node i can receive

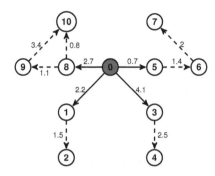

Figure 2.1 A directed graph $\mathcal{G}$ consists of 11 nodes, where the numbers around the edges represent the weights. Although $\mathcal{G}$ is not strongly connected because nodes 1–10 do not have directed paths to all other nodes, $\mathcal{G}$ contains a directed spanning tree with node 0 being the root.

the information from node j. When $(j, i) \in \mathcal{E}$, node j is said to be a neighbor of node i. Denote by $\mathcal{N}_i$ the set of neighbors of node i. If there exists a sequence of distinct nodes $i_1, \ldots, i_m$ such that $(i, i_1), (i_1, i_2), \ldots, (i_{m-1}, i_m), (i_m, k) \in \mathcal{E}$, then it is said that node i has a directed path to node k, or node k is reachable from node i. $\mathcal{G}$ is *strongly connected* if each node has at least one directed path to any other nodes. More generally, if there exists a node, called the root, which has at least one directed path to any other nodes, $\mathcal{G}$ is said to contain a directed *spanning tree*. Denote by a_{ij} the weight of the edge (j, i), $i, j = 1, \ldots, N$. It is assumed throughout this book that $a_{ij} \geq 0$, where $a_{ij} > 0$ if and only if $(j, i) \in \mathcal{E}$, and $a_{ij} = 0$, otherwise. In addition, it is assumed in this book that $a_{ii} = 0$, that is, self-loop is forbidden. $\mathcal{G}$ is called an undirected graph if $(i, j) \in \mathcal{E}$ whenever $(j, i) \in \mathcal{E}$ and $a_{ij} = a_{ji}$. An undirected graph is *connected* if there exists at least one undirected path between each pair of distinct nodes. For undirected graphs, the existence of an undirected spanning tree is equivalent to being connected. However, for directed graphs, the existence of a directed spanning tree is a weaker condition than being strongly connected. Please see Figure 2.1 for a directed graph which is not strongly connected but contains a directed spanning tree.

Let $\mathcal{A} = [a_{ij}] \in \mathbb{R}^{N \times N}$ be the adjacency matrix of the graph $\mathcal{G}$. Then the Laplacian matrix $\mathcal{L} = [l_{ij}] \in \mathbb{R}^{N \times N}$ is defined as

$$l_{ij} = \begin{cases} \sum_{j \in \mathcal{N}_i} a_{ij}, & \text{if } j = i, \\ -a_{ij}, & \text{if } j \neq i, \end{cases} \quad i = 1, \ldots, N. \tag{2.3}$$

If $\mathcal{G}$ is undirected, $\mathcal{L}$ is symmetric since $a_{ij} = a_{ji}$. However, when $\mathcal{G}$ is directed, $\mathcal{L}$ is not necessarily symmetric. No matter undirected or direct graphs, $\mathcal{L}$ has zero row sum. Hence 0 is an eigenvalue of $\mathcal{L}$ with an associated eigenvector $\mathbf{1}_N$. Note that $\mathcal{L}$ is diagonally dominant and has nonnegative diagonal entries. According to the Gershgorin's disc theorem (see Lemma 2.1), all nonzero eigenvalues of $\mathcal{L}$ have positive real parts if $\mathcal{G}$ is directed. If $\mathcal{G}$ is undirected, all nonzero eigenvalues of $\mathcal{L}$ is positive since $\mathcal{L}$ is symmetric which indicates that all the eigenvalues are real.

Lemma 2.14 *[127]* $\mathcal{L}$ *has a simple zero eigenvalue and all other eigenvalues have positive real parts (respectively, are positive) if and only if* $\mathcal{G}$ *has a directed spanning tree (respectively, is connected).*

Remark 2.1 *Let* $x = [x_1, \ldots, x_N]^T \in \mathbb{R}^N$. *It is not difficult to obtain that* $\mathcal{L}x = [\sum_{j=1}^{N} a_{1j}(x_1 - x_j), \ldots, \sum_{j=1}^{N} a_{Nj}(x_N - x_j)]^T$. *If* $\mathcal{G}$ *is undirected, then* $x^T \mathcal{L}x = \frac{1}{2}\sum_{i,j=1}^{N} a_{ij}(x_i - x_j)^2$. *Furthermore, when the undirected graph* $\mathcal{G}$ *is connected, it follows from Lemma 2.14 that* $x^T \mathcal{L}x = 0$ *if and only if* $x_i = x_j$ *for all* $i, j = 1, \ldots, N$.

Let $\mathcal{G}$ be a directed graph which has a directed spanning tree. Assume further that all possible directed spanning trees in such a directed graph $\mathcal{G}$ have a common root. Without loss of generality, assuming node 1 to be the common root. Then $\mathcal{L}$ can be rewritten as

$$
\mathcal{L} = \begin{bmatrix} 0 & \mathbf{0}_{N-1}^T \\ \mathbf{P} & \overline{\mathcal{L}} \end{bmatrix}, \quad \overline{\mathcal{L}} = \begin{bmatrix} \sum_{j \in \mathcal{N}_2} a_{2j} & -a_{23} & \cdots & -a_{2N} \\ -a_{32} & \sum_{j \in \mathcal{N}_3} a_{3j} & \cdots & -a_{3N} \\ \vdots & \vdots & \ddots & \vdots \\ -a_{N2} & -a_{N3} & \cdots & \sum_{j \in \mathcal{N}_N} a_{Nj} \end{bmatrix}, \quad (2.4)
$$

where $\mathbf{P} = -[a_{21}, \ldots, a_{N1}]^T$. Then Lemma 2.14 implies that all the eigenvalues of $\overline{\mathcal{L}}$ have positive real parts. On the other hand, all the non-diagonal entries of $\overline{\mathcal{L}}$ are non-positive. Then it can be got from Lemma 2.2 that $\overline{\mathcal{L}}$ is a nonsingular M-matrix and there exists a diagonal matrix $\Phi = \text{diag}\{\phi_2, \ldots, \phi_N\}$ such that $\overline{\mathcal{L}}^T \Phi + \Phi \overline{\mathcal{L}} > 0$, where $\phi_i > 0$, $i = 2, \ldots, N$. Unfortunately, Lemma 2.2 has not presented any methods for selecting appropriate diagonal entries ϕ_i which may be used for controller design.

Lemma 2.15 *[81] If* $\mathcal{G}$ *contains a directed spanning tree, then there exists a positive definite diagonal matrix* $\Phi = \text{diag}\{\phi_2, \ldots, \phi_N\}$ *such that* $\overline{\mathcal{L}}^T \Phi + \Phi \overline{\mathcal{L}} > 0$. *One such* $\phi = [\phi_2, \ldots, \phi_N]^T$ *can be obtained by solving the matrix equation* $\overline{\mathcal{L}}^T \phi = \mathbf{1}_{N-1}$.

Proof 2.3 *It suffices to show the second assertion. Since* $\overline{\mathcal{L}}$ *is a nonsingular M-matrix, it follows from Lemma 2.2 that* $(\overline{\mathcal{L}}^T)^{-1}$ *exists and is nonnegative, and thereby row sums cannot be all zero. Then it is easy to verify that* $\phi > \mathbf{0}_{N-1}$ *and* $\Phi \overline{\mathcal{L}} \mathbf{1}_N \geq \mathbf{0}_{N-1}$. *Noticing that* $\overline{\mathcal{L}}^T \Phi \mathbf{1}_{N-1} = \overline{\mathcal{L}}^T \phi = \mathbf{1}_{N-1}$, *we conclude that* $(\overline{\mathcal{L}}^T \Phi + \Phi \overline{\mathcal{L}})\mathbf{1}_{N-1} > 0$ *which implies* $\overline{\mathcal{L}}^T \Phi + \Phi \overline{\mathcal{L}}$ *is strictly diagonally dominant. Since the diagonal entries of* $\overline{\mathcal{L}}^T \Phi + \Phi \overline{\mathcal{L}}$ *are positive, it then follows from Lemma 2.1 that each eigenvalue of* $\overline{\mathcal{L}}^T \Phi + \Phi \overline{\mathcal{L}}$ *is positive, implying that* $\overline{\mathcal{L}}^T \Phi + \Phi \overline{\mathcal{L}}$ *is positive definite.*

2.4 SWITCHED SYSTEM THEORY

This section introduces the solutions of differential systems, MLFs, and stability theory under slow switching. For more detailed discussions, we refer the reader to Chapter 3 in [82].

2.4.1   Solutions of differential systems

Consider the system

$$\dot{x}(t) = f(t, x(t)), \ x(t) \in \mathbb{R}^n, \ t \in [t_0, +\infty), \tag{2.5}$$

where $f(t, x(t)) : [t_0, +\infty) \times \mathbb{R}^n \mapsto \mathbb{R}^n$. Denote by x_0 the initial value $x(t_0)$. A classical solution for the Cauchy problem of (2.5) with $x(t_0) = x_0$ on $[t_0, T]$ is a continuously differentiable map $x(t) : [t_0, T] \mapsto \mathbb{R}^n$ that satisfies (2.5). According to the well-known Peano's theorem, one knows that if the function f is continuous in a neighborhood of t_0, x_0, system (2.5) has at least one classical solution defined in a neighborhood of t_0, x_0. To proceed, the concept of Lipschitz condition is introduced.

Definition 2.2 *[27] A function $f(t, x(t)) : [t_0, +\infty) \times \mathbb{R}^n \mapsto \mathbb{R}^m$ is said to be globally Lipschitz in $x(t)$ uniformly over t if there exists a positive scalar L_0 such that*

$$\|f(t, x(t)) - f(t, y(t))\| \leq L_0 \|x(t) - y(t)\|, \tag{2.6}$$

for all $(t, x(t))$ and $(t, y(t))$.

Theorem 2.1 *[27] If $f(t, x(t)) : [t_0, +\infty) \times \mathbb{R}^n \mapsto \mathbb{R}^n$ is continuous in t and globally Lipschitz in $x(t)$ uniformly over t, then, for all $x_0 \in \mathbb{R}^n$, there exists a unique classical solution of (2.5) over the time interval $[t_0, +\infty)$ with initial condition x_0.*

However, since our view is toward systems with switching, the assumption that the function f is continuous in both t and $x(t)$ is too restrictive. The following example shows that, if the function is discontinuous, then classical solution of (2.5) might not exist.

Example 2.1 *[27] **Discontinuous Vector Field with Nonexistence of Classical Solutions:** Consider the function $f(t, x(t)) : [0, +\infty) \times \mathbb{R} \mapsto \mathbb{R}$ defined by*

$$f(t, x(t)) = \begin{cases} -1, & x(t) > 0, \\ 1, & x(t) \leq 0, \end{cases} \tag{2.7}$$

with initial value $x(0) = 0$. It is obviously that the function f is discontinuous at $x(t) = 0$. Suppose there exists a classical solution $x(t) : [0, T) \mapsto \mathbb{R}$ that satisfies (2.7). Then $\dot{x}(0) = f(0, x(0)) = f(0, 0) = 1$ which implies that, for sufficiently small $t > 0$, $x(t) > 0$, and hence $\dot{x}(t) = f(t, x(t)) = -1$. But this contradicts the fact that $t \mapsto \dot{x}(t)$ is continuous. Hence, there is no classical solution starting from zero.

It turns out that for the existence and uniqueness result to hold, it is sufficient to demand that f is piecewise continuous in t [82]. So we consider the Carathéodory's solution $x(\cdot)$ that is given by

$$x(t) = x_0 + \int_{t_0}^{t} f(s, x(s))ds. \tag{2.8}$$

Note that (2.8) satisfies the differential equation (2.5) almost everywhere.

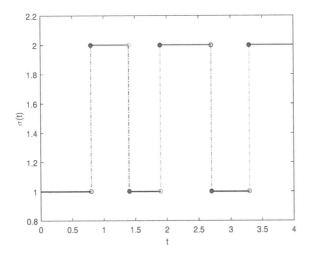

Figure 2.2 A switching signal $\sigma(t)$ with $\kappa = 2$.

2.4.2 Multiple Lyapunov functions

To proceed, the notion of time dependent switching is introduced.

As a special kind of hybrid dynamic system, switched system has been studied for quite some time by researchers from applied mathematics, systems and control fields. Roughly speaking, a switched system is a dynamic system that consists of a number of subsystems and a switching rule that determines switches among these subsystems. Suppose the switched system is generated by the following family of subsystems

$$\dot{x}(t) = f_p(t, x(t)), \ x(t) \in \mathbb{R}^n, \ p \in \{1, \ldots, \kappa\}, \tag{2.9}$$

together with a switching signal $\sigma(t) : [t_0, +\infty) \mapsto \{1, \ldots, \kappa\}$. Note that $\sigma(t)$ is a piecewise constant function that switches at the switching time instants $t_1, t_2, \ldots$, and is constant on the time interval $[t_k, t_{k+1})$, $k = 0, 1, \ldots$. In this book, we assume $\sigma(t)$ is right continuous, i.e., $\sigma(t) = \lim_{\iota \searrow t} \sigma(\iota)$, and $\inf_{k \in \mathbb{N}}(t_{k+1} - t_k) \geq \tau_m$ for some given positive scalar τ_m where inf represents the infimum. Please see Figure 2.2 for an example. Thus the switched systems with time-dependent switching signal $\sigma(t)$ can be described by the equation

$$\dot{x}(t) = f_{\sigma(t)}(t, x(t)). \tag{2.10}$$

According to Theorem 2.1, each subsystem has a unique solution over arbitrary interval $[t_k, t_{k+1})$, $k = 0, 1, \ldots$, with arbitrary initial value $x(t_k) \in \mathbb{R}^n$ if the function f_p, for each $p = 1, \ldots, \kappa$, is globally Lipschitz in $x(t)$ uniformly over t. Thus the switched system (2.10) is well defined for arbitrary switching signal $\sigma(t)$ defined above and any given initial value $x(t_0) \in \mathbb{R}^n$. Throughout this chapter, we assume that such a globally Lipschitz condition holds for the subsystems, and thus the well-definedness of the switched system is guaranteed. We further assume that $f_p(t, \mathbf{0}_n) = \mathbf{0}_n$ for each $p = 1, \ldots, \kappa$. Thus, the zero vector is an equilibrium point of the switched system (2.10). Next, some stability notions for the zero equilibrium point of switched systems are introduced.

Definition 2.3 *[153] Switched system (2.10) is said to be*

(1) stable *with respect to the switching signal $\sigma(t)$ if there exist positive scalar δ and a class $\mathcal{K}$ function α such that all solutions with $\|x(t_0)\| < \delta$ satisfying*

$$\|x(t)\| \leq \alpha(x(t_0)), \ \forall \ t \geq t_0. \tag{2.11}$$

(2) asymptotically stable *with respect to the switching signal $\sigma(t)$ if there exist a positive scalar δ and a class $\mathcal{KL}$ function β such that all solutions with $\|x(t_0)\| < \delta$ satisfying*

$$\|x(t)\| \leq \beta(x(t_0), t_0), \ \forall \ t \geq t_0. \tag{2.12}$$

(3) exponentially stable *with respect to the switching signal $\sigma(t)$ if there exist positive scalars δ, μ, and ν such that all solutions with $\|x(t_0)\| < \delta$ satisfying*

$$\|x(t)\| \leq \mu \cdot \exp(-\nu t)\|x(t_0)\|, \ \forall \ t \geq t_0. \tag{2.13}$$

Furthermore, switched system (2.10) is said to be globally stable, globally asymptotically stable, *and* globally exponentially stable *if the inequalities (2.11), (2.12), and (2.13) hold for all initial values, respectively.*

In the remainder of this section, MLFs based technique which is a useful tool for analyzing the stability of switched system is considered. Suppose that each system $\dot{x}(t) = f_p(t, x(t))$, $p \in \{1, \ldots, \kappa\}$, is (globally) asymptotically stable, and let $V_p(t, x(t))$ be their respective (radially unbounded) Lyapunov functions. If $V_{\sigma(t_{i-1})}(t_i, x(t_i)) = V_{\sigma(t_i)}(t_i, x(t_i))$ for all $i \geq 1$, then $V_{\sigma(t)}$ is continuous over time t (see Figure 2.3(a)), and thereby asymptotic stability follows. However, as depicted in Figure 2.3(b), although the value of the Lyapunov function decreases during $[t_k, t_{k+1})$, $k = 0, 1, \ldots$, it may increase at the switching time instant.

In the next subsection, we shall show how to apply MLFs based techniques to obtain stability criteria for switched systems under the constraint of slow switching.

2.4.3 Stability under slow switching

We firstly introduce the notion of dwell time. If there exist $\tau_M \geq \tau_m > 0$ such that $\tau_m \leq t_{i+1} - t_i \leq \tau_M < +\infty$ for $i = 0, 1, \ldots$, then τ_m is called the *dwell time* of the switching signal $\sigma(t)$ (see Figure 2.4 for a simple illustration). In the sequel, we assume that the switching signal $\sigma(t)$ always satisfies the condition that $\tau_m \leq t_{i+1} - t_i \leq \tau_M < +\infty$ for $i = 0, 1, \ldots$.

Theorem 2.2 *[82] Suppose all subsystems in the family (2.10) with $p \in \{1, \ldots, \kappa\}$ are globally exponentially stable, and there exists a Lyapunov function $V_p(t, x(t))$: $[t_0, +\infty) \times \mathbb{R}^n \mapsto [0, +\infty)$ for each $p \in \{1, \ldots, \kappa\}$ such that*

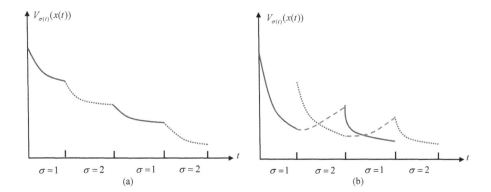

Figure 2.3 Two Lyapunov functions, where the solid lines correspond to V_1, and the dashed lines correspond to V_2. (a) Continuous $V_{\sigma(t)}$, (b) discontinuous $V_{\sigma(t)}$.

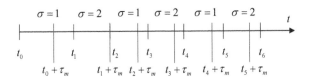

Figure 2.4 A switching signal having dwell time τ_m, where t_k, $k = 1, 2, \ldots$, are the switching instants and $\tau_m = t_2 - t_1$.

(1) $a_p\|x(t)\|^2 \le V_p(t, x(t)) \le b_p\|x(t)\|^2$;

(2) $\frac{\partial V_p(t,x(t))}{\partial t} + \frac{\partial V_p(t,x(t))}{x(t)} f_p(t, x(t)) \le -c_p\|x(t)\|^2$,

with a_p, b_p, and c_p being positive scalars. Then the switched system (2.10) is globally exponentially stable if the dwell time

$$\tau_m > \frac{\ln \gamma}{\rho}, \quad \gamma = \frac{\max_{p=1,\ldots,\kappa} b_p}{\min_{p=1,\ldots,\kappa} a_p}, \quad \rho = \min\left\{\frac{c_1}{b_1}, \ldots, \frac{c_\kappa}{b_\kappa}\right\}. \tag{2.14}$$

Proof 2.4 *Suppose that $\sigma(t) = p$, we may get from the conditions of the theorem that*

$$\dot{V}_p(t, x(t)) = \frac{\partial V_p(t, x(t))}{\partial t} + \frac{\partial V_p(t, x(t))}{\partial x(t)} f_p(t, x(t)) \le -\frac{c_p}{b_p} V_p(t, x(t)), \tag{2.15}$$

where the derivatives of the Lyapunov functions at switching time points should be considered as their right-hand derivatives. For each $t \in [t_i, t_{i+1})$ with $\sigma(t) = p$, $i = 0, 1, \ldots$, it follows from (2.15) that

$$V_p(t, x(t)) \le \exp\left(-\frac{c_p}{b_p}(t - t_i)\right) V_p(t, x(t_i)).$$

We now construct the following MLFs for the switched system (2.10):

$$V(t) = V_{\sigma(t)}(t, x(t)). \tag{2.16}$$

Noticing that $V(t_k) \le b_{\sigma(t_k)}\|x(t_k)\|^2 = b_{\sigma(t_k)}\|x(t_k^-)\|^2 \le \gamma V(t_k^-)$, for $k \ge 1$. We get from the above inequalities that

$$\begin{aligned} V(t_1) \le \gamma V(t_1^-) &\le \gamma \exp\left(-\rho(t_1 - t_0)\right) V(t_0) \\ &\le \gamma \exp\left(-\rho\tau_m\right) V(t_0), \end{aligned} \tag{2.17}$$

where the inequality $t_1 - t_0 \ge \tau_m$ was used to get the last inequality. Similarly, we have

$$\begin{aligned} V(t_2) \le \gamma V(t_2^-) &\le \gamma \exp\left(-\rho(t_2 - t_1)\right) V(t_1) \\ &\le \gamma \exp\left(-\rho\tau_m\right) V(t_1). \end{aligned} \tag{2.18}$$

We may thus conclude from (2.17) and (2.18) that

$$V(t_2) \le \gamma^2 \exp\left(-2\rho\tau_m\right) V(t_0).$$

By iteration, for each $k = 1, 2, \ldots$, we have

$$V(t_k) \le \gamma^k \exp\left(-k\rho\tau_m\right) V(t_0) = \exp\left(-k\rho\tau_m + k \ln \gamma\right) V(t_0). \tag{2.19}$$

As $\tau_m > (\ln \gamma)/\rho$, then $\mu = \rho - (\ln \gamma)/\tau_m > 0$. It follows from (2.19) that

$$V(t_k) \le \exp\left(-k\mu\tau_m\right) V(t_0). \tag{2.20}$$

On the other hand, for any given time instant t, there exists a $j \in \mathbb{N}$ such that $t_j \leq t < t_{j+1}$. For the case with $j = 0$, it can be verified that $V(t) \leq \exp(-\rho(t-t_0))V(t_0)$. For the case with $j \geq 1$, it is not difficult to get that

$$V(t) \leq \exp\left(-\rho(t-t_j)\right) V(t_j) \leq \exp\left(-\rho(t-t_j) - j\mu\tau_m\right) V(t_0)$$

$$\leq \exp\left(-j\mu\tau_m\right) V(t_0) \leq \exp\left(-\frac{j\tau_m}{(j+1)\tau_M}\mu t\right) V(t_0), \tag{2.21}$$

where we use the inequality $1 \geq t/[(j+1)\tau_M]$ to get the last inequality. By using the fact $j/(j+1) \geq 1/2$ when $j \geq 1$, we get from (2.21) that

$$V(t) \leq \exp\left(-\frac{\tau_m}{2\tau_M}\mu t\right) V(t_0). \tag{2.22}$$

By letting $\varrho_0 = \min\{\rho, \frac{\tau_m\mu}{2\tau_M}\}$, we may get from the above analysis that $V(t) \leq \exp(-\varrho_0(t-t_0))V(t_0)$. Then, according to the condition that $V(t) \geq a_{\min}\|x(t)\|^2$ with $a_{\min} = \min_{p \in \{1,\dots,\kappa\}}\{a_p\}$, it can be concluded that the switched system (2.10) is globally exponentially stable.

In the remainder of this section, a more general result will be presented which relies on the notion *average dwell time* (ADT).

In the context of dwell time switching, there can be no more switches for the next τ_m units of time after a switch occurs [82]. In some practical circumstances, however, the value of the switching signals may change fast during some time intervals, e.g., the communication topologies for a team of robots may switch quite fast at the some moments due to the quick change of some robots' positions which leads to some links failure or restoration. To relax the dwell time switching, the notion of ADT is introduced.

Definition 2.4 *[56] Let $N_\sigma(t,T)$ be the number of switches on an interval (t,T), $\forall$ $T > t \geq t_0$. If there exist two scalars $N_0 \geq 0$ and $\tau_a > 0$ such that*

$$N_\sigma(t,T) \leq N_0 + \frac{T-t}{\tau_a}, \tag{2.23}$$

then the switching signal σ is said to have ADT τ_a.

Remark 2.2 *The essence of ADT condition given in (2.23) is that there may exist some consecutive switches separated by the time intervals with length less than τ_a, but the length of the average time interval between consecutive switches should not be less than τ_a. More precisely, inequality (2.23) implies that, for $N_0 > 0$, the average length of time intervals between consecutive switches should not be less than τ_a by discarding the first $\lceil N_0 \rceil$ switches, where $\lceil N_0 \rceil$ represents the smallest integer larger than $\lceil N_0 \rceil$. Note also that $N_0 = 0$ means that there is no switching over any given time interval.*

We now extend the Theorem 2.2 to switching signals with ADT.

Theorem 2.3 *[82] Suppose all subsystems in the family (2.10) with $p \in \{1, \ldots, \kappa\}$ are globally exponentially stable and there exists a Lyapunov function $V_p(t, x(t))$: $[t_0, +\infty) \times \mathbb{R}^n \mapsto [0, +\infty)$ for each $p \in \{1, \ldots, \kappa\}$ such that*

(1) $a_p \|x(t)\|^2 \leq V_p(t, x(t)) \leq b_p \|x(t)\|^2$;

(2) $\frac{\partial V_p(t, x(t))}{\partial t} + \frac{\partial V_p(t, x(t))}{x(t)} f_p(t, x(t)) \leq -c_p \|x(t)\|^2$,

with a_p, b_p, and c_p being positive scalars. Then the switched system (2.10) is globally exponentially stable if the ADT

$$\tau_a > \frac{\ln \gamma}{\rho}, \quad \gamma = \frac{\max_{p=1,\ldots,\kappa} b_p}{\min_{p=1,\ldots,\kappa} a_p}, \quad \rho = \min \left\{ \frac{c_1}{b_1}, \ldots, \frac{c_\kappa}{b_\kappa} \right\}. \tag{2.24}$$

Proof 2.5 *Let $V(t)$ be the Lyapunov function with the same form as that given by (2.16). By using the same arguments as in (2.17) and (2.18), we have*

$$V(t_1) \leq \gamma \exp\left(-\rho(t_1 - t_0)\right) V(t_0), \tag{2.25}$$

and

$$V(t_2) \leq \gamma \exp\left(-\rho(t_2 - t_1)\right) V(t_1). \tag{2.26}$$

Combining (2.26) together with (2.25) gives

$$V(t_2) \leq \gamma^2 \exp\left(-\rho(t_2 - t_0)\right) V(t_0).$$

By iteration, it is not difficult to conclude that, for each $k = 1, 2, \ldots$,

$$V(t_k) \leq \gamma^k \exp\left(-\rho(t_k - t_0)\right) V(t_0). \tag{2.27}$$

For any given $t > t_0$, the number of switches on the interval (t_0, t) is $N_\sigma(t_0, t)$. It follows from (2.27) that

$$V(t) \leq \gamma^{N_\sigma(t_0, t)} \exp\left(-\rho(t - t_0)\right) V(t_0). \tag{2.28}$$

Substituting (2.23) into (2.28) gives

$$V(t) \leq \gamma^{N_0} \cdot \exp\left(-\rho(t - t_0) + \frac{\ln \gamma}{\tau_a}(t - t_0)\right) V(t_0)$$
$$= \gamma^{N_0} \cdot \exp\left(-\tilde{\mu}(t - t_0)\right) V(t_0), \tag{2.29}$$

where $\tilde{\mu} = \rho - (\ln \gamma)/\tau_a > 0$. Then the proof is completed by using some similar arguments as the last part of the proof of Theorem 2.2.

CHAPTER 3

Consensus of linear CNSs with directed switching topologies

This chapter studies the consensus problem of CNSs with linear dynamics and directed switching topologies. Section 3.1 studies the leaderless consensus problem. This section begins by introducing some previous works and by presenting our motivations. Then the linear CNSs model and the main theoretical results are presented. A simulation is also given to validate the obtained result. Compared to previous works, a main improvement of this section is that the switching topologies are allowed to have spanning trees rooted at different nodes. Section 3.2 studies the leader-following consensus problems for the case with an autonomous leader and the case with a nonautonomous leader, respectively. This section presents an iterative optimization algorithm to construct a class of novel MLFs such that a smaller (average) dwell time is required compared with that constructed by the traditional nonsingular M-matrix theory.

3.1 CONSENSUS OF LINEAR CNSs WITH DIRECTED SWITCHING TOPOLOGIES

3.1.1 Introduction

In the past decade, the consensus problem of general linear CNSs has received a lot of attention [76, 146, 162, 185, 186, 224]. Specifically, the consensus problem of linear CNSs under a directed fixed communication topology has been addressed in [76, 224]. In [162], the robust consensus of linear CNSs with additive perturbations of the transfer matrices of the nominal dynamics was studied. In [163] and a number of subsequent papers, the robust consensus was analyzed from the viewpoint of the $\mathcal{H}_\infty$ control theory. Among other relevant references, we mention [146] where, while assuming that the open loop systems are Lyapunov stable, the consensus problem of linear CNSs with undirected switching topologies has been investigated. In the situation where the CNS is equipped with a leader and the topology of the system belongs to the class of directed switching topologies, the consensus tracking problem has been studied in [185, 186]. One feature of the results in these references is that the open loop agents' dynamics do not have to be Lyapunov stable. Note that the presence of the leader in the CNSs considered in these references facilitate the derivations and the direct analyses of the consensus error system. However, when the open loop systems are not Lyapunov stable and/or there is no designated leader in the group, the consensus problem for linear CNSs with directed switching topologies remains challenging.

Motivated by the above discussion, this section aims to study the consensus problem for linear CNSs with directed switching topologies. Several aspects of the current study are worth mentioning. Firstly, some of the assumptions in the existing works are dismissed, e.g., the open loop dynamics of the agents do not have to be Lyapunov stable in this chapter. Furthermore, the CNSs under consideration are not required to have a leader. Compared with the consensus problems for linear CNSs with a designated leader, the point of difference here concerns the assumption on the system's communication topology. In the previous work on the consensus tracking of linear CNSs such as [185], each possible augmented system graph was required to contain a directed spanning tree rooted at the leader. Compared with that work, the switching topologies in this section are allowed to have spanning trees rooted at different nodes. This is a significant relaxation of the previous conditions since it enables the system to be reconfigured if necessary (e.g., to allow different nodes to serve as the formation leader). This also has a potential to make the system more reliable.

3.1.2 Problem formulation

Consider a CNS consists of N agents that are labelled as agents $1, \ldots, N$. The dynamics of agent i are described by

$$\dot{x}_i(t) = Ax_i(t) + Bu_i(t), \tag{3.1}$$

where $x_i(t) \in \mathbb{R}^n$ is the state, $u_i(t) \in \mathbb{R}^m$ is the control input, $A \in \mathbb{R}^{n \times n}$ and $B \in \mathbb{R}^{n \times m}$ are, respectively, the state matrix and control input matrix. It is assumed that the matrix pair (A, B) is stabilizable. And it is assumed that the communication topology of the CNS under consideration switches dynamically over a graph set $\widehat{\mathcal{G}}$, where $\widehat{\mathcal{G}} = \{\mathcal{G}^1, \ldots, \mathcal{G}^\kappa\}$, $\kappa \geq 1$, denotes the set of all possible directed topologies.

Suppose that $\mathcal{G}(t) \in \widehat{\mathcal{G}}$ for all t. To describe the time-varying property of communication topology, assume that there exists an infinite sequence of non-overlapping time intervals $[t_k, t_{k+1})$, $k = 0, 1, \ldots$, with $t_0 = 0$, $0 < \tau_m \leq t_{k+1} - t_k \leq \tau_M < +\infty$, over which the communication topology is fixed. Here, $\tau_M > \tau_m > 0$ and τ_m is called the dwell time. The introduction of the switching signal $\sigma(t) : [0, +\infty) \mapsto \{1, \ldots, \kappa\}$ makes the communication topology of CNS (3.1) well defined at every time instant $t \geq 0$. For notational convenience, we will describe this communication topology using the time-varying graph $\mathcal{G}^{\sigma(t)}$.

Within the context of CNSs, only relative information among neighboring agents can be used for coordination. For each agent i, the following distributed consensus protocol is proposed

$$u_i(t) = \alpha K \sum_{j=1}^{N} a_{ij}^{\sigma(t)}[x_j(t) - x_i(t)], \quad i = 1, \ldots, N, \tag{3.2}$$

where $\alpha > 0$ represents the coupling strength, $K \in \mathbb{R}^{m \times n}$ is the feedback gain matrix to be designed, and $\mathcal{A}^{\sigma(t)} = [a_{ij}^{\sigma(t)}]_{N \times N}$ is the adjacency matrix of graph $\mathcal{G}^{\sigma(t)}$. Then, it follows from (3.1) and (3.2) that

$$\dot{x}_i(t) = A x_i(t) + \alpha B K \sum_{j=1}^{N} a_{ij}^{\sigma(t)}[x_j(t) - x_i(t)], \tag{3.3}$$

where $i = 1, \ldots, N$.

Let $x(t) = [x_1^T(t), \ldots, x_N^T(t)]^T$, it thus follows from (3.3) that

$$\dot{x}(t) = \left[(I_N \otimes A) - \alpha \left(\mathcal{L}^{\sigma(t)} \otimes BK\right)\right] x(t), \tag{3.4}$$

where $\mathcal{L}^{\sigma(t)}$ is the Laplacian matrix of communication topology $\mathcal{G}^{\sigma(t)}$.

Before concluding this section, the following assumption is presented which will be used in the derivation of the main results.

Assumption 3.1 *For each $i \in \{1, \ldots, \kappa\}$, the graph $\mathcal{G}^i$ contains a directed spanning tree.*

Remark 3.1 *Note that we will not assume in the sequel that the directed spanning trees within the graphs $\mathcal{G}^i$, $i = 1, \ldots, \kappa$, share a common root node, though such an assumption is very common in the existing related literature [185]. Certainly, Assumption 3.1 holds in the special case considered in the above reference, where each possible topology $\mathcal{G}^i$, $i \in \{1, \ldots, \kappa\}$, contains a directed spanning tree, and all these trees are rooted at the same node. Furthermore, Assumption 3.1 holds if each possible topology is strongly connected.*

Note that all signals considered in this section are assumed to be differentiable on the right. Furthermore, for any given initial value $x(t_0) \in \mathbb{R}^{Nn}$, the switched systems (3.4) are assumed to have a unique and absolutely continuous solution $x(t)$ in the sense of Carathéodory.

3.1.3 Main results

Let $e(t) = [e_1^T(t), \ldots, e_{N-1}^T(t)]^T$ with $e_i(t) = x_i(t) - x_N(t)$ for $i = 1, \ldots, N-1$. It can then be obtained that $e(t) = (\Xi \otimes I_n)x(t)$, where $\Xi = [I_{N-1}, -\mathbf{1}_{N-1}] \in \mathbb{R}^{(N-1) \times N}$. Using this notation, it can thus be obtained from (3.4) that

$$\dot{e}(t) = (I_{N-1} \otimes A)e(t) - \alpha \left(\Xi \mathcal{L}^{\sigma(t)} \otimes BK\right) x(t). \tag{3.5}$$

Noticing that $\left(\mathcal{L}^{\sigma(t)} \otimes BK\right)(\mathbf{1}_N \otimes I_n)x_N(t) = \mathbf{0}$, one has that

$$\left(\Xi \mathcal{L}^{\sigma(t)} \otimes BK\right) x(t) = \left(\Xi \mathcal{L}^{\sigma(t)} \Pi \otimes BK\right) e(t), \tag{3.6}$$

where $\Pi = \begin{bmatrix} I_{N-1} \\ 0_{N-1}^T \end{bmatrix} \in \mathbb{R}^{N \times (N-1)}$. Substituting (3.6) into (3.5) gives that

$$\dot{e}(t) = \left[I_{N-1} \otimes A - \alpha \left(\Xi \mathcal{L}^{\sigma(t)} \Pi \otimes BK \right) \right] e(t). \tag{3.7}$$

Obviously, $\mathbf{0}$ is the equilibrium point of the switched system (3.7). Furthermore, by Definition 1.1, the CNS (3.4) achieves consensus if and only if the zero equilibrium point of the switched system (3.7) is globally attractive. Thus, to show that the CNS (3.4) achieves consensus, it is sufficient to establish that the zero equilibrium point of the switched system (3.7) is globally asymptotically stable.

According to Assumption 3.1, it can be obtained from Theorem 2.8 in [130] that, for each $i \in \{1, \ldots, \kappa\}$ and an arbitrarily given $\alpha > 0$, the linear time-invariant system

$$\dot{\zeta}(t) = -\alpha \left(\Xi \mathcal{L}^i \Pi \otimes I_n \right) \zeta(t)$$

is globally asymptotically stable at its zero equilibrium point, where $\zeta(t) \in \mathbb{R}^{(N-1)n}$. This implies that for each $i \in \{1, \ldots, \kappa\}$, the $(N-1) \times (N-1)$ matrix $\Xi \mathcal{L}^i \Pi$ is anti-stable. For notational convenience, let $\widehat{\mathcal{L}}^i = \Xi \mathcal{L}^i \Pi$, $i = 1, \ldots, \kappa$. Choose a positive scalar $c_i < \lambda_{\min}^i$, where $\lambda_{\min}^i = \min_{j=1,\ldots,N-1} \text{Re}(\lambda_j(\widehat{\mathcal{L}}^i))$, and $\lambda_j(\widehat{\mathcal{L}}^i)$, $j = 1, \ldots, N-1$, are the eigenvalues of $\widehat{\mathcal{L}}^i$. Then it is easy to verify that there exists a positive definite matrix Q^i such that

$$\left(\widehat{\mathcal{L}}^i \right)^T Q^i + Q^i \widehat{\mathcal{L}}^i > 2c_i Q^i. \tag{3.8}$$

Remark 3.2 *By introducing a linear transformation, the consensus problem for the CNS (3.3) is transformed into the problem of stabilizing globally the switched linear system (3.7). Note that the dynamics of e can be directly obtained when the CNSs have a common leader or each possible topology is strongly connected and balanced [185]. It is also worth noting that the transformation matrix Ξ in (3.7) is not unique [152, 215].*

The following theorem presents the design of the feedback gain matrix and the coupling strength for the protocol (3.2) to achieve consensus tracking by the closed-loop system (3.3).

This theorem summarizes the main theoretical results of this section.

Theorem 3.1 *Suppose that Assumption 3.1 holds and there exists $\beta > 0$ such that the following LMI*

$$AP + PA^T - BB^T + \beta P < 0 \tag{3.9}$$

has a feasible solution $P > 0$. Then, the CNS (3.3) with $K = (1/2)B^T P^{-1}$ achieves consensus if the following conditions hold:

(1) The coupling strength α satisfies $\alpha > 2/c_0$ where $c_0 = \min_{i=1,\ldots,\kappa} c_i$, and c_i, $i = 1, \ldots, \kappa$, are given in (3.8);

(2) *For some $\iota > 0$, the switching interconnection graph $\mathcal{G}^{\sigma(t)}$ satisfies the following condition*

$$\beta(t_{k+1} - t_k) - \ln\overline{\lambda}^k_{\max} > \iota, \tag{3.10}$$

where $\overline{\lambda}^k_{\max}$ is the largest eigenvalue of $\left(Q^{\sigma(t_k)}\right)^{-1} Q^{\sigma(t_{k+1})}$, $k \in \mathbb{N}$.

Proof 3.1 *Construct the following MLFs for the switched systems (3.5):*

$$V(t) = e^T(t) \left(Q^{\sigma(t)} \otimes P^{-1}\right) e(t), \tag{3.11}$$

where $Q^{\sigma(t)} \in \{Q^1, \ldots, Q^\kappa\}$, Q^i, $i = 1, \ldots, \kappa$, are defined in (3.8), $P > 0$ is the solution of LMI (3.9).

Taking the time derivative of $V(t)$ along the trajectories of system (3.5) gives

$$\dot{V}(t) = e^T(t) \left[Q^{\sigma(t)} \otimes \left(P^{-1}A + A^T P^{-1}\right)\right] e(t)$$
$$- 2\alpha e^T(t) \left[\left(Q^{\sigma(t)} \widehat{\mathcal{L}}^{\sigma(t)}\right) \otimes \left(P^{-1}BK\right)\right] e(t) \tag{3.12}$$

for all $t \neq t_k$. Substituting $K = \frac{1}{2}B^T P^{-1}$ into (3.12) yields

$$\dot{V}(t) = e^T(t) \left[Q^{\sigma(t)} \otimes \left(P^{-1}A + A^T P^{-1}\right)\right] e(t)$$
$$- \alpha e^T(t) \left[\left(Q^{\sigma(t)} \widehat{\mathcal{L}}^{\sigma(t)}\right) \otimes \left(P^{-1}BB^T P^{-1}\right)\right] e(t)$$
$$= e^T(t) \left[Q^{\sigma(t)} \otimes \left(P^{-1}A + A^T P^{-1}\right)\right] e(t)$$
$$- \frac{\alpha}{2} e^T(t) \left[\left(Q^{\sigma(t)} \widehat{\mathcal{L}}^{\sigma(t)} + \left(\widehat{\mathcal{L}}^{\sigma(t)}\right)^T Q^{\sigma(t)}\right) \otimes \left(P^{-1}BB^T P^{-1}\right)\right] e(t). \tag{3.13}$$

According to (3.8), it thus follows from (3.13) that

$$\dot{V}(t) \leq e^T(t) \left[Q^{\sigma(t)} \otimes \left(P^{-1}A + A^T P^{-1} - \frac{\alpha c_0}{2} P^{-1}BB^T P^{-1}\right)\right] e(t), \tag{3.14}$$

where $c_0 = \min_{i=1,\ldots,\kappa} c_i$. According to the condition $\alpha > 2/c_0$, one then has

$$\dot{V}(t) \leq e^T(t) \left[Q^{\sigma(t)} \otimes \left(P^{-1}A + A^T P^{-1} - P^{-1}BB^T P^{-1}\right)\right] e(t). \tag{3.15}$$

According to the fact $AP + PA^T - BB^T + \beta P < 0$, one has $P^{-1}A + A^T P^{-1} - P^{-1}BB^T P^{-1} + \beta P^{-1} < 0$. It thus follows from (3.15) that

$$\dot{V}(t) < -\beta e^T(t) \left(Q^{\sigma(t)} \otimes P^{-1}\right) e(t). \tag{3.16}$$

For an arbitrarily given $k \in \mathbb{N}$, one has

$$V(t^-_{k+1}) < \exp(-\beta(t_{k+1} - t_k))V(t_k). \tag{3.17}$$

Since $Q^{\sigma(t)}$ is positive definite for all $t \geq 0$, it can be obtained from Theorem 7.6.3 in [62] that all the eigenvalues of $\left(Q^{\sigma(t_k)}\right)^{-1} Q^{\sigma(t_{k+1})}$ are positive. From the generalized eigenvalue theory, since the topology switches at time points t_k, $k \in \mathbb{N}$, then the following fact holds:

$$V(t_{k+1}) < \overline{\lambda}_{\max}^k V(t_{k+1}^-), \quad \forall \, k \in \mathbb{N},$$

where $\overline{\lambda}_{\max}^k$ is the largest eigenvalue of $\left(Q^{\sigma(t_k)}\right)^{-1} Q^{\sigma(t_{k+1})}$. It can then be derived from the above analysis and (3.17) that

$$V(t_{k+1}) < \overline{\lambda}_{\max}^k V(t_{k+1}^-) < \exp\left(-\beta(t_{k+1} - t_k) + \ln \overline{\lambda}_{\max}^k\right) V(t_k). \tag{3.18}$$

Then it follows from (3.18) and (3.10) that

$$V(t_{k+1}) < \exp(-\iota)V(t_k), \quad \forall \, k \in \mathbb{N}. \tag{3.19}$$

In the sequel, the exponential convergence of $V(t)$ to 0 will be proved by recursion. For $t \in [t_0, t_1)$, we obtain from (3.18) that

$$V(t) < \exp(-\beta t)V(t_0). \tag{3.20}$$

Furthermore, since $t_0 = 0$ and $t_1 < \tau_M$, it is easy to verify that

$$V(t_1) < \exp\left(-\iota\right) V(0) < \exp\left(-\iota t_1/\tau_M\right) V(0). \tag{3.21}$$

Next we consider an arbitrarily given $t > t_1$. For any such $t > t_1$, there exists a positive integer $z \geq 1$ such that $t_z < t \leq t_{z+1}$. When $t \in (t_z, t_{z+1})$, a similar derivation using (3.19) yields

$$V(t) < \exp(-\beta(t - t_z))V(t_z) < \exp\left(-\left[\beta(t - t_z) + (z-1)\iota\right]\right) V(0)$$
$$< \exp\left(-(z-1)\iota\right) V(0) < \exp\left(-\frac{(z-1)\iota}{z\tau_M}t\right) V(0). \tag{3.22}$$

Since $z \geq 1$, it follows from (3.22) that

$$V(t) < \exp\left(-\frac{\iota}{2\tau_M}t\right) V(0), \quad t \in (t_z, t_{z+1}). \tag{3.23}$$

For the case of $t = t_{z+1}$, one also has that

$$V(t_{z+1}) < \exp\left(-\frac{\iota}{\tau_M}t_{z+1}\right) V(0). \tag{3.24}$$

It follows from conditions (3.20)–(3.24) that $\|e(t)\|$ converges to zero exponentially. Thus, the CNS (3.3) indeed achieves consensus.

Remark 3.3 It can be seen that the existence of the protocol (3.2) depends on the feasibility of the LMI (3.9). In the case where the pair (A, B) is controllable, the LMI (3.9) is feasible for any given $\beta > 0$. In the case of stabilizable but not completely controllable (A, B), denote by $\widetilde{\lambda}_i$, $i = 1, \dots, s$, all the uncontrollable modes of (A, B). Then the LMI (3.9) is feasible if and only if $\beta < \min_{i=1,\dots,s}\text{Re}(-\widetilde{\lambda}_i)$.

From Theorem 3.1, we can obtain the following corollary that provides a sufficient condition on the communication topology, in terms of its dwell time, for the system under consideration to achieve consensus.

Corollary 3.1 *Suppose that Assumption 3.1 holds and there exists $\beta > 0$ such that the LMI (3.9) has a feasible solution $P > 0$. Then, the CNS (3.3) with $K = (1/2)B^T P^{-1}$ achieves consensus if the coupling strength α satisfies the condition $\alpha > 2/c_0$ where c_0 was defined in Theorem 3.1, and the dwell time of the switching communication graph $\mathcal{G}^{\sigma(t)}$ satisfies the following condition*

$$\tau_m > \frac{\ln \overline{\lambda}_{max}}{\beta}, \tag{3.25}$$

where $\overline{\lambda}_{max} = \max_{i,j=1,\ldots,\kappa, i \neq j} \widehat{\lambda}_{i,j}$, where $\widehat{\lambda}_{i,j}$ is the largest eigenvalue of $(Q^i)^{-1} Q^j$.

Corollary 3.1 can then be proven by using the reasoning similar to that used in the proof of Theorem 3.1. We leave it as an exercise to the reader.

Remark 3.4 *Compared with Theorem 3.1, the consensus conditions given in Corollary 3.1 are more convenient to use in practical applications since one does not need to check the condition (3.10) for all time intervals. Corollary 3.1 tells that consensus in linear CNS with directed switching topologies with each possible topology containing a directed spanning tree can be achieved if the open-loop agent dynamics are stabilizable and the dwell time is larger than a threshold value given on the right-hand side of (3.25).*

Remark 3.5 *Suppose that (A, B) is controllable, it can be seen from Corollary 3.1 that the consensus problem of the CNS (3.3) with an arbitrarily given dwell time τ_m is solved by the protocol (3.2) designed in Theorem 3.1 with an appropriately selected β.*

Remark 3.6 *Since the underlying topology describing interactions between the agents is time-varying, even though the intrinsic dynamics of each agent are described by linear time-invariant systems, the closed-loop agent dynamics resulting from the application of the switching protocol proposed in the present section are indeed nonlinear. It is thus challenging or even impossible to predict the final state of consensus for such a closed-loop CNS. Clearly, the final consensus value depends on the intrinsic dynamics of each agent, the coupling strength α, the feedback gain matrix K, and the switching mode among different topologies.*

3.1.4 Numerical simulations

Consider the CNS (3.3) consisting of five agents, whose topology switches between the graphs $\mathcal{G}^1$ and $\mathcal{G}^2$ shown in Figure 3.1. For convenience, the weight of each edge is 1. Each agent represents a vertical take-off and landing (VTOL) aircraft. According to [109], the dynamics of the ith VTOL aircraft for a typical loading and

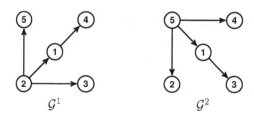

Figure 3.1 The communication graphs $\mathcal{G}^1$ and $\mathcal{G}^2$, where the weight of each edge is 1.

flight condition at the air speed of 135 kt can be described by the system (3.1), with $x_i(t) = [x_{i1}(t), x_{i2}(t), x_{i3}(t), x_{i4}(t)]^T \in \mathbb{R}^4$,

$$A = \begin{bmatrix} -0.0366 & 0.0271 & 0.0188 & -0.4555 \\ 0.0482 & -1.01 & 0.0024 & -4.0208 \\ 0.1002 & 0.3681 & -0.707 & 1.420 \\ 0.0 & 0.0 & 1.0 & 0.0 \end{bmatrix}, B = \begin{bmatrix} 0.4422 & 0.1761 \\ 3.5446 & -7.5922 \\ -5.52 & 4.49 \\ 0.0 & 0.0 \end{bmatrix},$$

where the state variables are defined as: $x_{i1}(t)$ is the horizontal velocity, $x_{i2}(t)$ is the vertical velocity, $x_{i3}(t)$ is the pitch rate, and $x_{i4}(t)$ is the pitch angle [109]. It can be seen from Figure 3.1 that $\mathcal{G}^1$ contains a directed spanning tree with node 2 as the leader, while $\mathcal{G}^2$ contains a directed spanning tree rooted at node 5.

The transformed Laplacian matrices $\widehat{\mathcal{L}}^1$, $\widehat{\mathcal{L}}^2$ in this example are

$$\widehat{\mathcal{L}}^1 = \begin{bmatrix} 1 & 0 & 0 & 0 \\ 0 & 1 & 0 & 0 \\ 0 & 0 & 1 & 0 \\ -1 & 1 & 0 & 1 \end{bmatrix}, \quad \widehat{\mathcal{L}}^2 = \begin{bmatrix} 1 & 0 & 0 & 0 \\ 0 & 1 & 0 & 0 \\ -1 & 0 & 1 & 0 \\ 0 & 0 & 0 & 1 \end{bmatrix}.$$

Set $c_1 = c_2 = 0.5$. Solving the LMI (3.8) gives that $\overline{\lambda}_{\max} = 2.5612$, where $\overline{\lambda}_{\max}$ is defined in Corollary 3.1. Let $\beta = 3$, solving LMI (3.9) gives that

$$K = \begin{bmatrix} 5.8206 & 0.2978 & -0.2615 & -2.7967 \\ -1.1646 & -0.4522 & 0.0530 & 2.0420 \end{bmatrix}.$$

Set $\alpha = 4.1 > 2/c_0 = 4.0$. Then, according to Corollary 3.1, one knows that consensus in the closed-loop CNS (3.3) can be achieved if the dwell time $\tau_m > 0.3135\,\text{s}$. In simulations, let the topology switches between graph $\mathcal{G}^1$ and $\mathcal{G}^2$ every 0.32 s. The state trajectories of the closed-loop CNS (3.3) are shown in Figs. 3.2 and 3.3. The evolution of $\|e(t)\|$ is shown in Figure 3.4, which confirms that the CNS (3.3) achieves consensus.

3.2 DISTRIBUTED CONSENSUS TRACKING FOR GENERAL LINEAR CNSS WITH DIRECTED SWITCHING TOPOLOGIES

3.2.1 Introduction

Despite the recent advances of stability analysis for switched linear systems and the aforementioned results on consensus tracking, there continues to be lack of an efficient approach to analyzing consensus of general linear CNSs with directed switching

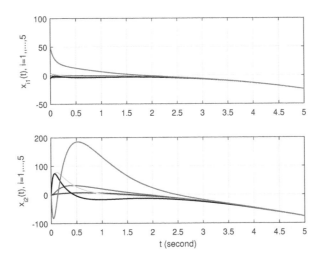

Figure 3.2 Trajectories of $x_{i1}(t)$ and $x_{i2}(t)$, $i = 1, \ldots, 5$.

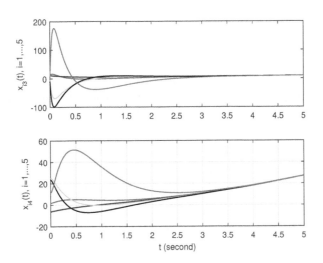

Figure 3.3 Trajectories of $x_{i3}(t)$ and $x_{i4}(t)$, $i = 1, \ldots, 5$.

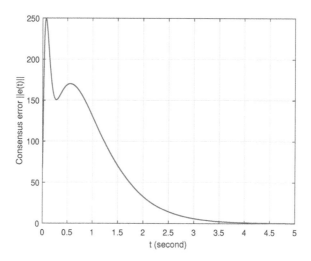

Figure 3.4 Trajectory of the Euclidean norm of consensus error $\|e(t)\|$.

topology, especially in the case with a nonautonomous leader. As shown in [123,188], the analysis approaches for stability of switched systems can not be directly applied to consensus of CNSs with switching topologies. These facts motivate us in this section to develop a new kind of MLFs by using the Lyapunov inequality approach for the tracking error systems of general linear CNSs with directed switching topologies. The primary purpose is to explore the excellent potential of topology-dependent MLFs insightfully, intending to arrive at less conservative criteria for consensus tracking and relax the requirement on topology graphs in terms of containing a directed spanning tree. On the other hand, it is desirable to provide some ADT based criteria on switching signals for consensus of CNSs with directed switching topology. Specifically, first, in the case where each possible topology graph contains a directed spanning tree rooted at the leader node and the dynamics of the leader are described by a linear autonomous system, some efficient criteria on the allowable ADT for reaching consensus tracking in the studied CNSs are derived and analyzed. The MLFs are obtained by performing an iterative optimization algorithm, which is more favorable in calculating the allowable ADT than those constructed using the M-matrix-based approach. The results are then further developed to accommodate the case where the topology graph only frequently contains a directed spanning tree as the CNSs evolve over time. At last, consensus tracking for general linear CNSs with a nonautonomous leader and directed switching topology is studied by developing a new class of nonlinear protocols and constructing a new kind of MLFs.

3.2.2 Model formulation

Consider a CNS consisting of a leader and N followers, where the leader is labelled as agent 0, and the followers are respectively labelled as agents $1, \ldots, N$. The dynamics of agent i, $i = 1, \ldots, N$, are described by

$$\dot{x}_i(t) = Ax_i(t) + Bu_i(t), \tag{3.26}$$

where $x_i(t) \in \mathbb{R}^n$ and $u_i(t) \in \mathbb{R}^m$ are respectively the state and the control input, $A \in \mathbb{R}^{n \times n}$ and $B \in \mathbb{R}^{n \times m}$ are respectively the system and input matrices. Assume that (A, B) is stabilizable. It is further assumed that the leader has no neighbors throughout this section, i.e., the leader's dynamics will not be affected by those of any followers. Then the dynamics of agent 0 are described by

$$\dot{x}_0(t) = Ax_0(t) + Bf(x_0(t), t), \tag{3.27}$$

where $x_0(t) \in \mathbb{R}^n$ is the leader's state, and $f(x_0(t), t) \in \mathbb{R}^m$ is an unknown nonlinear function describing external control inputs acting on the leader.

It is assumed that the communication graph for the $N + 1$ agents is switched within a finite graph set $\hat{\mathcal{G}} = \{\mathcal{G}^1, \ldots, \mathcal{G}^\kappa\}$, $\kappa > 1$ and $\kappa \in \mathbb{N}$. Since the leader has no neighbors, the associated Laplacian matrix can be rewritten as

$$\mathcal{L}^{\sigma(t)} = \begin{bmatrix} 0 & \mathbf{0}_N^T \\ \mathbf{P} & \overline{\mathcal{L}}^{\sigma(t)} \end{bmatrix},$$

$$\overline{\mathcal{L}}^{\sigma(t)} = \begin{bmatrix} \sum_{j \in \mathcal{N}_1} a_{1j}^{\sigma(t)} & -a_{12}^{\sigma(t)} & \cdots & -a_{1N}^{\sigma(t)} \\ -a_{21}^{\sigma(t)} & \sum_{j \in \mathcal{N}_2} a_{2j}^{\sigma(t)} & \cdots & -a_{2N}^{\sigma(t)} \\ \vdots & \vdots & \ddots & \vdots \\ -a_{N1}^{\sigma(t)} & -a_{N2}^{\sigma(t)} & \cdots & \sum_{j \in \mathcal{N}_N} a_{Nj}^{\sigma(t)} \end{bmatrix}, \tag{3.28}$$

where $\mathbf{P} = -[a_{10}^{\sigma(t)}, \ldots, a_{N0}^{\sigma(t)}]^T$, the piecewise constant function $\sigma(t) : [0, +\infty) \mapsto \{1, \ldots, \kappa\}$ represents the switching signal and satisfies the ADT condition (2.23).

Within the context of CNSs, each agent only communicates with its neighbors. Then the relative state information of other agents with respect to agent i is given by

$$\delta_i(t) = \sum_{j=0}^{N} a_{ij}^{\sigma(t)}(x_j(t) - x_i(t)), \quad i = 1, \ldots, N. \tag{3.29}$$

Note that $\delta_i(t)$ is a local information and thereby can be used for the controller design.

3.2.3 Main results for an autonomous leader case

In this subsection, the leader is assumed to be autonomous. That is, the dynamics of the leader are described by (3.27) with $f(x_0(t), t) = \mathbf{0}_m$. To achieve consensus tracking, a distributed protocol is proposed as follows

$$u_i(t) = cK\delta_i(t), \quad i = 1, \ldots, N, \tag{3.30}$$

where $\delta_i(t)$ is given in (3.29), $c > 0$ and $K \in \mathbb{R}^{m \times n}$ are the control parameters to be designed later.

Let $e(t) = [e_1^T(t), \ldots, e_N^T(t)]^T$, where $e_i(t) = x_i(t) - x_0(t)$, $i = 1, \ldots, N$. Combining (3.26), (3.27) together with (3.30) yields

$$\dot{e}(t) = \left[I_N \otimes A - c(\overline{\mathcal{L}}^{\sigma(t)} \otimes BK) \right] e(t). \tag{3.31}$$

Obviously, the consensus tracking problem is solved if and only if $\lim_{t \to +\infty} \| e(t) \| = 0$. That is, consensus tracking in the considered CNSs will be achieved if and only if the zero fixed point of switched systems (3.31) is globally attractive. Throughout the section, the derivatives of all signals at switching time instants should be considered as their right derivatives.

Before moving on, the following Assumption is made.

Assumption 3.2 *For each $i \in \{1, \ldots, \kappa\}$, the directed graph $\mathcal{G}^i$ contains a directed spanning tree rooted at node 0 (i.e., the leader).*

Under Assumption 3.2, it can be obtained from Lemma 2.14 that all the eigenvalues of $\overline{\mathcal{L}}^i$ have positive real parts, i.e., $\overline{\mathcal{L}}^i$ is anti-stable. Thus, the Lyapunov inequalities

$$Q^i \overline{\mathcal{L}}^i + (\overline{\mathcal{L}}^i)^T Q^i > 0 \tag{3.32}$$

are simultaneously feasible for some positive definite matrices Q^i, $i \in \{1, \ldots, \kappa\}$. Since the righ-hand side of (3.32) is homogeneous for Q^i for each $i \in \{1, \ldots, \kappa\}$, one gets that the matrix inequalities

$$\overline{Q}^i \overline{\mathcal{L}}^i + (\overline{\mathcal{L}}^i)^T \overline{Q}^i > 0, \ \ \overline{Q}^i \leq I_N, \ \text{and} \ \overline{Q}^i > 0, \tag{3.33}$$

are simultaneously feasible for some positive definite matrices $\overline{Q}^i$. To arrive at a less conservative estimation for the minimum allowable ADT for achieving consensus tracking, the following optimization algorithm is proposed.

Algorithm 3.1 *Suppose that Assumption 3.2 holds. Let ϵ_0 be a positive scalar such that $\epsilon_0 \ll 1$. Set $\epsilon_s = 0$ for all $s \in \{1, \ldots, \kappa\}$ and $i = 1$.*

(1) Check whether the following LMI: $\overline{\mathcal{L}}^i + (\overline{\mathcal{L}}^i)^T > 0$ holds. If yes, then set $\Phi^i = I_N$ and go to step (3); else, let $\Phi^i = \overline{Q}^i$ and go to step (2), where $\overline{Q}^i$ is a feasible solution of matrix inequalities (3.33).

(2) Check whether the following LMIs: $\Omega^i \overline{\mathcal{L}}^i + (\overline{\mathcal{L}}^i)^T \Omega^i > 0$, $\Omega^i > (\epsilon_i + \epsilon_0)I_N$, and $\Omega^i \leq I_N$ are simultaneously feasible for some positive definite matrices Ω^i. If yes, then solve these LMIs to get some feasible solutions, set $\Phi^i = \Omega^i$, $\epsilon_i = \epsilon_i + \epsilon_0$, and go back to the beginning of step (2); else, go to step (3).

(3) Let $i = i + 1$ and check whether the inequality: $i \leq \kappa$ holds. If yes, then go back to step (1); else, stop.

For notational convenience, let

$$\mu = \max_{i,j \in \{1,\ldots,\kappa\}, i \neq j} \{ \lambda_{\max}(\Phi^i) / \lambda_{\min}(\Phi^j) \}, \tag{3.34}$$

where matrices Φ^k, $k \in \{1, \ldots, \kappa\}$, are determined in Algorithm 3.1. Obviously, $\mu \geq 1$. According to the facts that $\Phi^i > 0$ and $\Phi^i \overline{\mathcal{L}}^i + (\overline{\mathcal{L}}^i)^T \Phi^i > 0$, for each $i \in \{1, \ldots, \kappa\}$, it is known that the eigenvalues of $\overline{\mathcal{L}}^i + (\Phi^i)^{-1}(\overline{\mathcal{L}}^i)^T \Phi^i$ are real and positive. Introduce

$$\underline{\lambda}_0 = \min_{i \in \{1, \ldots, \kappa\}} \{\lambda_{\min}(\overline{\mathcal{L}}^i + (\Phi^i)^{-1}(\overline{\mathcal{L}}^i)^T \Phi^i)\}. \tag{3.35}$$

Based on the above analysis, one may establish the following theorem.

Theorem 3.2 *Suppose that Assumption 3.2 holds and matrix pair (A, B) is stabilizable but not completely controllable with λ_{uc} being the largest real part of its uncontrollable eigenvalues. Then, consensus tracking in the CNSs consisting of followers (3.26) and a single leader (3.27) with $f(x_0(t), t) = \mathbf{0}_m$ will be ensured under protocol (3.30) with $K = B^T P^{-1}$ and $c > \alpha/\underline{\lambda}_0$, if*

$$\tau_a > \tau_{\text{th}}, \tag{3.36}$$

where $\tau_{\text{th}} = (\ln \mu)/\beta$, $P > 0$ and $\alpha > 0$ satisfy the LMI

$$AP + PA^T - \alpha BB^T + \beta P < 0 \tag{3.37}$$

for some $\beta \in (0, -2\lambda_{uc})$.

Proof 3.2 *Choose the following MLFs for the switched systems (3.31):*

$$V(t) = e^T(t)(\Phi^{\sigma(t)} \otimes P^{-1})e(t), \tag{3.38}$$

where $\Phi^{\sigma(t)} \in \{\Phi^1, \ldots, \Phi^\kappa\}$ for all $t \geq 0$, Φ^i, $i \in \{1, \ldots, \kappa\}$, are determined in Algorithm 3.1, and $P > 0$ is defined in (3.37). Taking the time derivative of $V(t)$ along the trajectories of systems (3.31) and invoking $K = B^T P^{-1}$ yield

$$\dot{V}(t) = e^T(t)[\Phi^{\sigma(t)} \otimes (P^{-1}A + A^T P^{-1})]e(t)$$
$$- ce^T(t)\{[\Phi^{\sigma(t)}\overline{\mathcal{L}}^{\sigma(t)} + (\overline{\mathcal{L}}^{\sigma(t)})^T \Phi^{\sigma(t)}] \otimes P^{-1}BB^T P^{-1}\}e(t). \tag{3.39}$$

Since $P^{-1}BB^T P^{-1}$ is positive semi-definite, it can be obtained from (3.39) and Lemma 2.9 that

$$\dot{V}(t) \leq e^T(t)[\Phi^{\sigma(t)} \otimes (P^{-1}A + A^T P^{-1} - c\underline{\lambda}_0 P^{-1}BB^T P^{-1})]e(t) < -\beta V(t), \tag{3.40}$$

where the last inequality is derived using the inequality (3.37) and the condition $c > \alpha/\underline{\lambda}_0$. For an arbitrarily given $\bar{t} > 0$, let $N_\sigma(0, \bar{t})$ be the number of discontinuities of switching signal $\sigma(t)$ over time interval $(0, \bar{t})$. It can be obtained from (3.40) that $V(\bar{t}) < \exp(-\beta\bar{t})V(0)$ when $N_\sigma(0, \bar{t}) = 0$. Suppose that there is at least one switching over time interval $(0, \bar{t})$, then let t_i, $i = 1, \ldots, N_\sigma(0, \bar{t})$, be the switching time points of $\sigma(t)$ over time interval $(0, \bar{t})$ with $t_1 < \ldots < t_{N_\sigma(0, \bar{t})}$. For the case with $N_\sigma(0, \bar{t}) = 1$, from (3.40) and the fact $V(t_i) \leq \mu V(t_i^-)$ it can be derived that

$V(\bar{t}) < \mu \exp(-\beta \bar{t})V(0)$. *For the case with* $N_\sigma(0, \bar{t}) \geq 2$, *it follows from (3.40) that*

$$V(\bar{t}) < \exp\left(-\beta(\bar{t} - t_{N_\sigma(0,\bar{t})})\right) V(t_{N_\sigma(0,\bar{t})})$$
$$< \mu \exp\left(-\beta(\bar{t} - t_{N_\sigma(0,\bar{t})-1})\right) V(t_{N_\sigma(0,\bar{t})-1}).$$

It can thus be derived by induction that

$$V(\bar{t}) < \mu^{N_\sigma(0,\bar{t})} \exp\left(-\beta \bar{t}\right) V(0). \tag{3.41}$$

By Definition 2.4, it can be obtained from (3.41) that

$$V(\bar{t}) < M_0 \exp\left(-(\beta - \frac{\ln \mu}{\tau_a})\bar{t}\right), \tag{3.42}$$

where $M_0 = V(0) \cdot \exp(N_0 \ln \mu)$. *Moreover, the condition* $\tau_a > (\ln \mu)/\beta$ *implies* $\beta > (\ln \mu)/\tau_a$. *Since* $\bar{t}$ *is arbitrarily chosen, it follows from (3.42) that consensus tracking in the considered CNSs is achieved.*

Remark 3.7 *It should be noted that, under the assumption that the matrix pair* (A, B) *is stabilizable but not completely controllable, LMI (3.37) is always feasible for some positive definite* $P > 0$ *as long as* $\beta < -2\lambda_{uc}$ *with* λ_{uc} *being the maximal real part of its uncontrollable eigenvalues. According to (3.36), it is known that the lower bound of the allowable ADT for consensus tracking is given by* $(\ln \mu)/(-2\lambda_{uc})$. *Here,* λ_{uc} *is determined by the intrinsic dynamics of followers. According to (3.34), one knows* $\mu \geq 1$. *Noticeably, the smaller the value of* μ, *the lower the minimum allowable ADT will be attained, which, in turn, reduces the conservatism of the criteria provided in Theorem 3.2. To this end, Algorithm 3.1 is given to select matrices* Φ^i, $i \in \{1, \ldots, \kappa\}$, *in* $V(t)$. *Suppose that Assumption 3.2 holds and the matrix pair* (A, B) *is completely controllable. Then, for any given* $\beta > 0$, *LMI (3.37) is always feasible for some* $P > 0$. *This indicates that, under Assumption 3.2 and the condition that* (A, B) *is completely controllable, consensus tracking in the considered CNSs with an arbitrarily given switching signal* $\sigma(t)$ *can be realized if the control parameters* c *and* K *are appropriately designed.*

Prior to the Lyapunov inequality based criteria provided in Theorem 3.2, the M-matrix based criteria on consensus tracking of CNSs with fixed and switching topologies have been derived in [185, 188, 225]. However, as pointed out in [81, 226], there is a flaw in those works due to the improper choice of the diagonal matrix when constructing topology-dependent quadratic Lyapunov functions for the tracking error systems. Under Assumption 3.2, it can be obtained from Lemma 2.15 that for each $t \geq 0$, $\overline{\mathcal{L}}^{\sigma(t)}$ in (3.28) is a nonsingular M-matrix. Since $\sigma(t) \in \{1, \ldots, \kappa\}$, one has $\overline{\mathcal{L}}^{\sigma(t)} \in \{\overline{\mathcal{L}}^1, \ldots, \overline{\mathcal{L}}^\kappa\}$ for each $t \geq 0$. Furthermore, it can be obtained from Theorem 1 in [226] that, for each $i \in \{1, \ldots, \kappa\}$, there exist positive definite diagonal matrices $\Theta^i = \text{diag}\{\theta_1^i, \ldots, \theta_N^i\}$, $i \in \{1, \ldots, \kappa\}$, such that

$$\Theta^i \overline{\mathcal{L}}^i + (\overline{\mathcal{L}}^i)^T \Theta^i > 0, \tag{3.43}$$

where $\theta_j^i = \widehat{\theta}_j^i/\widehat{\theta}_j^i$, $j \in \{1,\ldots,N\}$, $\widehat{\theta}^i = [\widehat{\theta}_1^i,\ldots,\widehat{\theta}_N^i]^T = (\overline{\mathcal{L}}^i)^{-T}\mathbf{1}_N$ and $\widetilde{\theta}^i = [\widehat{\theta}_1^i,\ldots,\widehat{\theta}_N^i]^T = (\overline{\mathcal{L}}^i)^{-1}\mathbf{1}_N$. Introduce

$$\widetilde{\kappa}_0 = \max_{i,j\in\{1,\ldots,\kappa\},i\neq j}\{\overline{\theta}^i/\underline{\theta}^j\},$$
$$\widetilde{\lambda}_0 = \min_{i\in\{1,\ldots,\kappa\}}\{\lambda_{\min}(\overline{\mathcal{L}}^i + (\Theta^i)^{-1}(\overline{\mathcal{L}}^i)^T\Theta^i)\}, \tag{3.44}$$

where $\overline{\theta}^i = \max_{j\in\{1,\ldots,N\}}\{\theta_j^i\}$, $\underline{\theta}^j = \min_{s\in\{1,\ldots,N\}}\{\theta_s^j\}$, $\Theta^i = [\theta_1^i,\ldots,\theta_N^i]^T$ is defined in (3.43). Based on the above analysis, the following theorem can be established.

Theorem 3.3 *Suppose that Assumption 3.2 holds and matrix pair (A, B) is stabilizable but not completely controllable with λ_{uc} being the largest real part of its uncontrollable eigenvalues. Then, consensus tracking in the CNSs consisting of followers (3.26) and a single leader (3.27) with $f(x_0(t),t) = \mathbf{0}_m$ will be ensured under protocol (3.30) with $K = B^T P^{-1}$ and $c > \alpha/\widetilde{\lambda}_0$, if*

$$\tau_a > \widetilde{\tau}_{th}, \tag{3.45}$$

where $\widetilde{\tau}_{th} = (\ln\widetilde{\kappa}_0)/\beta$, $\widetilde{\kappa}_0$ is defined in (3.44), the positive scalars α, β, and matrix P are defined as the same as those in Theorem 3.2, respectively.

Proof 3.3 *Construct the following MLFs for switched systems (3.31):*

$$V(t) = e^T(t)(\Theta^{\sigma(t)} \otimes P^{-1})e(t), \tag{3.46}$$

where $\Theta^{\sigma(t)} \in \{\Theta^1,\ldots,\Theta^\kappa\}$ for all $t \geq 0$, Θ^i, $i \in \{1,\ldots,\kappa\}$, are defined in (3.43), and $P > 0$ is defined as the same as that in Theorem 3.2. Thus, this theorem can be proven by following the steps in the proof of Theorem 3.2.

Noticeably, the lower bounds of the allowable ADT provided in Theorems 3.3 and 3.2 are different. Note also that the matrices of the quadratic MLFs (3.46) are assumed to be block diagonal while the ones in (3.38) do not need this hypothetical condition. Next, it will be theoretically verified that, compared with the criteria provided in Theorem 3.3, those given in Theorem 3.2 associated with an appropriately selected ϵ_0 in Algorithm 3.1 are less conservative. Let

$$\overline{\Theta}^i = \frac{\Theta^i}{\theta_{\max}}, \tag{3.47}$$

in which $\theta_{\max} = \max_{i\in\{1,\ldots,\kappa\},j\in\{1,\ldots,N\}}\{\theta_j^i\}$. According to (3.43), one may get that

$$\overline{\Theta}^i\overline{\mathcal{L}}^i + (\overline{\mathcal{L}}^i)^T\overline{\Theta}^i > 0,$$

$\overline{\Theta}^i > 0$, and $\overline{\Theta}^i \leq I_N$. Since Φ^i in (3.38) are determined by Algorithm 3.1 where the positive scalar ϵ_0 can be chosen as small as needed, one may get that $\mu \leq \widetilde{\kappa}_0$, where μ and $\widetilde{\kappa}_0$ are defined respectively in (3.34) and (3.44). Specifically, one may set

$\epsilon_0 = \overline{\theta}_{\min}/N_c$ in the above analysis, where $\overline{\theta}_{\min} = \min_{i\in\{1,\ldots,\kappa\},j\in\{1,\ldots,N\}}\{\theta^i_j/\theta_{\max}\}$ and N_c is an arbitrarily given positive integer. The larger the N_c, the less the conservatism. The above analysis indicates that, for a given CNS, the minimal allowable ADT for achieving consensus provided in Theorem 3.2 is less conservative than that given in Theorem 3.3.

Based upon Assumption 3.2, a few sufficient criteria on consensus tracking for CNSs with directed switching topologies are respectively provided in Theorems 3.2 and 3.3. However, Assumption 3.2 is a bit strong in some practical cases, as some possible topology graphs perhaps do not contain any directed spanning tree. Next, consensus tracking for general linear CNSs with directed switching topologies is further investigated when some possible topology graphs do not contain any directed spanning tree.

Assumption 3.3 *Suppose that $\widehat{\mathcal{G}} = \widehat{\mathcal{G}}_{\mathrm{ct}} \cup \widehat{\mathcal{G}}_{\mathrm{dt}}$ with $\widehat{\mathcal{G}}_{\mathrm{ct}} = \{\mathcal{G}^1,\ldots,\mathcal{G}^{\tilde{\kappa}}\}$ and $\widehat{\mathcal{G}}_{\mathrm{dt}} = \{\mathcal{G}^{\tilde{\kappa}+1},\ldots,\mathcal{G}^{\kappa}\}$, $\tilde{\kappa} \leq \kappa$, $\tilde{\kappa} \in \mathbb{N}$, where $\mathcal{G}^i$ contains a directed spanning tree for each $i \in \Upsilon_{\mathrm{ct}} = \{1,\ldots,\tilde{\kappa}\}$ and $\mathcal{G}^j$ does not contain any directed spanning tree for each $j \in \Upsilon_{\mathrm{dt}} = \{\tilde{\kappa}+1,\ldots,\kappa\}$.*

Let $T^{\mathrm{ct}}_{t_0}(t)$ and $T^{\mathrm{dt}}_{t_0}(t)$ respectively be the total time lengths of activation time for $\sigma(s) \in \Upsilon_{\mathrm{ct}}$ and $\sigma(s) \in \Upsilon_{\mathrm{dt}}$ for $s \in (t_0, t)$ with $t \geq t_0 \geq 0$. To facilitate the subsequent analysis, the following assumption is made.

Assumption 3.4 *There exist two nonnegative scalars δ_0 and t_0 such that $T^{\mathrm{dt}}_{t_0}(t) \leq \delta_0 T^{\mathrm{ct}}_{t_0}(t)$ for all $t \geq t_0$.*

According to (3.28), it can be obtained from Algorithm 3.1 that, for each $i \in \Upsilon_{\mathrm{ct}}$, there exists $\widehat{\Phi}^i > 0$ such that

$$\widehat{\Phi}^i\overline{\mathcal{L}}^i + (\overline{\mathcal{L}}^i)^T\widehat{\Phi}^i > 0 \tag{3.48}$$

and $\widehat{\Phi}^i \leq I_N$. Let

$$\widehat{\kappa}_0 = \begin{cases} \max\limits_{i,j\in\Upsilon_{\mathrm{ct}};i\neq j}\{\overline{\mu}^i/\underline{\mu}^j\}, & \text{if } \tilde{\kappa} > 1, \\ 1, & \text{if } \tilde{\kappa} = 1, \end{cases} \tag{3.49}$$

where $\overline{\mu}^i = \lambda_{\max}(\widehat{\Phi}^i)$, $\underline{\mu}^j = \lambda_{\min}(\widehat{\Phi}^i)$, and matrices $\widehat{\Phi}^i$, $i \in \Upsilon_{\mathrm{ct}}$, are given in (3.48). Furthermore, let

$$\widehat{\lambda}_0 = \min_{i\in\Upsilon_{\mathrm{ct}}}\{\lambda_{\min}(\overline{\mathcal{L}}^i + (\widehat{\Phi}^i)^{-1}(\overline{\mathcal{L}}^i)^T\widehat{\Phi}^i)\},$$
$$\widehat{\widehat{\lambda}}_0 = \min_{j\in\Upsilon_{\mathrm{dt}}}\{\lambda_{\min}(\overline{\mathcal{L}}^j + (\Psi)^{-1}(\overline{\mathcal{L}}^j)^T\Psi)\}, \tag{3.50}$$

where $\Psi = (1/\tilde{\kappa})\sum_{i=1}^{\tilde{p}}\widehat{\Phi}^i$. Obviously, $\Psi > 0$ and $\Psi \leq \widehat{\kappa}_0\widehat{\Phi}^i$, for each $i \in \Upsilon_{\mathrm{ct}}$, where $\widehat{\kappa}_0$ is defined by (3.49).

Theorem 3.4 *Suppose that Assumptions 3.3 and 3.4 hold, and matrix pair (A, B) is stabilizable but not completely controllable with λ_{uc} being the largest real part of its uncontrollable eigenvalues. Then, consensus tracking in the CNSs consisting of* followers (3.26) and a single leader (3.27) with $f(x_0(t), t) = \mathbf{0}_m$ will be ensured under protocol (3.30) with $K = B^T P^{-1}$ and $c > \alpha/\widehat{\lambda}_0$, if

(1) $\tau_a > \widehat{\tau}_{\text{th}}$,

$$(2) \ \delta_0 < \frac{\beta - (\ln \widehat{\kappa}_0)/\tau_a}{\gamma + (\ln \widehat{\kappa}_0)/\tau_a}, \tag{3.51}$$

where δ_0 is defined in Assumption 3.4, $\widehat{\tau}_{\text{th}} = (\ln \widehat{\kappa}_0)/\beta$, $\gamma \geq 0$ and $P > 0$ are solutions of the following optimization problem:

$$\text{minimize } \gamma \geq 0,$$
$$\text{s.t. } \begin{cases} AP + PA^T - \alpha BB^T + \beta P < 0, \\ AP + PA^T - c\widehat{\lambda}_0 BB^T - \gamma P < 0, \end{cases} \tag{3.52}$$

for some $\beta \in (0, -2\lambda_{uc})$ and $\alpha > 0$.

Proof 3.4 *For brevity of expression, introduce the following piecewise constant function of time $\overline{\sigma}(t) : [t_0, +\infty) \mapsto \{1, \ldots, \tilde{\kappa} + 1\}$ such that $\overline{\sigma}(t) = \sigma(t)$ when $\sigma(t) \in \Upsilon_{\text{ct}}$ and $\overline{\sigma}(t) = \tilde{\kappa} + 1$ otherwise. Moreover, let $\widehat{\Phi}^{\tilde{\kappa}+1} = \Psi$. For $t \geq t_0$, construct the following MLFs for the switched systems (3.31):*

$$V(t) = e^T(t)(\widehat{\Phi}^{\overline{\sigma}(t)} \otimes P^{-1})e(t), \tag{3.53}$$

where $P > 0$ is defined in (3.52). Note that for any given $t \geq t_0$, there exists a positive integer k such that $t \in [t_{k-1}, t_k)$.

For the case with $\sigma(t_{k-1}) \in \Upsilon_{\text{ct}}$, the following inequality can be derived by using some similar analysis to that in (3.39)–(3.40):

$$V(t) \leq \exp\left(-\beta(t - t_{k-1})\right) V(t_{k-1}), \ t \in [t_{k-1}, t_k). \tag{3.54}$$

For the case with $\sigma(t_{k-1}) \in \Upsilon_{\text{dt}}$, one gets

$$\dot{V}(t) = e^T(t)\left[\Psi \otimes \left(P^{-1}A + A^T P^{-1}\right)\right]e(t)$$
$$- ce^T(t)\left\{[\Psi\overline{\mathcal{L}}^{\sigma(t)} + (\overline{\mathcal{L}}^{\sigma(t)})^T\Psi] \otimes P^{-1}BB^T P^{-1}\right\}e(t) \tag{3.55}$$

for $t \in [t_{k-1}, t_k)$. Since $P^{-1}BB^T P^{-1}$ is positive semi-definite, it can be derived from Lemma 2.9 that

$$- ce^T(t)\left\{[\Psi\overline{\mathcal{L}}^{\sigma(t)} + (\overline{\mathcal{L}}^{\sigma(t)})^T\Psi] \otimes P^{-1}BB^T P^{-1}\right\}e(t)$$
$$\leq -\widehat{\lambda}_0 ce^T(t)(\Psi \otimes P^{-1}BB^T P^{-1})e(t),$$

where $\widehat{\lambda}_0$ is defined in (3.50). This together with (3.55) yields

$$\dot{V}(t) \leq e^T(t)\left[\Psi \otimes (P^{-1}A + A^T P^{-1} - c\widehat{\lambda}_0 P^{-1}BB^T P^{-1})\right]e(t),$$
$$< \gamma V(t), \ t \in [t_{k-1}, t_k), \tag{3.56}$$

where the last inequality is obtained by using (3.52).

According to (3.53), one knows that $V(t_k) \leq \widehat{\kappa}_0 \cdot \lim_{t \nearrow t_k} V(t)$ for all $k \in \mathbb{N}$, where $\widehat{\kappa}_0$ is defined in (3.49). By Definition 2.4, one gets that $N_\sigma(t_0, t) \leq N_0 + (t - t_0)/\tau_a$ for all $t \geq t_0$. Based on the above analysis, combining (3.54) and (3.56) together yields

$$V(t) \leq \widehat{\kappa}_0^{[N_0 + (t - t_0)/\tau_a]} \exp\left(\gamma T_{t_0}^{\mathrm{dt}}(t) - \beta T_{t_0}^{\mathrm{ct}}(t)\right) V(t_0), \tag{3.57}$$

for any given $t \geq t_0$. By Assumption 3.4 and (3.51), it can be obtained from (3.57) that

$$V(t) \leq M_1 \exp\left(-\varsigma T_{t_0}^{\mathrm{ct}}(t)\right) \leq M_1 \exp\left(-\frac{\varsigma}{1 + \delta_0}(t - t_0)\right), \tag{3.58}$$

for some $\varsigma > 0$, where $M_1 = V(t_0) \cdot \exp\left(N_0 \ln \widehat{\kappa}_0\right)$. Thus, it follows from (3.58) that consensus tracking in the considered CNSs is achieved.

3.2.4   Main results for a nonautonomous leader case

In this subsection, the leader is assumed to be a nonautonomous agent. To achieve consensus tracking, a distributed protocol is proposed as follows

$$u_i(t) = d_1 F \delta_i(t) + d_2 \mathbf{sgn}(F \delta_i(t)), \quad i = 1, \ldots, N, \tag{3.59}$$

where $\delta_i(t)$ is given in (3.29), $d_1 > 0$, $d_2 > 0$, and $F \in \mathbb{R}^{m \times n}$ are the control parameters to be designed later, $\mathbf{sgn}(\cdot)$ denotes the element-wise sign function. For brevity, let $e(t) = [e_1^T(t), \ldots, e_N^T(t)]^T$ with $e_i(t) = x_i(t) - x_0(t)$. The tracking error system for the CNS (3.26) under protocol (3.59) with a nonautonomous leader (3.27) can be found to be

$$\dot{e}(t) = \left[I_N \otimes A - d_1\left(\overline{\mathcal{L}}^{\sigma(t)} \otimes BF\right)\right] e(t)$$
$$- d_2(I_N \otimes B) \cdot \mathbf{sgn}\left(\left(\overline{\mathcal{L}}^{\sigma(t)} \otimes F\right)e(t)\right) - (\mathbf{1}_N \otimes B)f(x_0(t), t), \tag{3.60}$$

where $\overline{\mathcal{L}}^{\sigma(t)}$ is given in (3.28). Note that the subsequent analysis is performed based on Assumption 3.2 and the following assumption.

Assumption 3.5 *There exists a positive scalar d_0 such that $\|f(x_0(t), t)\|_\infty \leq d_0$ for all $t \geq 0$.*

It is worth noticing that Assumption 3.5 provides an assurance preventing the actuators from blowing up physically. Note also that the explicit form of nonlinear function $f(x_0(t), t)$ is unknown to any follower. For notational brevity, let

$$\overline{\kappa}_0 = \max_{i,j \in \{1, \ldots, \kappa\}, i \neq j} \{\overline{\chi}^i / \underline{\chi}^j\}, \tag{3.61}$$

where $\overline{\chi}^i = \lambda_{\max}((\overline{\mathcal{L}}^i)^T \Phi^i \overline{\mathcal{L}}^i)$, $\underline{\chi}^j = \lambda_{\min}((\overline{\mathcal{L}}^i)^T \Phi^i \overline{\mathcal{L}}^i)$, and matrices Φ^i, $i \in \{1, \ldots, \kappa\}$, are determined in Algorithm 3.1 by restricting $\overline{Q}^i$ and Ω^i to be positive definite and diagonal matrices. In this case, one knows that Φ^i, $i \in \{1, \ldots, \kappa\}$, are all positive definite and diagonal matrices. Obviously, $\overline{\kappa}_0 \geq 1$. Furthermore, introduce

$$\chi_0 = \min_{i \in \{1, \ldots, \kappa\}} \{\lambda_{\min}(\overline{\mathcal{L}}^i + (\Phi^i)^{-1}(\overline{\mathcal{L}}^i)^T \Phi^i)\}. \tag{3.62}$$

Theorem 3.5 *Suppose that Assumptions 3.2 and 3.5 hold, and matrix pair (A, B) is stabilizable but not completely controllable with λ_{uc} being the largest real part of its uncontrollable eigenvalues. Then, consensus tracking in the CNS consisting of followers (3.26) and a nonautonomous leader (3.27) will be ensured under protocol (3.59) with $F = B^T P^{-1}$, $d_1 > \alpha/\chi_0$, and $d_2 \geq d_0$, if*

$$\tau_a > \bar{\tau}_{\text{th}}, \tag{3.63}$$

where $\bar{\tau}_{\text{th}} = (\ln \bar{\kappa}_0)/\beta$, $P > 0$ and $\alpha > 0$ satisfy the LMI

$$AP + PA^T - \alpha BB^T + \beta P < 0 \tag{3.64}$$

for some $\beta \in (0, -2\lambda_{uc})$, and χ_0 is defined in (3.62).

Proof 3.5 *Under Assumption 3.2, one gets that $\overline{\mathcal{L}}^{\sigma(t)}$ is anti-stable and thus non-singular. Then, we choose the following MLFs for the switched systems (3.60):*

$$V(t) = e^T(t)\left[\left(\left(\overline{\mathcal{L}}^{\sigma(t)}\right)^T \Phi^{\sigma(t)} \overline{\mathcal{L}}^{\sigma(t)}\right) \otimes P^{-1}\right]e(t), \tag{3.65}$$

where $\Phi^{\sigma(t)} \in \{\Phi^1, \ldots, \Phi^\kappa\}$ for all $t \geq 0$, Φ^i, $i \in \{1, \ldots, \kappa\}$, are determined by Algorithm 3.1 by restricting $\overline{Q}^i$ and Ω^i to be positive definite and diagonal matrices, and $P > 0$ is defined in (3.64). Taking the time derivative of $V(t)$ along the trajectories of systems (3.60) and invoking $F = B^T P^{-1}$ yield

$$\begin{aligned}
\dot{V}(t) =\ & \tilde{e}^T(t)\left[\Phi^{\sigma(t)} \otimes \left(P^{-1}A + A^T P^{-1}\right)\right]\tilde{e}(t) \\
& - d_1 \tilde{e}^T(t)\left\{\left[\Phi^{\sigma(t)}\overline{\mathcal{L}}^{\sigma(t)} + (\overline{\mathcal{L}}^{\sigma(t)})^T \Phi^{\sigma(t)}\right] \otimes P^{-1}BB^T P^{-1}\right\}\tilde{e}(t) \\
& - 2d_2 \tilde{e}^T(t)(\Phi^{\sigma(t)}\widehat{\mathcal{L}}^{\sigma(t)} \otimes P^{-1}B)\mathbf{sgn}((I_N \otimes B^T P^{-1})\tilde{e}(t)) \\
& - 2d_2 \tilde{e}^T(t)(\Phi^{\sigma(t)}\text{diag}\{\mathbf{P}^{\sigma(t)}\} \otimes P^{-1}B)\mathbf{sgn}((I_N \otimes B^T P^{-1})\tilde{e}(t)) \\
& - 2\tilde{e}^T(t)(\Phi^{\sigma(t)}\text{diag}\{\mathbf{P}^{\sigma(t)}\}\mathbf{1}_N \otimes P^{-1}B)f(x_0(t), t), \tag{3.66}
\end{aligned}$$

where $\tilde{e}(t) = (\overline{\mathcal{L}}^{\sigma(t)} \otimes I_n)e(t) = [\tilde{e}_1^T(t), \ldots, \tilde{e}_N^T(t)]^T$ with $\tilde{e}_i(t) \in \mathbb{R}^n$ for each $i = 1, \ldots, N$, $\widehat{\mathcal{L}}^{\sigma(t)} = \overline{\mathcal{L}}^{\sigma(t)} + \text{diag}\{\mathbf{P}^{\sigma(t)}\}$, and $\mathbf{P}^{\sigma(t)}$ is defined in (3.28). Since the off-diagonal elements of $\Phi^{\sigma(t)}\widehat{\mathcal{L}}^{\sigma(t)}$ are non-positive and the sum of elements in each row of this matrix is equal to 0, we obtain $\tilde{e}^T(t)(\Phi^{\sigma(t)}\widehat{\mathcal{L}}^{\sigma(t)} \otimes P^{-1}B)\mathbf{sgn}((I_N \otimes B^T P^{-1})\tilde{e}(t)) = \tilde{e}^T(t)(I_N \otimes P^{-1}B)(\Phi^{\sigma(t)}\widehat{\mathcal{L}}^{\sigma(t)} \otimes I)\mathbf{sgn}((I_N \otimes B^T P^{-1})\tilde{e}(t)) \geq 0$. Without loss of generality, let $\Phi^{\sigma(t)} = \text{diag}\{\phi_1^{\sigma(t)}, \ldots, \phi_N^{\sigma(t)}\}$. Since $\|a\|_1 = a \cdot \mathbf{sgn}(a)$, we have

$$\begin{aligned}
& 2d_2 \tilde{e}^T(t)(\Phi^{\sigma(t)}\text{diag}\{\mathbf{P}^{\sigma(t)}\} \otimes P^{-1}B)\mathbf{sgn}((I_N \otimes B^T P^{-1})\tilde{e}(t)) \\
& = 2d_2 \tilde{e}^T(t)(I_N \otimes P^{-1}B)(\Phi^{\sigma(t)}\text{diag}\{\mathbf{P}^{\sigma(t)}\} \otimes I_n)\mathbf{sgn}((I_N \otimes B^T P^{-1})\tilde{e}(t)) \\
& = 2d_2 \sum\nolimits_{i=1}^N a_{i0}^{\sigma(t)}\phi_i^{\sigma(t)}\|B^T P^{-1}\tilde{e}_i(t)\|_1.
\end{aligned}$$

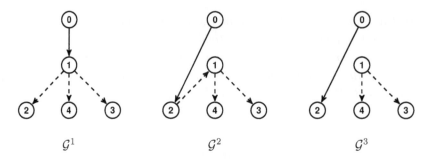

Figure 3.5 The communication graphs $\mathcal{G}^1$, $\mathcal{G}^2$, and $\mathcal{G}^3$. The solid lines indicate the pinning links from the leader to the follower agent.

In addition, according to the Hölder's inequality, we have

$$- 2\tilde{e}^T(t)(\Phi^{\sigma(t)}\text{diag}\{\mathbf{P}^{\sigma(t)}\}\mathbf{1}_N\otimes P^{-1}B)f(x_0(t),t)$$

$$\leq 2\|\tilde{e}^T(t)(\Phi^{\sigma(t)}\text{diag}\{\mathbf{P}^{\sigma(t)}\}\mathbf{1}_N\otimes P^{-1}B)\|_1 \cdot \|f(x_0(t),t)\|_\infty$$

$$\leq 2d_0 \sum_{i=1}^{N} a_{i0}^{\sigma(t)}\phi_i^{\sigma(t)}\|B^T P^{-1}\tilde{e}_i(t)\|_1,$$

where the last inequality is derived based on Assumption 3.5. According to the above analysis and the condition $d_2 \geq d_0$, we have

$$\dot{V}(t) \leq \tilde{e}^T(t)[\Phi^{\sigma(t)} \otimes \left(P^{-1}A+A^T P^{-1}-d_1\chi_0 P^{-1}BB^T P^{-1}\right)]\tilde{e}(t)$$

$$< -\beta V(t),$$

where the last inequality is obtained by using $d_1 > \alpha/\chi_0$ and LMI (3.64). Thus, this theorem can be proven by following the steps in the proof of Theorem 3.2.

3.2.5 Numerical simulations

Consider the consensus tracking problem of CNS with followers' dynamics given by (3.26) and leader's dynamics given by (3.27). Figure 3.5 indicates three possible switching topologies of the considered CNS, where topology $\mathcal{G}^1$ and $\mathcal{G}^2$ both contain a directed spanning trees with the leader agent being the root, while no spanning tree is involved in $\mathcal{G}^3$. The associated Laplacian matrix among the followers $\overline{\mathcal{L}}^{\sigma(t)}$ are given by

$$\overline{\mathcal{L}}^1 = \begin{bmatrix} 1 & 0 & 0 & 0 \\ -1 & 1 & 0 & 0 \\ -1 & 0 & 1 & 0 \\ -1 & 0 & 0 & 1 \end{bmatrix}, \quad \overline{\mathcal{L}}^2 = \begin{bmatrix} 1 & -1 & 0 & 0 \\ 0 & 1 & 0 & 0 \\ -1 & 0 & 1 & 0 \\ -1 & 0 & 0 & 1 \end{bmatrix}, \quad \overline{\mathcal{L}}^3 = \begin{bmatrix} 0 & 0 & 0 & 0 \\ 0 & 1 & 0 & 0 \\ -1 & 0 & 1 & 0 \\ -1 & 0 & 0 & 1 \end{bmatrix}.$$

Suppose that the states of agent i in the CNS are represented by $x_i(t) = [x_{i1}(t), x_{i2}(t)]^T \in \mathbb{R}^2$, and their system's matrices are given by

$$A = \begin{bmatrix} -4 & 1 \\ 0 & -1 \end{bmatrix}, \quad B = \begin{bmatrix} 1 \\ 0 \end{bmatrix}.$$

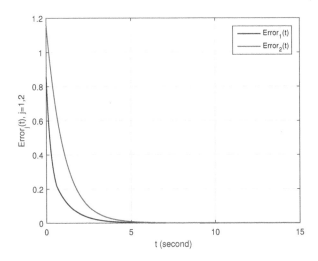

Figure 3.6 Trajectories of the consensus tracking errors $\text{Error}_1(t)$ and $\text{Error}_2(t)$ in Case 1.

It could be checked that $\text{rank}[B\ AB] < 2$, thus matrix pair (A, B) is not controllable. However, the system is stabilizable with the uncontrollable eigenvalue $\lambda_{uc} = -1$. Next, three typical cases will be investigated corresponding to the results derived in Theorems 3.3, 3.4, and 3.5, respectively.

$Case\ 1$: Suppose that the communication topology of the CNS switches between graphs $\mathcal{G}^1$ and $\mathcal{G}^2$, and $f(x_0(t), t) = 0$. It could be calculated that $\tilde{\kappa}_0 = 4$ and $\tilde{\lambda}_0 = 0.8098$. Then, allowable τ_a could be determined with $\tau_a > \tilde{\tau}_{\text{th}} = 1.3863$, where we select $\beta = 1 \in (0, 2)$. By solving LMI (3.37) and selecting proper $c = \alpha/\tilde{\lambda}_0 + 1$, one could get

$$cBK = \begin{bmatrix} 0.2512 & -0.0628 \\ 0 & 0 \end{bmatrix}.$$

Then, according to Theorem 3.3, one knows that consensus tracking can be achieved if the dwell time is longer than 1.3865 s. In this simulations, we set $\tau_a = 1.5$ s. Define the tracking errors $\text{Error}_j(t) = \sum_{i=1}^{4} \|x_{ij}(t) - x_{0j}(t)\|$, $j = 1, 2$. As can be seen in Fig. 3.6, both $\text{Error}_1(t)$ and $\text{Error}_2(t)$ approach to zero as time tends to infinity, which validates the effectiveness of Theorem 3.3.

$Case\ 2$: Suppose that the communication topology of the CNS switches between graphs $\mathcal{G}^1$, $\mathcal{G}^2$, and $\mathcal{G}^3$, and $f(x_0(t), t) = 0$. It could be calculated from (3.50) that $\tilde{\lambda}_0 = 0.2679$, and $\hat{\lambda}_0 = -0.7321$. Select $\alpha = 1$ and $\beta = 1$, then the allowable τ_a could be selected as $\tau_a = 1.5$ s. Solve the optimization problem (3.52) to get $\gamma = 0$ and

$$cBK = \begin{bmatrix} 0.701 & -0.153 \\ 0 & 0 \end{bmatrix}.$$

Thus, parameter δ_0 could be calculated as $\delta_0 < 0.0820$. We simulate graph $\mathcal{G}^3$ appears only once in every 16 switching times thus $\delta_0 < 0.0820$ could be satisfied. As shown in Fig. 3.7, both $\text{Error}_1(t)$ and $\text{Error}_2(t)$ approach to zero as time approaches infinity, which validates the effectiveness of Theorem 3.4.

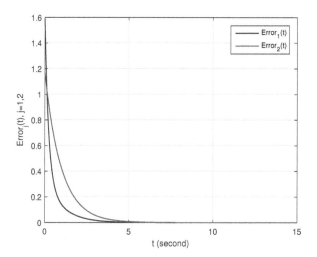

Figure 3.7 Trajectories of the consensus tracking errors $Error_1(t)$ and $Error_2(t)$ in Case 2.

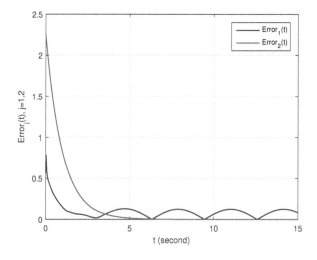

Figure 3.8 Trajectories of the consensus tracking errors $Error_1(t)$ and $Error_2(t)$ in Case 3 without **sgn** term.

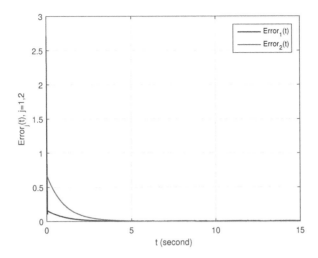

Figure 3.9 Trajectories of the consensus tracking errors $\mathrm{Error}_1(t)$ and $\mathrm{Error}_2(t)$ in Case 3 with controllers (3.59).

Case 3: Suppose that the communication topology of the CNS switches between graphs $\mathcal{G}^1$, $\mathcal{G}^2$, and $f(x_0(t), t) = \sin(t)$. One could check that Assumption (3.5) is satisfied with $d_0 = 1$. It could be calculated that $\bar{\kappa}_0 = 22.9578$ and $\bar{\tau}_{\mathrm{th}} = 3.1333$ with $\beta = 1$. Similar to Case 1, LMI (3.64) could be solved to get cBK. For illustrating the importance of **sgn** term in (3.59), Fig. 3.8 depicts the consensus tracking errors $\mathrm{Error}_1(t)$ and $\mathrm{Error}_2(t)$ in Case 3 without **sgn** term, and one could see that the consensus tracking errors will not approach to zero. By utilizing the control protocol (3.59), Fig. 3.9 shows that the consensus tracking with the nonautonomous leader could indeed be achieved, which validates the correctness of Theorem 3.5.

3.3 CONCLUSIONS

This chapter has studied the consensus problems for linear CNSs with directed switching topologies. For consensus of CNSs without a leader, we assumed that each possible switching topology contains a directed spanning tree. Requirements for switching signals have been derived, under which the criteria for selecting efficient feedback gain matrix and coupling strength have also been given. Compared to previous works with a fixed designated leader, the switching topologies considered in this chapter could have directed spanning trees with various leaders. This could facilitate a more flexible and more resilient configuration of the CNSs when confronted with complex scenarios. For leader following consensus with a linear autonomous system, by assuming each possible topology graph contains a directed spanning tree rooted at the leader node, different MLFs have been constructed by performing an iterative optimization algorithm and the M-matrix-based approach, respectively. We have further extended the results to the case where underline switching graphs only frequently contain a directed spanning. Last, consensus tracking for general linear CNSs with a nonautonomous leader has also been successfully explored by applying nonlinear control protocols.

CHAPTER 4

Consensus disturbance rejection of MIMO linear CNSs with directed switching topologies

This chapter studies the consensus disturbance rejection problem for multiple-input multiple-output linear CNSs subject to nonvanishing disturbances. This Chapter begins by overviewing some previous works and by indicating our motivations. Section 4.2 presents the models and proposes an unknown input observer (UIO) based on the relative outputs among neighboring agents. Section 4.3 studies the case with static coupling and directed switching communication topologies. By using the MLFs based technique, it is shown that consensus is achieved and the disturbances are fully rejected. Section 4.4 studies the case with dynamic couplings and directed fixed topology. As the control parameters do not depend on any global information, so the obtained consensus disturbance rejection is fully distributed. Finally, some simulations are given to validate the obtained theoretical results.

4.1 INTRODUCTION

As an interesting issue continued from single systems [21,31], disturbance rejection of CNSs has received more and more attention recently [14,34,57,148,170,220]. In [14], the authors solved the consensus problem for MIMO linear CNSs with undirected fixed topology subject to unknown disturbances which were assumed to have steady state values. Furthermore, consensus problem was solved for CNSs in the presence of harmonic nonvanishing disturbances [34]. Later, Sun *et al.* [148] proposed a fully distributed approach for achieving consensus disturbance rejection in linear CNSs with a directed fixed topology. More recently, the authors [170] applied the state predictor feedback method to settle the consensus disturbance rejection problem for CNSs with input delays as well as output delays.

The aforementioned literature and some references therein have broadened our knowledge on consensus control for CNSs under external disturbances. So we study the consensus disturbance rejection problem for MIMO linear CNSs under deterministic but nonvanishing disturbances, where both directed fixed and switching topologies are considered. Since only the agents' outputs are available, a UIO is designed for each follower based upon the relative outputs to estimate the consensus error between each follower and its neighbors. With the aid of this UIO, a state estimator with static coupling strength and a disturbance estimator are designed. Based on these two estimators, a controller is designed for the considered CNSs with directed switching topologies. We show that consensus disturbance rejection can be achieved by choosing suitable control parameters if the ADT is greater than a positive constant. Furthermore, a controller which incorporates a state estimator with an adaptive coupling law and a disturbance estimator is designed for the considered CNSs with directed fixed topology. We show that consensus disturbance rejection can be achieved in a fully distributed manner.

4.2 MODEL FORMULATION AND UNKNOWN INPUT OBSERVER

The CNSs under consideration have a leader and N followers. For illustration convenience, we label the leader as agent 0, and label the N followers as agents $1, \ldots, N$. The dynamics of the agent i, $i = 1, \ldots, N$, are described by:

$$
\begin{aligned}
\dot{x}_i(t) &= Ax_i(t) + Bu_i(t) + Dd_i(t), \\
y_i(t) &= Cx_i(t),
\end{aligned}
\tag{4.1}
$$

where $x_i(t) \in \mathbb{R}^n$, $u_i(t) \in \mathbb{R}^m$, and $y_i(t) \in \mathbb{R}^q$ are, respectively, the state, the control input, and the output, $A \in \mathbb{R}^{n \times n}$, $B \in \mathbb{R}^{n \times m}$, and $C \in \mathbb{R}^{q \times n}$ represent, respectively, the state matrix, the control input matrix, and the output matrix, $D \in \mathbb{R}^{n \times p}$ is a constant matrix, $d_i(t) \in \mathbb{R}^p$ is the external disturbance generated by the exogenous system:

$$
\dot{d}_i(t) = W d_i(t),
\tag{4.2}
$$

where $W \in \mathbb{R}^{p \times p}$ represents a known exosystem matrix. In some practical applications, the leader acts as a reference generator which provides desired trajectory for the following agents to track. So the dynamics of the agent 0 are described by:

$$
\begin{aligned}
\dot{x}_0(t) &= Ax_0(t), \\
y_0(t) &= Cx_0(t),
\end{aligned}
\tag{4.3}
$$

where $x_0(t) \in \mathbb{R}^n$ and $y_0(t) \in \mathbb{R}^q$ are, respectively, the leader's state and output.

One goal of this chapter is that the disturbance $d_i(t)$ can be completely rejected, and the other is to make each follower evolves along the trajectory provided by the leader finally. To achieve these goals, we make the following assumptions.

Assumption 4.1 *There is a constant matrix $E \in \mathbb{R}^{m \times p}$ such that $D = BE$.*

Assumption 4.2 *The exosystem matrix W has p distinct eigenvalues with real part being zero, and the matrix pair (W, D) is observable.*

Assumption 4.3 *Each graph $\mathcal{G}^k$ contains a directed spanning tree with agent 0 (i.e., the leader) being the root, $k = 1, \ldots, \kappa$.*

As the leader does not receive any information from the followers, the Laplacian matrix among the $N + 1$ agents can be rewritten as:

$$\mathcal{L} = \begin{bmatrix} 0 & \mathbf{0}_N^T \\ \mathbf{P} & \overline{\mathcal{L}}^{\sigma(t)} \end{bmatrix},$$

$$\overline{\mathcal{L}}^{\sigma(t)} = \begin{bmatrix} \sum_{j \in \mathcal{N}_1} a_{1j}^{\sigma(t)} & -a_{12}^{\sigma(t)} & \cdots & -a_{1N}^{\sigma(t)} \\ -a_{21}^{\sigma(t)} & \sum_{j \in \mathcal{N}_2} a_{2j}^{\sigma(t)} & \cdots & -a_{2N}^{\sigma(t)} \\ \vdots & \vdots & \ddots & \vdots \\ -a_{N1}^{\sigma(t)} & -a_{N2}^{\sigma(t)} & \cdots & \sum_{j \in \mathcal{N}_N} a_{Nj}^{\sigma(t)} \end{bmatrix},$$

where $\mathbf{P} = -[a_{10}^{\sigma(t)}, \ldots, a_{N0}^{\sigma(t)}]^T$, $\overline{\mathcal{L}}^{\sigma(t)} \in \mathbb{R}^{N \times N}$. Under Assumption 4.3, it can be got from Lemma 2.15 that there exists a positive definite matrix $\Phi^k = \text{diag}\{\phi_1^k, \ldots, \phi_N^k\}$ such that

$$\left(\overline{\mathcal{L}}^k\right)^T \Phi^k + \Phi^k \overline{\mathcal{L}}^k > 0,$$

where $[\phi_1^k, \ldots, \phi_N^k]^T = \left(\overline{\mathcal{L}}^k\right)^{-T} \cdot \mathbf{1}_N$. For convenience, we denote

$$\lambda_0 = \min_{k=1,\ldots,\kappa} \left\{ \lambda_{\min} \left(\left(\Phi^k\right)^{-1} \left(\overline{\mathcal{L}}^k\right)^T \Phi^k + \overline{\mathcal{L}}^k \right) \right\}. \tag{4.4}$$

Remark 4.1 *Assumption 4.1 presents a matching condition under which the disturbance effects can be compensated through the control action. A sufficient criterion for the existence of the matrix E is that $\text{rank}(B, D) = \text{rank}(B)$. Since E may not equal to I_m, the disturbances may be imposed on some channels other than the control input channels. Assumption 4.2 provides a standard requirement which is frequently employed in the study of output regulation and disturbance rejection [34]. Although Assumption 4.2 provides some requirements on the eigenvalues of W, it covers a wide range of periodic disturbances [21] such as the trigonometric functions upon which many other functions can be approximated with a bias. Assumption 4.3 gives a necessary condition for achieving consensus in CNSs with directed switching communication topologies [187].*

Let $e_i(t) = x_i(t) - x_0(t)$, $i = 1, \ldots, N$. Define $\delta_i(t) = \sum_{j=1}^N a_{ij}^{\sigma(t)}(x_i(t) - x_j(t)) + a_{i0}^{\sigma(t)}(x_i(t) - x_0(t))$. Letting $e(t) = [e_1^T(t), \ldots, e_N^T(t)]^T$ and $\delta(t) = [\delta_1^T(t), \ldots, \delta_N^T(t)]^T$. It is not difficult to get $\delta(t) = (\overline{\mathcal{L}}^{\sigma(t)} \otimes I_n)e(t)$. Due to the invertibility of $\overline{\mathcal{L}}^k$, $k = 1, \ldots, \kappa$, which can be got if Assumption 4.3 holds, $e(t) \to \mathbf{0}_{Nn}$ if and only if $\delta(t) \to \mathbf{0}_{Nn}$ when

$t \to \infty$. Since only a few followers can receive the leaders' output information, neither $e(t)$ nor $\delta(t)$ is available for protocols design. Based on the relative output information among neighboring agents, the UIO (4.5) is designed for each follower to estimate $\delta(t)$.

$$\dot{v}_i(t) = (GA - FC)v_i(t) + (F(I_q + CH) - GAH)$$

$$\cdot \left(\sum_{j=1}^{N} a_{ij}^{\sigma(t)}(y_i(t) - y_j(t)) + a_{i0}^{\sigma(t)}(y_i(t) - y_0(t)) \right),$$

$$\hat{\delta}_i(t) = v_i(t) - H \left(\sum_{j=1}^{N} a_{ij}^{\sigma(t)}(y_i(t) - y_j(t)) + a_{i0}^{\sigma(t)}(y_i(t) - y_0(t)) \right), \qquad (4.5)$$

where $v_i(t) \in \mathbb{R}^n$, $\hat{\delta}_i(t) \in \mathbb{R}^n$ is the estimate of $\delta_i(t)$, $H = -B[(CB)^T(CB)]^{-1}(CB)^T$, $G = I_n + HC$, and $F \in \mathbb{R}^{n \times q}$ is chosen such that $GA - FC$ is stable. In the sequel, we firstly show $\tilde{\delta}_i(t) \triangleq \hat{\delta}_i(t) - \delta_i(t)$ will asymptotically converge to the zero vector $\mathbf{0}_n$ whatever the controller is designed.

Theorem 4.1 *If $GA - FC$ is stable, then the UIO $\hat{\delta}_i(t)$ given by (4.5) can asymptotically estimate the consensus error $\delta_i(t)$.*

Proof 4.1 *Noticing $C\delta_i(t) = \sum_{j=1}^{N} a_{ij}^{\sigma(t)}(y_i(t) - y_j(t)) + a_{i0}^{\sigma(t)}(y_i(t) - y_0(t))$, we get from (4.5) that $\hat{\delta}_i(t) = v_i(t) - HC\delta_i(t)$. This together with (4.1), (4.5) gives that*

$$\dot{\tilde{\delta}}_i(t) = \dot{v}_i(t) - G\dot{\delta}_i(t)$$
$$= (GA - FC)v_i(t) + (F(I_q + CH) - GAH)C\delta_i(t)$$
$$- GA\delta_i(t) - GB \left[\sum_{j=1}^{N} a_{ij}^{\sigma(t)}(u_i(t) - u_j(t)) + a_{i0}^{\sigma(t)}u_i(t) \right]$$
$$- GD \left[\sum_{j=1}^{N} a_{ij}^{\sigma(t)}(d_i(t) - d_j(t)) + a_{i0}^{\sigma(t)}d_i(t) \right]. \qquad (4.6)$$

Observing the structure of H, it is not difficult to show that

$$GB = (I_n + HC)B = 0,$$
$$(F(I_q + CH) - GAH)C - GA = (FC - GA)G.$$

Substituting these with $D = BE$ into (4.6) yield

$$\dot{\tilde{\delta}}_i(t) = (GA - FC)(v_i(t) - G\delta_i(t))$$
$$= (GA - FC)\tilde{\delta}_i(t). \qquad (4.7)$$

Then the proof is completed since $GA - FC$ is stable.

Remark 4.2 *A critical issue is the existence of the observer (4.5), i.e., whether the matrix H is well defined, and whether the feedback gain matrix F exists or not.*

Now we give some necessary and sufficient conditions. First, H is well defined if and only if $\mathrm{rank}(CB) = \mathrm{rank}(B) = m$. *Noticing F exists if the matrix pair (GA, C) is detectable. According to Theorem 3 of [31], (GA, C) is detectable if and only if*

$$\mathrm{rank} \begin{bmatrix} sI_n - A & B \\ C & 0 \end{bmatrix} = n + m \ for \ \forall \ s \in \mathbb{C} \ with \ \mathrm{Re}(s) \geq 0.$$ *Consequently, a necessary and sufficient condition for the existence of (4.5) is*

(1) $\mathrm{rank}(CB) = \mathrm{rank}(B) = m$;

(2) $\mathrm{rank} \begin{bmatrix} sI - A & B \\ C & 0 \end{bmatrix} = n + m, \ \forall \ s \in \mathbb{C}, \ \mathrm{Re}(s) \geq 0.$

If conditions (1) and (2) hold, then we can get H and G according to the definitions. Since (GA, C) is detectable, there exist some $U > 0$ such that

$$(GA)^T U + U(GA) - C^T C < 0. \tag{4.8}$$

Then, we can select $F = U^{-1}C^T$.

With the aid of (4.5), the following state estimator is designed for each follower.

$$\dot{\xi}_i(t) = A\xi_i(t) + \alpha BK\xi_i(t) + \rho(1 + \varrho_i(t))BK(\zeta_i(t) - \hat{\delta}_i(t)),$$
$$\varrho_i(t) = (\zeta_i(t) - \hat{\delta}_i(t))^T P^{-1}(\zeta_i(t) - \hat{\delta}_i(t)), \tag{4.9}$$

where $\alpha > 0$ and $K \in \mathbb{R}^{m \times n}$ are, respectively, the coupling strength and the feedback gain matrix to be designed, ρ and $P \in \mathbb{R}^{n \times n}$ are, respectively, a positive constant and a positive definite matrix to be determined, $\zeta_i(t) = \sum_{j=1}^{N} a_{ij}^{\sigma(t)}(\xi_i(t) - \xi_j(t)) + a_{i0}^{\sigma(t)}\xi_i(t)$. Based on (4.5) and (4.9), the following disturbance observer is designed.

$$\hat{d}_i(t) = z_i(t) + Q\hat{\delta}_i(t),$$
$$\dot{z}_i(t) = Wz_i(t) + (WQ - QA)\hat{\delta}_i(t) - \alpha QBK\zeta_i(t), \tag{4.10}$$

where $\hat{d}_i(t) \in \mathbb{R}^p$ and $z_i(t) \in \mathbb{R}^p$ are, respectively, the state and the internal state of the disturbance observer [21], $Q \in \mathbb{R}^{p \times n}$ will be given later. Now, the controller (4.11) is designed.

$$u_i(t) = \alpha K\xi_i(t) - E\hat{d}_i(t). \tag{4.11}$$

Substituting (4.11) into (4.1) gives that

$$\dot{e}_i(t) = Ae_i(t) + \alpha BK\xi_i(t) - D\tilde{d}_i(t), \tag{4.12}$$

where $\tilde{d}_i(t) = \hat{d}_i(t) - d_i(t)$, $i = 1, \ldots, N$. Let $\zeta(t) = [\zeta_1^T(t), \ldots, \zeta_N^T(t)]^T$ and $\tilde{d}(t) = [\tilde{d}_1^T(t), \ldots, \tilde{d}_N^T(t)]^T$. We get from (4.12) that

$$\dot{\delta}(t) = (I_N \otimes A)\delta(t) + \alpha(I_N \otimes BK)\zeta(t) - (\overline{\mathcal{L}}^{\sigma(t)} \otimes D)\tilde{d}(t).$$

This together with (4.7) gives that

$$\dot{\hat{\delta}}(t) = (I_N \otimes A)\hat{\delta}(t) + \alpha(I_N \otimes BK)\zeta(t) \tag{4.13}$$
$$- (\overline{\mathcal{L}}^{\sigma(t)} \otimes D)\tilde{d}(t) + [I_N \otimes (GA - FC - A)]\tilde{\delta}(t),$$

where $\hat{\delta}(t) = [\hat{\delta}_1^T(t), \ldots, \hat{\delta}_N^T(t)]^T$ and $\tilde{\delta}(t) = [\tilde{\delta}_1^T(t), \ldots, \tilde{\delta}_N^T(t)]^T$. And it follows from

(4.7) that

$$\dot{\tilde{\delta}}(t) = [I_N \otimes (GA - FC)]\tilde{\delta}(t). \tag{4.14}$$

Combining (4.2), (4.10), and (4.13), we have

$$\begin{aligned}\dot{\tilde{d}}(t) =&(I_N \otimes W)\tilde{d}(t) - (\overline{\mathcal{L}}^{\sigma(t)} \otimes QD)\tilde{d}(t) \\&+ [I_N \otimes Q(GA - FC - A)]\tilde{\delta}(t).\end{aligned} \tag{4.15}$$

Let $\varrho(t) = \text{diag}\{\varrho_1(t), \ldots, \varrho_N(t)\}$. We get from (4.9) that

$$\dot{\zeta}(t) = [I_N \otimes (A + \alpha BK)]\zeta(t) + [\rho(I_N + \varrho(t))\overline{\mathcal{L}}^{\sigma(t)} \otimes BK](\zeta(t) - \hat{\delta}(t)). \tag{4.16}$$

4.3 CNSS WITH STATIC COUPLING AND SWITCHING TOPOLOGIES

This section studies the consensus disturbance rejection problem for CNSs under directed switching topologies. Before moving forward, the definition of consensus disturbance rejection is given.

Definition 4.1 *The consensus disturbance rejection of CNSs (4.1) and (4.3) with disturbances generated by (4.2) is said to be achieved if*

$$\lim_{t \to \infty} \|x_i(t) - x_0(t)\| = 0, \quad \lim_{t \to \infty} \|\hat{d}_i(t) - d_i(t)\| = 0, \tag{4.17}$$

hold for arbitrary initial values $x_i(t_0)$, $x_0(t_0)$, $\hat{d}_i(t_0)$, $d_i(t_0)$, $i = 1, \ldots, N$.

Theorem 4.2 *Suppose Assumptions 4.1–4.3 hold. If the ADT $\tau_a > \ln \nu$, then the consensus disturbance rejection of CNSs (4.1) and (4.3) with the disturbances generated by (4.2) can be achieved by adopting the consensus error estimator (4.5), the state estimator (4.9), and the disturbance observer (4.10) based controller (4.11) with $K = -B^T P^{-1}$, $Q = \mu R^{-1}D^T$, $\rho \geq 4\alpha/\lambda_0$, $\mu \geq 4/\lambda_0$, where α is a positive constant, λ_0 is given by (4.4), $P > 0$ and $R > 0$ are, respectively, obtained by solving the LMIs (4.18) and (4.19),*

$$AP + PA^T - \alpha BB^T + P < 0, \tag{4.18}$$

$$W^T R + RW - D^T D + 2R < 0. \tag{4.19}$$

Proof 4.2 *For any $t \in [t_j, t_{j+1})$, $j = 0, 1, 2, \ldots$, we construct the following MLFs*

$$V_1(t) = V_{11}(t) + V_{12}(t) + V_{13}(t) + V_{14}(t), \tag{4.20}$$

where

$$V_{11}(t) = \zeta^T(t)(I_N \otimes P^{-1})\zeta(t),$$

$$V_{12}(t) = \frac{\gamma_1}{2} \sum_{i=1}^{N} \phi_i^{\sigma(t)}(2 + \varrho_i(t))\varrho_i(t),$$

$$V_{13}(t) = \gamma_1\gamma_2\tilde{d}^T(t)(\Phi^{\sigma(t)} \otimes R)\tilde{d}(t),$$

$$V_{14}(t) = \gamma_1\gamma_3\tilde{\delta}^T(t)(\Phi^{\sigma(t)} \otimes S)\tilde{\delta}(t),$$

where $\gamma_1 > 0$, $\gamma_2 > 0$, $\gamma_3 > 0$ will be given later, $\Phi^{\sigma(t)} \in \{\Phi^1, \ldots, \Phi^\kappa\}$, $S > 0$ satisfies

$$(GA - FC)^T S + S(GA - FC) + 2S < 0. \tag{4.21}$$

Calculating the derivative of $V_{11}(t)$ along the trajectory (4.16), we have

$$
\begin{aligned}
\dot{V}_{11}(t) &= \zeta^T(t) \left[I_N \otimes (A^T P^{-1} + P^{-1} A - 2\alpha\Theta) \right] \zeta(t) \\
&\quad - 2\zeta^T(t) \left[\rho(I_N + \varrho(t)) \overline{\mathcal{L}}^{\sigma(t)} \otimes \Theta \right] \left(\zeta(t) - \hat{\delta}(t) \right) \\
&\leq \zeta^T(t) \left[I_N \otimes (A^T P^{-1} + P^{-1} A - 2\alpha\Theta) \right] \zeta(t) + \alpha \zeta^T(t)(I_N \otimes \Theta)\zeta(t) \\
&\quad + \frac{\rho^2 \iota_1}{\alpha} \left(\zeta(t) - \hat{\delta}(t) \right)^T \left[(I_N + \varrho(t))^2 \otimes \Theta \right] \left(\zeta(t) - \hat{\delta}(t) \right) \\
&\leq -\zeta^T(t)(I_N \otimes P^{-1})\zeta(t) \\
&\quad + \frac{\rho^2 \iota_1}{\alpha} \left(\zeta(t) - \hat{\delta}(t) \right)^T \left[(I_N + \varrho(t))^2 \otimes \Theta \right] \left(\zeta(t) - \hat{\delta}(t) \right),
\end{aligned} \tag{4.22}
$$

where $\iota_1 = \max_{k=1,2,\ldots,\kappa} \lambda_{\max}((\overline{\mathcal{L}}^k)^T \overline{\mathcal{L}}^k)$, $\Theta = P^{-1} B B^T P^{-1}$, and the last inequality follows from (4.18).

Calculating the derivative of $V_{12}(t)$ along the trajectories (4.9) and (4.16), we have

$$
\begin{aligned}
\dot{V}_{12}(t) &= \gamma_1 \sum_{i=1}^{N} \phi_i^{\sigma(t)}(1 + \varrho_i(t))\dot{\varrho}_i(t) \\
&= 2\gamma_1 \left(\zeta(t) - \hat{\delta}(t) \right)^T \left[\Phi^{\sigma(t)}(I_N + \varrho(t)) \otimes P^{-1} \right] \left(\dot{\zeta}(t) - \dot{\hat{\delta}}(t) \right) \\
&= \gamma_1 \left(\zeta(t) - \hat{\delta}(t) \right)^T \left\{ \Phi^{\sigma(t)}(I_N + \varrho(t)) \otimes (A^T P^{-1} + P^{-1} A) \right. \\
&\quad \left. - \rho(I_N + \varrho(t)) \left[\Phi^{\sigma(t)} \overline{\mathcal{L}}^{\sigma(t)} + (\overline{\mathcal{L}}^{\sigma(t)})^T \Phi^{\sigma(t)} \right] (I_N + \varrho(t)) \otimes \Theta \right\} \left(\zeta(t) - \hat{\delta}(t) \right) \\
&\quad + 2\gamma_1 \left(\zeta(t) - \hat{\delta}(t) \right)^T \left[\Phi^{\sigma(t)}(I_N + \varrho(t)) \overline{\mathcal{L}}^{\sigma(t)} \otimes P^{-1} D \right] \tilde{d}(t) \\
&\quad - 2\gamma_1 \left(\zeta(t) - \hat{\delta}(t) \right)^T \left[\Phi^{\sigma(t)}(I_N + \varrho(t)) \otimes P^{-1}(GA - FC - A) \right] \hat{\delta}(t). \tag{4.23}
\end{aligned}
$$

Let ι_2 be the smallest eigenvalue of $B^T B$. Note that $\iota_2 > 0$ since B is full column rank under Assumption 4.1, then

$$
\begin{aligned}
& 2 \left(\zeta(t) - \hat{\delta}(t) \right)^T \left[\Phi^{\sigma(t)}(I_N + \varrho(t)) \overline{\mathcal{L}}^{\sigma(t)} \otimes P^{-1} D \right] \tilde{d}(t) \\
&\leq \frac{\rho \lambda_0}{4} \left(\zeta(t) - \hat{\delta}(t) \right)^T \left[\Phi^{\sigma(t)}(I_N + \varrho(t))^2 \otimes P^{-1} B B^T P^{-1} \right] \left(\zeta(t) - \hat{\delta}(t) \right) \\
&\quad + \frac{4\iota_1}{\rho \lambda_0} \tilde{d}^T(t) \left(\Phi^{\sigma(t)} \otimes E^T E \right) \tilde{d}(t) \\
&\leq \frac{\rho \lambda_0}{4} \left(\zeta(t) - \hat{\delta}(t) \right)^T \left[\Phi^{\sigma(t)}(I_N + \varrho(t))^2 \otimes \Theta \right] \left(\zeta(t) - \hat{\delta}(t) \right) \\
&\quad + \frac{4\iota_1}{\rho \lambda_0 \iota_2} \tilde{d}^T(t) \left(\Phi^{\sigma(t)} \otimes D^T D \right) \tilde{d}(t). \tag{4.24}
\end{aligned}
$$

Let $\iota_3 = \lambda_{\max}((GA - FC - A)^T(GA - FC - A))$. Direct calculation gives that

$$
\begin{aligned}
&- 2\left(\zeta(t) - \hat{\delta}(t)\right)^T\left[\Phi^{\sigma(t)}(I_N + \varrho(t)) \otimes P^{-1}(GA - FC - A)\right]\tilde{\delta}(t) \\
&\leq \frac{\rho\lambda_0\iota_2}{4}\left(\zeta(t) - \hat{\delta}(t)\right)^T\left[\Phi^{\sigma(t)}(I_N + \varrho(t))^2 \otimes P^{-2}\right]\left(\zeta(t) - \hat{\delta}(t)\right) \\
&\quad + \frac{4\iota_3}{\rho\lambda_0\iota_2}\tilde{\delta}^T(t)\left(\Phi^{\sigma(t)} \otimes I_n\right)\tilde{\delta}(t) \\
&\leq \frac{\rho\lambda_0}{4}\left(\zeta(t) - \hat{\delta}(t)\right)^T\left[\Phi^{\sigma(t)}(I_N + \varrho(t))^2 \otimes \Theta\right]\left(\zeta(t) - \hat{\delta}(t)\right) \\
&\quad + \frac{4\iota_3}{\rho\lambda_0\iota_2\lambda_{\min}(S)}\tilde{\delta}^T(t)\left(\Phi^{\sigma(t)} \otimes S\right)\tilde{\delta}(t).
\end{aligned}
\tag{4.25}
$$

By using the inequalities $\rho \geq 4\alpha/\lambda_0$ and $(I_N + \varrho(t)) \geq I_N$, we get

$$
\begin{aligned}
&- \frac{\rho\lambda_0}{4}\left(\zeta(t) - \hat{\delta}(t)\right)^T\left[\Phi^{\sigma(t)}(I_N + \varrho(t))^2 \otimes \Theta\right]\left(\zeta(t) - \hat{\delta}(t)\right) \\
&\leq - \alpha\left(\zeta(t) - \hat{\delta}(t)\right)^T\left[\Phi^{\sigma(t)}(I_N + \varrho(t)) \otimes \Theta\right]\left(\zeta(t) - \hat{\delta}(t)\right).
\end{aligned}
\tag{4.26}
$$

Substituting the inequalities (4.24)–(4.26) into (4.23) gives that

$$
\begin{aligned}
\dot{V}_{12}(t) \leq &\gamma_1(\zeta(t) - \hat{\delta}(t))^T[\Phi^{\sigma(t)}(I_N + \varrho(t)) \otimes (A^T P^{-1} + P^{-1} - \alpha\Theta)](\zeta(t) - \hat{\delta}(t)) \\
&- \frac{\gamma_1\rho\lambda_0}{4}(\zeta(t) - \hat{\delta}(t))^T[\Phi^{\sigma(t)}(I_N + \varrho(t))^2 \otimes \Theta](\zeta(t) - \hat{\delta}(t)) \\
&+ \frac{4\gamma_1\iota_1}{\rho\lambda_0\iota_2}\tilde{d}^T(t)(\Phi^{\sigma(t)} \otimes D^T D)\tilde{d}(t) \\
&+ \frac{4\gamma_1\iota_3}{\rho\lambda_0\iota_2\lambda_{\min}(S)}\tilde{\delta}^T(t)(\Phi^{\sigma(t)} \otimes S)\tilde{\delta}(t) \\
\leq &- \gamma_1\sum_{i=1}^{N}\phi_i^{\sigma(t)}(1 + \varrho_i(t))\varrho_i(t) \\
&- \frac{\gamma_1\rho\lambda_0}{4}(\zeta(t) - \hat{\delta}(t))^T[\Phi^{\sigma(t)}(I_N + \varrho(t))^2 \otimes \Theta](\zeta(t) - \hat{\delta}(t)) \\
&+ \frac{4\gamma_1\iota_1}{\rho\lambda_0\iota_2}\tilde{d}^T(t)(\Phi^{\sigma(t)} \otimes D^T D)\tilde{d}(t) \\
&+ \frac{4\gamma_1\iota_3}{\rho\lambda_0\iota_2\lambda_{\min}(S)}\tilde{\delta}^T(t)(\Phi^{\sigma(t)} \otimes S)\tilde{\delta}(t).
\end{aligned}
\tag{4.27}
$$

Calculating the derivative of $V_{13}(t)$ along the trajectory (4.15), we have

$$
\begin{aligned}
\dot{V}_{13}(t) = &\gamma_1\gamma_2\tilde{d}^T(t)[\Phi^{\sigma(t)} \otimes (W^T R + RW) - (\Phi^{\sigma(t)}\overline{\mathcal{L}}^{\sigma(t)} + (\overline{\mathcal{L}}^{\sigma(t)})^T\Phi^{\sigma(t)}) \otimes RQD]\tilde{d}(t) \\
&+ 2\gamma_1\gamma_2\tilde{d}^T(t)[\Phi^{\sigma(t)} \otimes RQ(GA - FC - A)]\tilde{\delta}(t) \\
\leq &\gamma_1\gamma_2\tilde{d}^T(t)[\Phi^{\sigma(t)} \otimes (W^T R + RW - \lambda_0\mu D^T D)]\tilde{d}(t) \\
&+ \frac{\gamma_1\gamma_2\mu\lambda_0}{4}\tilde{d}^T(t)(\Phi^{\sigma(t)} \otimes D^T D)\tilde{d}(t) + \frac{4\gamma_1\gamma_2\mu\iota_3}{\lambda_0}\tilde{\delta}^T(t)(\Phi^{\sigma(t)} \otimes I_n)\tilde{\delta}(t) \\
\leq &- \gamma_1\gamma_2\tilde{d}^T(t)\left[\Phi^{\sigma(t)} \otimes R\right]\tilde{d}(t) - \gamma_1\gamma_2\tilde{d}^T(t)\left[\Phi^{\sigma(t)} \otimes D^T D\right]\tilde{d}(t)
\end{aligned}
$$

$$+ \frac{4\gamma_1\gamma_2\mu\iota_3}{\lambda_0\lambda_{\min}(S)}\tilde{\delta}^T(t)(\Phi^{\sigma(t)} \otimes S)\tilde{\delta}(t), \tag{4.28}$$

where we use $\mu \geq 4/\lambda_0$ and (4.19) to get the last inequality.

Calculating the derivative of $V_{14}(t)$ along the trajectory (4.14), we have

$$\dot{V}_{14}(t) = \gamma_1\gamma_3\tilde{\delta}^T(t)[\Phi^{\sigma(t)} \otimes (GA - FC)^T S + S(GA - FC)]\tilde{\delta}(t)$$
$$\leq -2\gamma_1\gamma_3\tilde{\delta}^T(t)(\Phi^{\sigma(t)} \otimes S)\tilde{\delta}(t), \tag{4.29}$$

where the inequality follows from (4.21).

We now conclude from (4.22), (4.27), (4.28), and (4.29) that

$$\dot{V}_1(t) \leq -\zeta^T(t)(I_N \otimes P^{-1})\zeta(t) - \gamma_1\sum_{i=1}^N \phi_i^{\sigma(t)}(1 + \varrho_i(t))\varrho_i(t)$$

$$- \gamma_1\gamma_2\tilde{d}^T(t)(\Phi^{\sigma(t)} \otimes R)\tilde{d}(t) - \gamma_1\gamma_3\tilde{\delta}^T(t)(\Phi^{\sigma(t)} \otimes S)\tilde{\delta}(t)$$

$$- \left(\frac{\gamma_1\rho\lambda_0\underline{\phi}}{4} - \frac{\rho^2\iota_1}{\alpha}\right)\left(\zeta(t) - \hat{\delta}(t)\right)^T\left[(I_N + \varrho(t))^2 \otimes \Theta\right]\left(\zeta(t) - \hat{\delta}(t)\right)$$

$$- \gamma_1\left(\gamma_2 - \frac{4\gamma_1\iota_1}{\rho\lambda_0\iota_2}\right)\tilde{d}^T(t)\left(\Phi^{\sigma(t)} \otimes D^T D\right)\tilde{d}(t)$$

$$- \gamma_1\tilde{\delta}^T(t)\left[\Phi^{\sigma(t)} \otimes \left(\gamma_3 - \frac{4\gamma_2\mu\iota_3}{\lambda_0\lambda_{\min}(S)} - \frac{4\iota_3}{\rho\lambda_0\iota_2\lambda_{\min}(S)}\right)S\right]\tilde{\delta}(t). \tag{4.30}$$

We now choose $\gamma_1 \geq \frac{4\rho\iota_1}{\alpha\lambda_0\underline{\phi}}$, $\gamma_2 \geq \frac{4\gamma_1\iota_1}{\rho\lambda_0\iota_2}$, $\gamma_3 \geq \frac{4\gamma_2\mu\iota_3}{\lambda_0\lambda_{\min}(S)} + \frac{4\iota_3}{\rho\lambda_0\iota_2\lambda_{\min}(S)}$, and insert these inequalities into (4.30) gives that

$$\dot{V}_1(t) \leq -\zeta^T(t)(I_N \otimes P^{-1})\zeta(t) - \frac{\gamma_1}{2}\sum_{i=1}^N \phi_i^{\sigma(t)}(2 + 2\varrho_i(t))\varrho_i(t)$$

$$- \gamma_1\gamma_2\tilde{d}^T(t)(\Phi^{\sigma(t)} \otimes R)\tilde{d}(t) - \gamma_1\gamma_3\tilde{\delta}^T(t)(\Phi^{\sigma(t)} \otimes S)\tilde{\delta}(t)$$

$$\leq -V_{11}(t) - V_{12}(t) - V_{13}(t) - V_{14}(t) = -V_1(t), \tag{4.31}$$

where we use the fact $2 + 2\varrho_i(t) \geq 2 + \varrho_i(t) > 1$ to get the second inequality.

For any $t \in [t_j, t_{j+1})$, we get from (4.31) that

$$V_1(t) \leq \exp(-(t - t_j))V_1(t_j) \leq \nu \cdot \exp(-(t - t_j))V_1(t_j^-), \tag{4.32}$$

where we use $V_1(t_j) \leq \nu \cdot V_1(t_j^-)$ with $\nu = \max_{i=1,2,\ldots,N}^{k=1,2,\ldots,\kappa} \phi_i^k \ / \min_{i=1,2,\ldots,N}^{k=1,2,\ldots,\kappa} \phi_i^k$ and $V_1(t_j^-) = \lim_{t \nearrow t_j} V_1(t)$ to get the second inequality. For any time instant $t > t_0$, let $N_\sigma[t_0, t)$ be the number of switchings during the interval $[t_0, t)$. It is not difficult to obtain from (4.32) that

$$V_1(t) \leq \nu^{N_\sigma[t_0,t)}\exp(-(t - t_0))V_1(t_0)$$
$$\leq \nu^{N_0}\exp\left(-(1 - \frac{\ln\nu}{\tau_a})(t - t_0)\right)V_1(t_0). \tag{4.33}$$

Since $\tau_a > \ln\nu$, (4.33) implies $V_1(t) \to 0$ when $t \to \infty$. This combines with (4.20) gives that $\|\zeta(t)\|$, $\|\tilde{d}(t)\|$, $\|\zeta(t) - \hat{\delta}(t)\|$, and $\|\tilde{\delta}(t)\|$ approach 0. Therefore, $\|\delta(t)\| \to 0$ when $t \to \infty$ which further implies the consensus tracking is achieved.

Remark 4.3 *We learn from* $\|\tilde{\delta}(t)\| \to 0$ *that the UIO* (4.5) *can estimate the exact consensus error whatever the controller is designed. Since* $\|\tilde{d}(t)\| \to 0$, *the external disturbance that generated by the exogenous system* (4.2) *can be fully rejected by the designed disturbance observer* (4.10) *which is designed upon the information of the UIO* (4.5) *as well as the state estimator* (4.9). *The LMI* (4.18) *is feasible for some positive constants* α *since the matrix pair* (A, B) *is completely controllable which can be obtained from the assumption that* $\mathrm{rank} \begin{bmatrix} sI - A & B \end{bmatrix} = n, \forall s \in \mathbb{C}, \mathrm{Re}(s) \geq 0$. *In addition, Assumption 4.2 provides a sufficient criterion for the feasibility of the LMI* (4.19). *And the LMI* (4.21) *is feasible since* $GA - FC$ *is stable. The LMIs* (4.18) *and* (4.19) *are independently of any global information such as the total number of agents and the smallest eigenvalue of the Laplacian matrices, etc. The coupling strength* ρ *and the ADT condition* τ_a, *however, depend on* λ_0 *which is a global parameter and thus limits the application of Theorem 4.2 to some large scale networks.*

4.4 CNSS WITH DYNAMIC COUPLING AND FIXED TOPOLOGY

We could learn from Theorem 4.2 that the coupling strength ρ depends on the smallest eigenvalue λ_0 which is a global information associated with all the possible communication graphs. Consequently, the controller (4.11) can not be implemented in a distributed way. Motivated by this observation, we give a new state estimator with dynamic coupling strengths upon which a fully distributed controller can be reconstructed. While, unlike the last subsection, the directed topology of the CNSs considered in this subsection is assumed to be fixed. The state estimator is given as follows.

$$\dot{\hat{\xi}}_i(t) = A\hat{\xi}_i(t) + \alpha BK\hat{\xi}_i(t) + (\rho_i + \varrho_i)BK\left(\hat{\zeta}_i(t) - \hat{\delta}_i(t)\right),$$

$$\dot{\rho}_i = \left(\hat{\zeta}_i(t) - \hat{\delta}_i(t)\right)^T \Theta \left(\hat{\zeta}_i(t) - \hat{\delta}_i(t)\right),$$

$$\varrho_i = \left(\hat{\zeta}_i(t) - \hat{\delta}_i(t)\right)^T P^{-1} \left(\hat{\zeta}_i(t) - \hat{\delta}_i(t)\right), \quad (4.34)$$

where $\hat{\zeta}_i(t) = \sum_{j=1}^{N} a_{ij}\left(\hat{\xi}_i(t) - \hat{\xi}_j(t)\right) + a_{i0}\hat{\xi}_i(t)$, $\Theta = P^{-1}BB^T P^{-1}$, $P > 0$ will be given later, and the initial value $\rho_i(t_0) > 0$. Based on the estimator (4.34), the disturbance observer and the controller are then given by (4.35) and (4.36), respectively.

$$\hat{d}_i(t) = z_i(t) + Q\hat{\delta}_i(t),$$

$$\dot{z}_i(t) = Wz_i(t) + (WQ - QA)\hat{\delta}_i(t) - \alpha QBK\hat{\zeta}_i(t), \quad (4.35)$$

$$u_i(t) = \alpha K\hat{\xi}_i(t) - E\hat{d}_i(t). \quad (4.36)$$

By using the same analyses to those presented in Section 4.2, we get

$$\dot{\hat{\delta}}(t) = (I_N \otimes A)\hat{\delta}(t) + \alpha(I_N \otimes BK)\hat{\zeta}(t)$$
$$- (\overline{\mathcal{L}} \otimes D)\tilde{d}(t) + [I_N \otimes (GA - FC - A)]\tilde{\delta}(t), \quad (4.37)$$

$$\dot{\tilde{\delta}}(t) = [I_N \otimes (GA - FC)]\tilde{\delta}(t), \tag{4.38}$$

$$\dot{\tilde{d}}(t) = (I_N \otimes W)\tilde{d}(t) - (\overline{\mathcal{L}} \otimes QD)\tilde{d}(t) + [I_N \otimes Q(GA - FC - A)]\tilde{\delta}(t), \tag{4.39}$$

$$\dot{\hat{\zeta}}(t) = [I_N \otimes (A + \alpha BK)]\hat{\zeta}(t) + [\overline{\mathcal{L}}(\rho + \varrho) \otimes BK]\left(\hat{\zeta}(t) - \hat{\delta}(t)\right), \tag{4.40}$$

where $\hat{\zeta}(t) = \left[\hat{\zeta}_1^T(t), \ldots, \hat{\zeta}_2^T(t)\right]^T$, $\rho = \mathrm{diag}\{\rho_1, \ldots, \rho_N\}$, and the other symbols are the same as those defined in Section 4.2.

Theorem 4.3 *Suppose Assumptions 4.1–4.3 hold. Then the fully distributed consensus disturbance rejection problem of CNSs (4.1) and (4.3) with the disturbances generated by (4.2) can be solved by adopting the consensus error estimator (4.5), the state estimator (4.34), and the disturbance observer (4.35) based controller (4.36) with $\alpha > 0$, $K = -B^T P^{-1}$, where Q is chosen such that QD is positive definite, $P > 0$ satisfies the LMI*

$$AP + PA^T - 2\alpha BB^T < 0. \tag{4.41}$$

Proof 4.3 *Since the graph $\mathcal{G}$ satisfies Assumption 4.3, there exists a positive vector $\phi = [\phi_1, \ldots, \phi_N]^T$ such that $\overline{\mathcal{L}}^T \phi = \mathbf{1}_N$. Let $\Phi = \mathrm{diag}\{\phi_1, \ldots, \phi_N\}$. We construct the following Lyapunov function:*

$$V_2(t) = V_{21}(t) + V_{22}(t) + V_{23}(t) + V_{24}(t), \tag{4.42}$$

where

$$V_{21}(t) = \hat{\zeta}^T(t)\left(I_N \otimes P^{-1}\right)\hat{\zeta}(t),$$

$$V_{22}(t) = \frac{\gamma_1}{2}\sum_{i=1}^N \phi_i(2\rho_i + \varrho_i)\varrho_i + \frac{\gamma_1}{2}\sum_{i=1}^N (\rho_i - c)^2,$$

$$V_{23}(t) = \gamma_1\gamma_2 \tilde{d}^T(t)(\Phi \otimes I_p)\tilde{d}(t),$$

$$V_{24}(t) = \gamma_1\gamma_3 \tilde{\delta}^T(t)(\Phi \otimes S)\tilde{\delta}(t),$$

where γ_1, γ_2, γ_3, c are positive constants to be given later, $S > 0$ satisfies $(GA - FC)^T S + S(GA - FC) < 0$.

Calculating the derivative of $V_{21}(t)$ along the trajectory (4.40), we have

$$\dot{V}_{21}(t) = \hat{\zeta}^T(t)\left[I_N \otimes (A^T P^{-1} + P^{-1}A - 2\alpha\Theta)\right]\hat{\zeta}(t)$$
$$\quad - 2\hat{\zeta}^T(t)\left[\overline{\mathcal{L}}(\rho + \varrho) \otimes \Theta\right]\left(\hat{\zeta}(t) - \hat{\delta}(t)\right)$$
$$\leq -\hat{\zeta}^T(t)(I_N \otimes \Lambda)\hat{\zeta}(t) + \frac{1}{2}\hat{\zeta}^T(t)(I_N \otimes \Lambda)\hat{\zeta}(t)$$
$$\quad + \frac{2\iota_4\lambda_{\max}(\Theta)}{\lambda_{\min}(\Lambda)}\left(\hat{\zeta}(t) - \hat{\delta}(t)\right)^T\left[(\rho + \varrho)^2 \otimes \Theta\right](\hat{\zeta}(t) - \hat{\delta}(t))$$

$$\leq -\frac{1}{2}\hat{\zeta}^T(t)(I_N \otimes \Lambda)\hat{\zeta}(t)$$

$$+ \frac{2\iota_4 \lambda_{\max}(\Theta)}{\lambda_{\min}(\Lambda)}\left(\hat{\zeta}(t) - \hat{\delta}(t)\right)^T \left[(\rho + \varrho)^2 \otimes \Theta\right]\left(\hat{\zeta}(t) - \hat{\delta}(t)\right), \qquad (4.43)$$

where $\iota_4 = \lambda_{\max}\left(\overline{\mathcal{L}}^T\overline{\mathcal{L}}\right)$ and $\Lambda = -(A^TP^{-1} + P^{-1}A - 2\alpha\Theta)$. Since P satisfies (4.41), *Λ is positive definite.*

Calculating the derivative of $V_{22}(t)$ along the trajectories (4.37) *and* (4.40), *we have*

$$\dot{V}_{22}(t) = \gamma_1 \sum_{i=1}^{N}\left[\phi_i(\rho_i + \varrho_i)\dot{\varrho}_i + \phi_i\dot{\rho}_i\varrho_i + (\rho_i - c)\dot{\rho}_i\right]$$

$$= 2\gamma_1\left(\hat{\zeta}(t) - \hat{\delta}(t)\right)^T\left[\Phi(\rho + \varrho) \otimes P^{-1}\right]\left(\dot{\hat{\zeta}}(t) - \dot{\hat{\delta}}(t)\right)$$

$$+ \gamma_1\left(\hat{\zeta}(t) - \hat{\delta}(t)\right)^T\left[(\Phi\varrho + \rho - cI_N) \otimes \Theta\right]\left(\hat{\zeta}(t) - \hat{\delta}(t)\right)$$

$$= \gamma_1\left(\hat{\zeta}(t) - \hat{\delta}(t)\right)^T\left[\Phi(\rho + \varrho) \otimes (A^TP^{-1} + P^{-1}A) - (\rho + \varrho)\right.$$

$$\cdot \left.\left(\overline{\mathcal{L}}^T\Phi + \Phi\overline{\mathcal{L}}\right)(\rho + \varrho) \otimes \Theta + (\Phi\varrho + \rho - cI_N) \otimes \Theta\right]\left(\hat{\zeta}(t) - \hat{\delta}(t)\right)$$

$$+ 2\gamma_1\left(\hat{\zeta}(t) - \hat{\delta}(t)\right)^T\left[(\rho + \varrho)\Phi\overline{\mathcal{L}} \otimes P^{-1}D\right]\tilde{d}(t)$$

$$- 2\gamma_1\left(\hat{\zeta}(t) - \hat{\delta}(t)\right)^T\left[\Phi(\rho + \varrho) \otimes P^{-1}(GA - FC - A)\right]\tilde{\delta}(t)$$

$$\leq \gamma_1\left(\hat{\zeta}(t) - \hat{\delta}(t)\right)^T\left\{\Phi(\rho + \varrho) \otimes (A^TP^{-1} + P^{-1}A)\right.$$

$$- \left.\left[\lambda_0\Phi(\rho + \varrho)^2 - (\Phi\varrho + \rho - cI_N)\right] \otimes\Theta\right\}\left(\hat{\zeta}(t) - \hat{\delta}(t)\right)$$

$$+ 2\gamma_1\left(\hat{\zeta}(t) - \hat{\delta}(t)\right)^T\left[(\rho + \varrho)\Phi\overline{\mathcal{L}} \otimes P^{-1}D\right]\tilde{d}(t)$$

$$- 2\gamma_1\left(\hat{\zeta}(t) - \hat{\delta}(t)\right)^T\left[\Phi(\rho + \varrho) \otimes P^{-1}(GA - FC - A)\right]\tilde{\delta}(t), \qquad (4.44)$$

where $\lambda_0 = \lambda_{\min}(\overline{\mathcal{L}} + \Phi^{-1}\overline{\mathcal{L}}^T\Phi)$. By using the similar arguments made in (4.24) *and* (4.25), *we get*

$$2\left(\zeta(t) - \hat{\delta}(t)\right)^T\left[(\rho + \varrho)\Phi\overline{\mathcal{L}} \otimes P^{-1}D\right]\tilde{d}(t)$$

$$\leq \frac{\lambda_0}{4}\left(\zeta(t) - \hat{\delta}(t)\right)^T\left[\Phi(\rho + \varrho)^2 \otimes \Theta\right]\left(\zeta(t) - \hat{\delta}(t)\right)$$

$$+ \frac{4\iota_4\lambda_{\max}(D^TD)}{\lambda_0\iota_2}\tilde{d}^T(t)\left(\Phi \otimes I_p\right)\tilde{d}(t), \qquad (4.45)$$

$$- 2\left(\zeta(t) - \hat{\delta}(t)\right)^T\left[\Phi(\rho + \varrho) \otimes P^{-1}(GA - FC - A)\right]\tilde{\delta}(t)$$

$$\leq \frac{\lambda_0}{4}\left(\zeta(t) - \hat{\delta}(t)\right)^T\left[\Phi(\rho + \varrho)^2 \otimes \Theta\right]\left(\zeta(t) - \hat{\delta}(t)\right)$$

$$+ \frac{4\iota_3}{\lambda_0\iota_2}\tilde{\delta}^T(t)\left(\Phi \otimes I_n\right)\tilde{\delta}(t). \qquad (4.46)$$

Moreover,

$$- \left(\hat{\zeta}(t) - \hat{\delta}(t)\right)^T \left\{\left[\frac{\lambda_0}{4}\Phi(\rho+\varrho)^2 - (\Phi(\varrho+\rho) - cI_N)\right] \otimes \Theta\right\} \left(\hat{\zeta}(t) - \hat{\delta}(t)\right)$$

$$\leq - \left(\hat{\zeta}(t) - \hat{\delta}(t)\right)^T \left\{\left[\frac{\lambda_0}{4}\Phi(\rho+\varrho)^2 + \frac{\lambda_0}{4}\Phi\rho^2 + \frac{\lambda_0}{4}\Phi\varrho^2\right.\right.$$
$$\left.\left. - \frac{1}{2}\left(\frac{2\Phi}{\lambda_0} + \frac{\lambda_0}{2}\Phi\varrho^2\right) - \rho + cI_N\right] \otimes \Theta\right\} \left(\hat{\zeta}(t) - \hat{\delta}(t)\right)$$

$$\leq - \left(\hat{\zeta}(t) - \hat{\delta}(t)\right)^T \left\{\left[\frac{\lambda_0}{4}\Phi(\rho+\varrho)^2 - \frac{\Phi}{\lambda_0} + \left(\frac{\sqrt{\lambda_0}}{2}\Phi^{\frac{1}{2}}\rho\right.\right.\right.$$
$$\left.\left.\left. - \frac{1}{\sqrt{\lambda_0}}\Phi^{-\frac{1}{2}}\right)^2 - \frac{1}{\lambda_0}\Phi^{-1} + cI_N\right] \otimes \Theta\right\} \left(\hat{\zeta}(t) - \hat{\delta}(t)\right)$$

$$\leq - \left(\hat{\zeta}(t) - \hat{\delta}(t)\right)^T \left\{\left[\frac{\lambda_0}{4}\Phi(\rho+\varrho)^2 + \bar{c}I_N\right] \otimes \Theta\right\} \left(\hat{\zeta}(t) - \hat{\delta}(t)\right)$$

$$\leq - \left(\hat{\zeta}(t) - \hat{\delta}(t)\right)^T \left[\sqrt{\bar{c}\lambda_0}\Phi^{\frac{1}{2}}(\rho+\varrho) \otimes \Theta\right] \left(\hat{\zeta}(t) - \hat{\delta}(t)\right)$$

$$\leq - \left(\hat{\zeta}(t) - \hat{\delta}(t)\right)^T \left[\Phi(\rho+\varrho) \otimes 2\alpha\Theta\right] \left(\hat{\zeta}(t) - \hat{\delta}(t)\right), \tag{4.47}$$

where we choose $c \geq \bar{c} + \bar{\phi}/\lambda_0 + 1/(\lambda_0\underline{\phi})$ with $\sqrt{\bar{c}\lambda_0}/\bar{\phi}^{\frac{1}{2}} \geq 2\alpha$ to get the last two inequalities, here $\bar{\phi} = \max_{i=1,\dots,N} \phi_i$ and $\underline{\phi} = \min_{i=1,\dots,N} \phi_i$. Substituting (4.45)– (4.47) into (4.44) gives that

$$\dot{V}_{22}(t) \leq - \gamma_1 \left(\hat{\zeta}(t) - \hat{\delta}(t)\right)^T \left[\Phi(\rho+\varrho) \otimes \Lambda + \frac{\lambda_0}{4}\Phi(\rho+\varrho)^2 \otimes \Theta\right] \left(\hat{\zeta}(t) - \hat{\delta}(t)\right)$$
$$+ \frac{4\gamma_1\iota_4\lambda_{\max}(D^T D)}{\lambda_0\iota_2}\tilde{d}^T(t)(\Phi \otimes I_p)\tilde{d}(t) + \frac{4\gamma_1\iota_3}{\lambda_0\iota_2}\tilde{\delta}^T(t)(\Phi \otimes I_n)\tilde{\delta}(t). \tag{4.48}$$

We choose appropriate Q such that $\bar{\Lambda} = \frac{\lambda_0}{2}QD - W^T - W$ is positive definite. Calculating the derivative of $V_{23}(t)$ along the trajectory (4.39), we have

$$\dot{V}_{23}(t) = \gamma_1\gamma_2\tilde{d}^T(t)\left[\Phi \otimes (W^T + W) - (\bar{\mathcal{L}}^T\Phi + \Phi\bar{\mathcal{L}}) \otimes QD\right]\tilde{d}(t)$$
$$+ 2\gamma_1\gamma_2\tilde{d}^T(t)\left[\Phi \otimes Q(GA - FC - A)\right]\tilde{\delta}(t)$$
$$\leq \gamma_1\gamma_2\tilde{d}^T(t)\left[\Phi \otimes (W^T + W - \lambda_0 QD)\right]\tilde{d}(t)$$
$$+ \gamma_1\gamma_2\left[\frac{\lambda_0\lambda_{\min}(QD)}{2\lambda_{\max}(QQ^T)}\tilde{d}^T(t)\left(\Phi \otimes QQ^T\right)\tilde{d}(t)\right.$$
$$\left. + \frac{2\iota_3\lambda_{\max}(QQ^T)}{\lambda_0\lambda_{\min}(QD)}\tilde{\delta}^T(t)(\Phi \otimes I_n)\tilde{\delta}(t)\right]$$
$$\leq \gamma_1\gamma_2\tilde{d}^T(t)\left[\Phi \otimes \left(W^T + W - \frac{\lambda_0}{2}QD\right)\right]\tilde{d}(t)$$
$$+ \frac{2\gamma_1\gamma_2\iota_3\lambda_{\max}(QQ^T)}{\lambda_0\lambda_{\min}(QD)}\tilde{\delta}^T(t)(\Phi \otimes I_n)\tilde{\delta}(t)$$
$$\leq - \gamma_1\gamma_2\tilde{d}^T(t)(\Phi \otimes \bar{\Lambda})\tilde{d}(t)$$

$$+ \frac{2\gamma_1\gamma_2\iota_3\lambda_{\max}(QQ^T)}{\lambda_0\lambda_{\min}(QD)}\tilde{\delta}^T(t)(\Phi \otimes I_n)\tilde{\delta}(t). \tag{4.49}$$

Calculating the derivative of $V_{24}(t)$ along the trajectory (4.38), we have

$$\dot{V}_{24}(t) = \gamma_1\gamma_3\tilde{\delta}^T(t)\left\{\Phi \otimes \left[(GA - FC)^T S + S(GA - FC)\right]\right\}\tilde{\delta}(t)$$

$$= -\gamma_1\gamma_3\tilde{\delta}^T(t)\left(\Phi \otimes \tilde{\Lambda}\right)\tilde{\delta}(t), \tag{4.50}$$

where $\tilde{\Lambda} = -(GA - FC)^T S - S(GA - FC)$ is positive definite.
We then conclude from (4.43) to (4.50) that

$$\dot{V}_2(t) \leq -\frac{1}{2}\hat{\zeta}^T(t)(I_N \otimes \Lambda)\hat{\zeta}(t) - \gamma_1\left(\hat{\zeta}(t) - \hat{\delta}(t)\right)^T[\Phi(\rho + \varrho) \otimes \Lambda]\left(\hat{\zeta}(t) - \hat{\delta}(t)\right)$$

$$- \left(\frac{\gamma_1\lambda_0\phi}{4} - \frac{2\iota_4\lambda_{\max}(\Theta)}{\lambda_{\min}(\Lambda)}\right)\left(\hat{\zeta}(t) - \hat{\delta}(t)\right)^T\left[(\rho + \varrho)^2 \otimes \Theta\right]\left(\hat{\zeta}(t) - \hat{\delta}(t)\right)$$

$$- \gamma_1\left[\gamma_2\tilde{d}^T(t)(\Phi \otimes \bar{\Lambda})\tilde{d}(t) - \frac{4\iota_4\lambda_{\max}(D^T D)}{\lambda_0\iota_2}\tilde{d}^T(t)(\Phi \otimes I_p)\tilde{d}(t)\right]$$

$$- \gamma_1\left[\tilde{\delta}^T(t)\left(\gamma_3(\Phi \otimes \tilde{\Lambda}) - \left(\frac{2\gamma_2\iota_3\lambda_{\max}(QQ^T)}{\lambda_0\lambda_{\min}(QD)} + \frac{4\iota_3}{\lambda_0\iota_2}\right)(\Phi \otimes I_n)\right)\tilde{\delta}(t)\right]. \tag{4.51}$$

Now we choose

$$\gamma_1 \geq \frac{8\iota_4\lambda_{\max}(\Theta)}{\lambda_{\min}(\Lambda)\lambda_0\phi},$$

$$\gamma_2 \geq \frac{4\iota_4\lambda_{\max}(D^T D)}{\lambda_0\iota_2\lambda_{\min}(\bar{\Lambda})} + \frac{1}{\lambda_{\min}(\bar{\Lambda})},$$

$$\gamma_3 \geq \frac{2\gamma_2\iota_3\lambda_{\max}(QQ^T)}{\lambda_0\lambda_{\min}(QD)\lambda_{\min}(\tilde{\Lambda})} + \frac{4\iota_3}{\lambda_0\iota_2\lambda_{\min}(\tilde{\Lambda})} + \frac{1}{\lambda_{\min}(\tilde{\Lambda})}.$$

This together with (4.51) gives

$$\dot{V}_2(t) \leq -\frac{1}{2}\hat{\zeta}^T(t)(I_N \otimes \Lambda)\hat{\zeta}(t) - \gamma_1\left(\hat{\zeta}(t) - \hat{\delta}(t)\right)^T[\Phi(\rho + \varrho) \otimes \Lambda]\left(\hat{\zeta}(t) - \hat{\delta}(t)\right)$$

$$- \gamma_1\tilde{d}^T(t)(\Phi \otimes I_n)\tilde{d}(t) - \gamma_1\tilde{\delta}^T(t)(\Phi \otimes I_n)\tilde{\delta}(t), \tag{4.52}$$

which implies $\dot{V}_2(t) \leq 0$ holds for arbitrary time instant $t \geq t_0$. We then get from $\dot{V}_2(t) \leq 0$ and $V_2(t) \geq 0$ that the limit of $V_2(t)$ as $t \to \infty$ exists, and we denote the limit by $V_2(\infty)$. It is certainly that $0 \leq V_2(\infty) \leq V_2(t) \leq V_2(t_0)$. This together with (4.42) shows $\hat{\zeta}(t)$, $\hat{\zeta}(t) - \hat{\delta}(t)$, $\tilde{d}(t)$, $\tilde{\delta}(t) \in \mathbb{L}_\infty$, and ρ, ϱ are bounded. We then learn from (4.37) to (4.40) that $\dot{\hat{\zeta}}(t)$, $\dot{\hat{\zeta}}(t) - \dot{\hat{\delta}}(t)$, $\dot{\tilde{d}}(t)$, $\dot{\tilde{\delta}}(t) \in \mathbb{L}_\infty$. On the other hand, by integrating both sides of (4.52), we have

$$\int_{t_0}^\infty \frac{1}{2}\left\{\hat{\zeta}^T(t)(I_N \otimes \Lambda)\hat{\zeta}(t) + \gamma_1\left(\hat{\zeta}(t) - \hat{\delta}(t)\right)^T[\Phi(\rho + \varrho) \otimes \Lambda](\hat{\zeta}(t) - \hat{\delta}(t))\right.$$

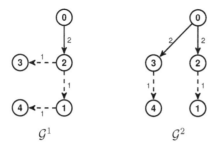

Figure 4.1 The communication graphs $\mathcal{G}^1$ and $\mathcal{G}^2$.

$$+ \gamma_1 \tilde{d}^T(t)(\Phi \otimes I_n)\tilde{d}(t) + \gamma_1 \tilde{\delta}^T(t)(\Phi \otimes I_n)\tilde{\delta}(t)\Big\}$$
$$\leq V_2(t_0) - V_2(\infty),$$

which implies $\hat{\zeta}(t)$, $\hat{\zeta}(t) - \hat{\delta}(t)$, $\tilde{d}(t)$, $\tilde{\delta}(t) \in \mathbb{L}_2$ as well. According to the well-known Barbălat lemma, the norm $\|\hat{\zeta}(t)\|$, $\|\hat{\zeta}(t) - \hat{\delta}(t)\|$, $\|\tilde{d}(t)\|$, $\|\tilde{\delta}(t)\|$ approach 0 as $t \to \infty$. Since $\delta(t) = \hat{\delta}(t) - \tilde{\delta}(t)$, $\|\delta(t)\| \to 0$ which further implies consensus tracking is achieved. In addition, $\|\tilde{d}(t)\| \to 0$ implies the disturbances are fully rejected. And the coupling strengths ρ_i converge to some positive constants since $\dot{\rho}_i \geq 0$ and ρ_i is bounded.

Remark 4.4 *We learn from the preceding analysis that the coupling strengths α in (4.36), ρ_i in (4.34) are independent of any global information and the feedback gain matrices K in (4.36), F in (4.5) only depend on the agents' inherent dynamics. So the consensus disturbance rejection can be achieved in a fully distributed way. Unlike Theorem 4.2, Theorem 4.3 aims at presenting some sufficient criteria for achieving fully distributed consensus disturbance rejection in CNSs with directed fixed communication topologies. Nevertheless, it is much more interesting but more challenging to investigate the fully distributed consensus disturbance rejection problem for CNSs with directed switching communication topologies. However, it is still an outstanding issue even for consensus problem of CNSs without any uncertainties or disturbances as far as we know.*

Remark 4.5 *In contrast to the consensus disturbance controllers given in [34, 148] which require the relative states among neighboring agents, the controller (4.36) uses the information of the state estimator as well as the disturbance observer which are designed upon the relative output information among neighboring agents. Hence, our results are much better for practical applications.*

4.5 NUMERICAL SIMULATIONS

We perform two examples to validate Theorems 4.2 and 4.3, respectively. The CNSs under consideration consist of five YF-22 research UAVs [34] whose longitudinal dynamics satisfy (4.1) with

$$A = \begin{bmatrix} -0.284 & -23.096 & 2.420 & 9.913 \\ 0 & -4.117 & 0.843 & 0.272 \\ 0 & -33.884 & -8.263 & -19.543 \\ 0 & 0 & 1 & 0 \end{bmatrix},$$

$$B = \begin{bmatrix} 20.168 \\ 0.544 \\ -39.085 \\ 0 \end{bmatrix}, \quad D = B \begin{bmatrix} 1 & 0 \end{bmatrix}, \quad C = \begin{bmatrix} 1 & 1 & 0 & 0 \\ 0 & 0 & 1 & 1 \end{bmatrix},$$

where $x_i(t) = [x_{i1}(t), x_{i2}(t), x_{i3}(t), x_{i4}(t)]^T$ and $x_{i1}(t)$, $x_{i2}(t)$, $x_{i3}(t)$, $x_{i4}(t)$ represent, respectively, the speed, the attack angle, the pitch rate, and the pitch angle, $i = 0, 1, \ldots, 4$. The harmonic disturbances are generated by (4.2) with $d_i(t) = [d_{i1}(t), d_{i2}(t)]^T$ and

$$W = \begin{bmatrix} 0 & 1.5 \\ -1.5 & 0 \end{bmatrix}.$$

It is not difficult to verify that Assumptions 4.1, 4.2 and the conditions (1) and (2) in Remark 4.2 hold. Then, we get

$$H = \begin{bmatrix} -0.2135 & -0.0058 & 0.4137 & 0 \\ 0.4029 & 0.0109 & -0.7808 & 0 \end{bmatrix}^T,$$

$$G = \begin{bmatrix} 0.7865 & -0.2135 & 0.4029 & 0.4029 \\ -0.0058 & 0.9942 & 0.0109 & 0.0109 \\ 0.4137 & 0.4137 & 0.2192 & -0.7808 \\ 0 & 0 & 0 & 1 \end{bmatrix}.$$

Solving the LMI (4.8) gives that

$$F = \begin{bmatrix} 11.3220 & 1.5912 & 5.9677 & 0.0839 \\ 5.2910 & 0.7606 & 3.1369 & 0.3498 \end{bmatrix}^T.$$

Example 1: This example is given to validate Theorem 4.2. The communication topologies $\mathcal{G}^1$ and $\mathcal{G}^2$ are plotted in Figure 4.1, where the numbers around the edges represent the weights. It is obviously that Assumption 4.3 hold. Direct calculation gives $\lambda_0 = 1.0561$ and $\nu = 2$. Set $\alpha = 1$, we can choose $\rho = \mu = 4 > 3.7876$. By solving the LMIs (4.18) and (4.19), we get

$$P^{-1} = \begin{bmatrix} 0.0166 & -0.1772 & 0.0089 & 0.1785 \\ -0.1772 & 1.9042 & -0.0963 & -1.9156 \\ 0.0089 & -0.0963 & 0.0053 & 0.0983 \\ 0.1785 & -1.9156 & 0.0983 & 1.9343 \end{bmatrix},$$

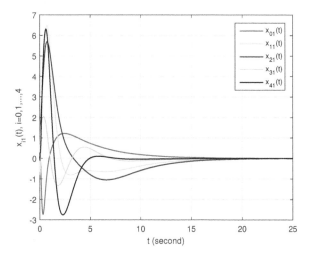

Figure 4.2 The agents' state trajectories $x_{i1}(t)$, $i = 0, 1, \ldots, 4$, in Example 1.

$$K = \begin{bmatrix} 0.1108 & -1.2277 & 0.0807 & 1.2849 \end{bmatrix},$$

$$Q = \begin{bmatrix} 0.4620 & 0.0125 & -0.8953 & 0 \\ 0.6013 & 0.0162 & -1.1652 & 0 \end{bmatrix}.$$

According to Theorem 4.2, the consensus disturbances rejection is achieved if $\tau_a = 0.7 > 0.6931$. Suppose $N_0 = 4$ and the communication topologies switch between $\mathcal{G}^1$ and $\mathcal{G}^2$ at the time instants $t = 0.2\,\mathrm{s}, 0.5\,\mathrm{s}, 0.8\,\mathrm{s}, 1\,\mathrm{s}$, and switch periodically with period $0.7\,\mathrm{s}$ after $t = 1\,\mathrm{s}$. The evolution of all the five agents are plotted in Figs. 4.2–4.5 which show the consensus tracking is achieved. The evolution of the disturbances (4.2) and the disturbance observer (4.10) are plotted in Figure 4.6 which shows the harmonic disturbances are fully rejected. Hence, this example validates Theorem 4.2 very well.

Example 2: This example is given to validate Theorem 4.3. The communication topology $\mathcal{G}^1$ is plotted in Figure 4.1 which clearly satisfies Assumption 4.3. Let $\alpha = 1$. Solving the LMI (4.41) gives that

$$P^{-1} = \begin{bmatrix} 0 & -0.0001 & 0 & 0 \\ -0.0001 & 0.0028 & -0.0001 & -0.0019 \\ 0 & -0.0001 & 0.0001 & 0.0004 \\ 0 & -0.0019 & 0.0004 & 0.0027 \end{bmatrix},$$

$$K = \begin{bmatrix} -0.0008 & -0.0059 & 0.0054 & 0.0172 \end{bmatrix}.$$

By solving the LMI (4.19) and setting $Q = R^{-1}D^T$, we have

$$Q = \begin{bmatrix} 0.0608 & 0.0016 & -0.1178 & 0 \\ 0.0791 & 0.0021 & -0.1533 & 0 \end{bmatrix}.$$

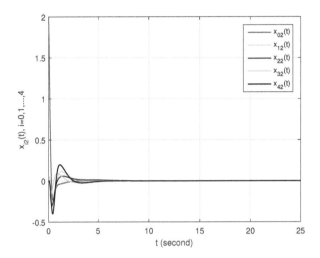

Figure 4.3 The agents' state trajectories $x_{i2}(t)$, $i = 0, 1, \ldots, 4$, in Example 1.

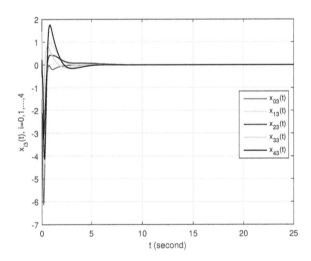

Figure 4.4 The agents' state trajectories $x_{i3}(t)$, $i = 0, 1, \ldots, 4$, in Example 1.

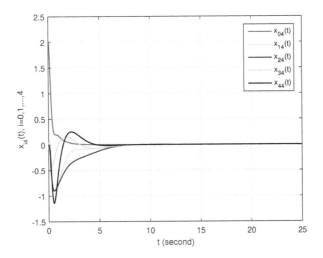

Figure 4.5 The agents' state trajectories $x_{i4}(t)$, $i = 0, 1, \ldots, 4$, in Example 1.

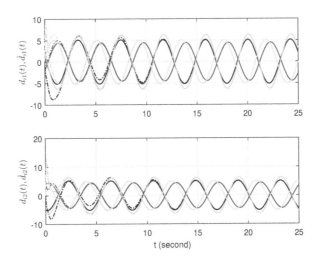

Figure 4.6 The evolution of matching disturbances, where the solid lines and the dashed lines represent, respectively, the disturbances $d_{ij}(t)$ and $\hat{d}_{ij}(t)$, $i = 1, \ldots, 4$, $j = 1, 2$, in Example 1.

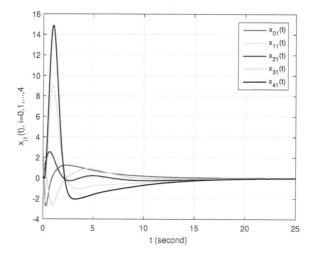

Figure 4.7 The agents' state trajectories $x_{i1}(t)$, $i = 0, 1, \ldots, 4$, in Example 2.

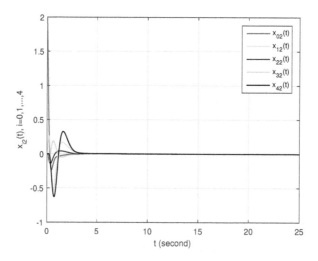

Figure 4.8 The agents' state trajectories $x_{i2}(t)$, $i = 0, 1, \ldots, 4$, in Example 2.

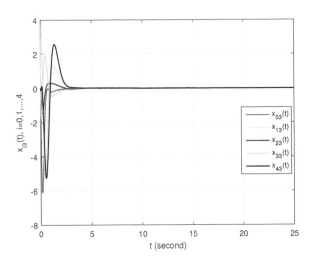

Figure 4.9 The agents' state trajectories $x_{i3}(t)$, $i = 0, 1, \ldots, 4$, in Example 2.

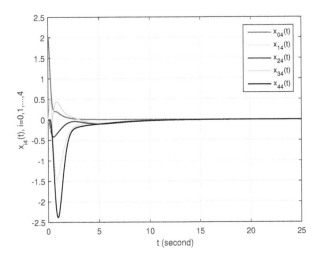

Figure 4.10 The agents' state trajectories $x_{i4}(t)$, $i = 0, 1, \ldots, 4$, in Example 2.

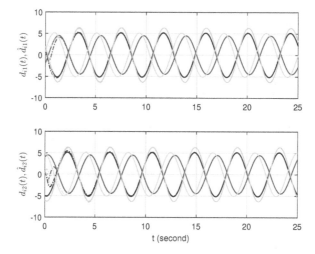

Figure 4.11 The evolution of matching disturbances, where the solid lines and the dashed lines represent, respectively, the disturbances $d_{ij}(t)$ and $\hat{d}_{ij}(t)$, $i = 1, \ldots, 4$, $j = 1, 2$, in Example 2.

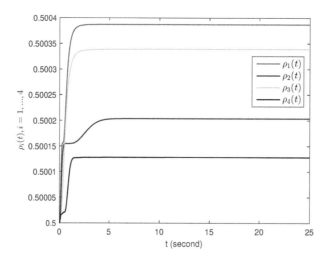

Figure 4.12 The evolution of the coupling strengths $\rho_i(t)$, $i = 1, \ldots, 4$, in Example 2.

According to Theorem 4.3, the fully distributed consensus disturbance rejection is achieved. The evolution of all the five agents are plotted in Figs. 4.7–4.10 which show the consensus tracking is achieved. The evolution of the disturbances (4.2) and the disturbance observer (4.35) are plotted in Figure 4.11 which shows the harmonic disturbances are fully rejected. Figure 4.12 shows the coupling strengths converge to some positive constants finally. Hence, this example validates Theorem 4.3 very well.

4.6 CONCLUSIONS

We have solved the consensus disturbance rejection problem for MIMO linear CNSs subject to a class of harmonic nonvanishing disturbances under directed fixed topology as well as directed switching topologies. Based on the relative output information among neighboring agents, each follower is equipped with a newly-designed UIO that could estimate the exact consensus error. With the aid of this UIO, a state estimator with static coupling and a disturbance observer were developed upon which the controller was designed for CNSs with directed switching communication topologies. Furthermore, a state estimator with adaptive coupling law was proposed for CNSs with directed fixed communication topology such that the designed controller could be implemented in a fully distributed manner. The simulations of UAVs indicate the prospect of the proposed controllers being used in engineering practices. In the future, we shall study the fully distributed consensus problem of CNSs with directed switching communication topologies which is a more challenging issue.

The main contributions of this chapter are listed as follows: (**I**) Although consensus disturbance rejection was studied in [34, 148, 170] for CNSs with fixed topology, this problem is solved for CNSs with directed switching topologies for the first time. As the technical analysis made for achieving our goal is much more challenging. Moreover, in contrast to the general linear CNSs studied in [34, 148], MIMO linear CNSs are investigated in this chapter which pose more challenge on the controller design since only the agents' output information are available. (**II**) In contrast to the recent work [193], disturbance rejection problem is further considered in this chapter. As disturbances are inevitable in real applications, thus the obtained results are much more suitable for practical applications.

CHAPTER 5

Consensus tracking of CNSs with first-order nonlinear dynamics and directed switching topologies

This chapter studies the consensus tracking of CNSs with first-order nonlinear dynamics and directed switching topologies. This chapter begins by overviewing some previous works and indicating our motivations. Section 5.2 studies the case with Lipschitz type nonlinear dynamics without assuming that each possible network topology has a directed spanning tree. Specifically, this section proposes an algorithm for selecting the pinned nodes such that the graph contains a directed spanning tree. Section 5.3 studies the case with Lorenz type nonlinear dynamics under directed fixed topology as well as directed switching topologies, where the Lorenz systems include the Chen and Lü systems. Finally, some simulations are given to validate the obtained theoretical results.

5.1 INTRODUCTION

According to whether the final synchronization states depend on the initial value or not, synchronization in CNSs can be generally categorized into local synchronization [98, 102] and global synchronization [96]. Compared with the local synchronization, the global synchronization means that the network synchronization can be achieved under any given initial conditions, thus is more favorable in practical applications. In [96], a distance between the nodes' states and the synchronization manifold was introduced, based on which a new methodology was proposed to investigate the global synchronization of coupled systems. Later, general algebraic connectivity was proposed in [218] to study the global synchronization as well as local synchronization problems in strongly connected networks. Global synchronization for coupled linear systems via state or output feedback control was studied in [224]. In [179, 204], global synchronization for a class of CNSs with sampling-data coupling was addressed. For the case that the considered networks are not strongly connected or even do not contain any directed spanning tree, the pinning synchronization problem arises [74, 176, 178, 216]. Pinning synchronization in scale-free and small-world complex networks were addressed in [178] and [176], respectively. Later, local and global pinning synchronization in random and scale-free networks were studied in [74]. It is worth noting that global synchronization is actually consensus tracking by regarding the target system in the network as a leader. In [216], pinning synchronization of undirected CNSs was further addressed. Without assuming the network topology is undirected or strongly connected, it was proved in [20] that a single controller can pin a coupled CNS to its homogeneous trajectory under some suitable conditions. Global pinning synchronization for a class of CNSs has been investigated in [66] under a V-stability framework. However, it is previously assumed in the aforementioned literature that each possible network topology contains a directed spanning tree with the leader being the root node. This indicates that each agent in the considered network can be influenced by the leader directly or indirectly. In some real cases, the aforementioned condition is hard or even impossible to verify.

Motivated by the aforementioned works on consensus tracking (i.e., global pinning synchronization) of CNSs, this chapter aims to solve the consensus tracking problem for a class of switched CNSs where some possible network topologies may not contain any directed spanning tree. By using a combined tool from M-matrix theory and stability analysis of switched systems, a new kind of topology-dependent MLFs for the switched networks is constructed. Theoretical analysis indicates that the consensus tracking in such a CNS can be achieved if some carefully selected nodes are pinned such that the network topology contains a directed spanning tree rooted at the target node frequently enough as the network evolves with time. Without causing any confusion, global pinning synchronization is referred as consensus tracking in the subsequent analysis in this chapter.

5.2 CONSENSUS TRACKING OF CNS WITH LIPSCHITZ TYPE DYNAMICS

5.2.1 Model formulation

Suppose that the considered CNS consists of N nodes, the dynamics of agent i are given by

$$\dot{x}_i(t) = f(x_i(t), t) + \alpha \sum_{j=1}^{N} a_{ij}(t)(x_j(t) - x_i(t)), \tag{5.1}$$

where $x_i(t) = [x_{i1}(t), \ldots, x_{in}(t)]^T \in \mathbb{R}^n$ for $i = 1, \ldots, N$ represent the states of agent i, $\alpha > 0$ is the coupling strength, and $\mathcal{A}(t) = [a_{ij}(t)]_{N \times N}$ is the adjacency matrix of graph $\mathcal{G}(\mathcal{A}(t))$ at time t. Throughout this chapter, the derivatives of all functions at switching time points should be considered as their right-hand derivatives. According to the definition of Laplacian matrix for a graph, it follows from (5.1) that

$$\dot{x}_i(t) = f(x_i(t), t) - \alpha \sum_{j=1}^{N} l_{ij}(t)x_j(t), \tag{5.2}$$

where $\mathcal{L}(t) = [l_{ij}(t)]_{N \times N}$ is the Laplacian matrix of graph $\mathcal{G}(\mathcal{A}(t))$.

The control goal in this section is to design pinning controllers for some appropriately selected agents in (5.2) such that the states of each agent in the considered network will approach $s(t)$ when t approaches $+\infty$, i.e., $\lim_{t\to\infty}\|x_i(t) - s(t)\| = 0$, for all $i = 1, \dots, N$ and arbitrarily given initial conditions, where

$$\dot{s}(t) = f(s(t), t). \tag{5.3}$$

Here, $s(t)$ may be an equilibrium point, a periodic orbit, or even a chaotic orbit. Motivated by the works in [74], pinning network (5.2) by using linear controllers $-\alpha c_i(t)(x_i(t) - s(t))$ to agent i leads to

$$\dot{x}_i(t) = f(x_i(t), t) - \alpha \sum_{j=1}^{N} l_{ij}(t)x_j(t) - \alpha c_i(t)(x_i(t) - s(t)), \tag{5.4}$$

where $c_i(t) \in \{0, 1\}$ and $c_i(t) = 1$ if and only if agent i of (5.2) is pinned at time t. Let $e_i(t) = x_i(t) - s(t)$, $i = 1, \dots, N$. It thus follows from (5.4) that

$$\dot{e}_i(t) = f(x_i(t), t) - f(s(t), t) - \alpha \sum_{j=1}^{N} l_{ij}(t)e_j(t) - \alpha c_i(t)e_i(t). \tag{5.5}$$

By taking the target system (5.3) as a virtual leader of the CNS under consideration, one may get the augmented network topology $\mathcal{G}(\widetilde{\mathcal{A}}(t))$ consisting of $N+1$ agents. Labeling the index of the virtual agent as 0, the Laplacian matrix $\widetilde{L}(t)$ of the augmented network topology $\mathcal{G}(\widetilde{\mathcal{A}}(t))$ can be written as:

$$\widetilde{\mathcal{L}}(t) = \begin{bmatrix} 0 & \mathbf{0}_N^T \\ \mathbf{P}(t) & \overline{\mathcal{L}}(t) \end{bmatrix} \in \mathbb{R}^{(N+1)\times(N+1)}, \tag{5.6}$$

$$\overline{\mathcal{L}}(t) = \begin{bmatrix} \sum_{j\in\mathcal{N}_1} a_{1j}(t) & -a_{12}(t) & \cdots & -a_{1N}(t) \\ -a_{21}(t) & \sum_{j\in\mathcal{N}_2} a_{2j}(t) & \cdots & -a_{2N}(t) \\ \vdots & \vdots & \ddots & \vdots \\ -a_{N1}(t) & -a_{N2}(t) & \cdots & \sum_{j\in\mathcal{N}_N} a_{Nj}(t) \end{bmatrix}, \tag{5.7}$$

where $\mathbf{P}(t) = -[a_{10}(t), \dots, a_{N0}(t)]^T$ with $a_{i0}(t) = c_i(t)$, $i = 1, \dots, N$. In the present section, the augmented network $\mathcal{G}(\widetilde{\mathcal{A}}(t))$ does not need to contain a directed spanning tree all the time. Note that the initial network topologies of practical CNSs always meet some connectivity conditions. For example, in the flocking control problem, as studied in [182], the initial network topology is assumed to be connected. In the context of consensus tracking, the condition that the augmented network topology $\mathcal{G}(\widetilde{\mathcal{A}}(0))$ contains a directed spanning tree can be ensured by appropriately pinning some agents in network $\mathcal{G}(\mathcal{A}(0))$ selected by using the following linear time complexity algorithm.

Algorithm 5.1 *Find the strongly connected components and the agent with zero in-degree in $\mathcal{G}(\mathcal{A}(0))$ by using Tarjan's algorithm [157]. Suppose that there are $\widetilde{\omega}$ ($\widetilde{\omega} \geq 0$) strongly connected components, labeled as $\mathcal{G}_1, \dots, \mathcal{G}_{\widetilde{\omega}}$, and $\widehat{\omega}$ ($\widehat{\omega} \geq 0$) agents with zero in-degree, labeled as $\nu_1, \dots, \nu_{\widehat{\omega}}$, in $\mathcal{G}(\mathcal{A}(0))$. Set $\kappa_0 = 0$ and $m = 1$.*

(1) Check whether the following condition holds: $\widehat{\omega} > 0$. If it does not hold, go to step (2); else, pin all the $\widehat{\omega}$ agents with zero in-degree and update the value of κ_0 by $\kappa_0 = \kappa_0 + \widehat{\omega}$;

(2) Check whether the following condition holds: $\widetilde{\omega} > 1$. If it holds, go to step (3); else stop.

(3) Check whether there exists at least one agent belonging to $\mathcal{G}_m$ which is reachable from at least one agent belonging to $\mathcal{G}_j$, $j = 1, \ldots, \widetilde{\omega}, j \neq m$. If it holds, go to step (4); if it does not hold, go to step (5).

(4) Check whether the following condition holds: $m < \widetilde{\omega}$. If it holds, let $m = m + 1$ and re-perform step (3); else stop.

(5) Arbitrarily select one agent in $\mathcal{G}_m$ to be pinned. Let $\kappa_0 = \kappa_0 + 1$. Check whether the following condition holds: $m < \widetilde{\omega}$. If it holds, let $m = m + 1$ and re-perform step (3); else stop.

Remark 5.1 *Note that the time complexity of Tarjan's algorithm is $\mathcal{O}(N + |\mathcal{E}(0)|)$, where N and $|\mathcal{E}(0)|$ are respectively the order and size of $\mathcal{G}(\mathcal{A}(0))$, and $\mathcal{O}(\cdot)$ is a Landau symbol representing the complexity function. It is not difficult to verify that $\mathcal{G}(\widetilde{A}(0))$ will contain a directed spanning tree if the κ_0 agents selected by Algorithm 5.1 are pinned. It can also be concluded that κ_0 is the minimal number of pinned vertices such that the augmented network topology contains a directed spanning tree.*

However, some links may be lost as the networked systems evolve with time. To solve such a consensus tracking problem, it is previously assumed in some existing works that each possible network topology contains at least a directed spanning tree or the agents could synchronously discard the incoming links when the network topology does not contain any directed spanning tree. However, it is sometimes difficult or even impossible to implement such an information discarding scheme for CNSs especially for those with huge size.

In this section, the consensus tracking in switched CNSs will be addressed without assuming that each possible network topology has a directed spanning tree. Furthermore, the dynamic agents in the considered switched networks do not need to discard the incoming links synchronously as the networks evolve with time. To proceed the analysis, it is assumed that the switching among the different topologies is triggered by communication links' loss or recovery and there is no agent that will be deleted from the network. Specifically, it is assumed that the augmented network contains a directed spanning tree at the beginning and some links will be lost as the network evolves with time. The above statements indicate that there may exist some time intervals over which the augmented network $\mathcal{G}(\widetilde{A}(t))$ does not contain any directed spanning tree. However, the augmented network may contain a directed spanning tree again by some repairing efforts, although it may take a certain period of time in practice.

For the convenience of analysis, suppose that there exists an infinite sequence of uniformly bounded non-overlapping time intervals $[t_k, t_{k+1})$, $k \in \mathbb{N}$, with $t_1 = 0$,

$0 < \tau_m \leq t_{k+1} - t_k \leq \tau_M < +\infty$, across which the interaction graph is time-invariant. Here, the positive constant τ_m is called the dwell time. It can be verified that the Zeno behavior is excluded during the network's evolution, i.e., $t_k \to +\infty$ as $k \to +\infty$. The time sequence $t_1, t_2, \ldots,$ is called the switching sequence, at which the network topology changes. For the convenience of expression, one may introduce a switching signal $\sigma(t) : [0, +\infty) \mapsto \{1, \ldots, q\}$. Then, let $\mathcal{G}(\tilde{\mathcal{A}}(t)) = \mathcal{G}(\tilde{\mathcal{A}}^{\sigma(t)})$ be the augmented interaction graph of network (5.4) at time t. Obviously, $\mathcal{G}(\tilde{\mathcal{A}}^{\sigma(t)}) \in \hat{\mathcal{G}}$ for all $t \geq 0$, where $\hat{\mathcal{G}} = \{\mathcal{G}(\tilde{\mathcal{A}}^1), \ldots, \mathcal{G}(\tilde{\mathcal{A}}^q)\}$, $q > 1$, denotes the set of all possible augmented directed interaction graphs. Furthermore, let $\overline{\mathcal{G}} = \{\mathcal{G}(\tilde{\mathcal{A}}^{\vartheta_1}), \ldots, \mathcal{G}(\tilde{\mathcal{A}}^{\vartheta_p})\}$ be the set of all possible augmented interaction graphs containing at least one directed spanning tree with $\{\vartheta_1, \ldots, \vartheta_p\} \subseteq \{1, \ldots, q\}$ and $\{\vartheta_1, \ldots, \vartheta_p\} \neq \varnothing$. One then has that $\{\vartheta_1, \ldots, \vartheta_p\} = \{1, \ldots, q\}$ if and only if each possible augmented interaction graph contains at least one directed spanning tree rooted at the target agent. According to the above analysis, assume that there exists an infinite sequence of uniformly bounded non-overlapping time intervals $[\bar{t}_\rho, \bar{t}_{\rho+1})$, $\rho \in \mathbb{N}$, with $\bar{t}_1 = 0$, $\overline{\tau}_M > \bar{t}_{\rho+1} - \bar{t}_\rho > \overline{\tau}_m$, such that $\sigma(\bar{t}_\rho) \in \overline{\mathcal{G}}$ for all $\rho \in \mathbb{N}$. Note that, for each $\rho \in \mathbb{N}$, there exists a $k \in \mathbb{N}$ such that $t_k = \bar{t}_\rho$. Obviously, the augmented network $\mathcal{G}(\tilde{\mathcal{A}}^{\sigma(\bar{t}_\rho)})$ contains at least one directed spanning tree with the agent $N + 1$ being the root. For the convenience of expression, let $t^\rho_{\min} = \min_{s \in \mathbb{N}} t_s$ subject to $t_s > \bar{t}_\rho$, $\sigma(t_s) \neq \sigma(\bar{t}_\rho)$. Note that the time points $\bar{t}_\rho$ and $t^\rho_{\min}$ can be designed offline or determined online as the network evolves with time (see Fig. 5.1 for illustration where it is assumed that $q = 4$ and $\overline{\mathcal{G}} = \{\mathcal{G}(\tilde{\mathcal{A}}^2), \mathcal{G}(\tilde{\mathcal{A}}^4)\}$. It can be thus obtained from Fig. 5.1 that $\bar{t}_1 = 0$, $t^1_{\min} = 0.5$, $\bar{t}_2 = 2$, and $t^2_{\min} = 2.8$).

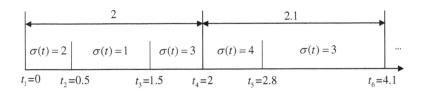

Figure 5.1 Time points t_k, $k \in \mathbb{N}$.

For the case of $t^\rho_{\min} < \bar{t}_{\rho+1}$, it follows from the above analysis that $\sigma(t) \in \mathcal{P}$, for $t \in [\bar{t}_\rho, t^\rho_{\min})$ and $\sigma(t) \in \mathcal{Q} \setminus \mathcal{P}$, for $t \in [t^\rho_{\min}, \bar{t}_{\rho+1})$, where $\mathcal{P} = \{\vartheta_1, \ldots, \vartheta_p\}$, $\mathcal{Q} = \{1, \ldots, q\}$. It thus follows from (5.5) that

$$\begin{aligned}
\dot{e}(t) &= F(e(t), t) - \alpha(\overline{\mathcal{L}}^{\sigma(\bar{t}_\rho)} \otimes I_n)e(t), \quad t \in [\bar{t}_\rho, t^\rho_{\min}), \\
\dot{e}(t) &= F(e(t), t) - \alpha(\overline{\mathcal{L}}^{\sigma(t)} \otimes I_n)e(t), \quad t \in [t^\rho_{\min}, \bar{t}_{\rho+1}),
\end{aligned} \tag{5.8}$$

for the case of $t^\rho_{\min} < \bar{t}_{\rho+1}$, and

$$\dot{e}(t) = F(e(t), t) - \alpha(\overline{\mathcal{L}}^{\sigma(\bar{t}_\rho)} \otimes I_n)e(t), \quad t \in [\bar{t}_\rho, \bar{t}_{\rho+1}), \tag{5.9}$$

for the case of $t^\rho_{\min} = \bar{t}_{\rho+1}$, where $e(t) = [e_1^T(t), \ldots, e_N^T(t)]^T$, $F(e(t), t) = [(f(x_1(t), t) -$

$f(s(t), t))^T, \ldots, (f(x_N(t), t) - f(s(t), t))^T]^T$, $\tilde{\mathcal{L}}^{\sigma(t)} = \tilde{\mathcal{L}}(t)$ is given in (5.6). Obviously, $e(t) = \mathbf{0}_{Nn}$ is a fixed point of switched systems (5.8). Furthermore, it can be verified that consensus tracking in CNS (5.2) with target trajectory generated by (5.3) will be achieved if and only if the zero equilibrium point of (5.5) is globally attractive. Thus, it is sufficient to show that consensus tracking in CNS (5.2) with target trajectory given in (5.3) could be ensured if the zero equilibrium point of (5.8) is globally asymptotically stable. It should be noted that the right hand of systems (5.9) is discontinuous due the time-dependent switching over different topologies. However, it can be verified that $e(t)$ is differentiable on the right. Furthermore, for any given initial value $e(0)$, the switched systems (5.9) possess a unique and absolutely continuous solution $e(t)$ in the sense of *Carathéodory*.

To derive the main results, the following assumption is needed.

Assumption 5.1 *There exists a semi-positive definite matrix* $\Gamma \in \mathbb{R}^{n \times n}$ *such that*

$$(x(t) - \tilde{x}(t))^T (f(x(t), t) - f(\tilde{x}(t), t)) \leq (x(t) - \tilde{x}(t))^T \Gamma (x(t) - \tilde{x}(t)), \quad (5.10)$$

for all $x(t)$, $\tilde{x}(t) \in \mathbb{R}^n$, *and* $t \geq 0$.

Note that Assumption 5.1 is satisfied by many well-known systems such as the Chua's circuit systems and the Lorenz systems. Furthermore, it is not hard to verify that Assumption 5.1 holds if the nonlinear function $f(\cdot, \cdot) : \mathbb{R}^n \times [0, +\infty) \mapsto \mathbb{R}^n$ satisfies the global Lipschitz condition.

5.2.2 Main results

Based on the analysis in the last section, one has that, for each $s \in \mathcal{P}$, $\mathcal{G}(\tilde{A}^s)$ contains a directed spanning tree rooted at agent 0. Denote the Laplacian matrix of $\mathcal{G}(\tilde{A}^s)$ by $\tilde{\mathcal{L}}^s$. Without loss of generality, let

$$\tilde{\mathcal{L}}^s = \begin{bmatrix} 0 & \mathbf{0}_N^T \\ \mathbf{P}^s & \overline{\mathcal{L}}^s \end{bmatrix} \in \mathbb{R}^{(N+1) \times (N+1)},$$

$$\overline{\mathcal{L}}^s = \begin{bmatrix} \sum_{j \in \mathcal{N}_1} a_{1j}^s & -a_{12}^s & \cdots & -a_{1N}^s \\ -a_{21}^s & \sum_{j \in \mathcal{N}_2} a_{2j}^s & \cdots & -a_{2N}^s \\ \vdots & \vdots & \ddots & \vdots \\ -a_{N1}^s & -a_{N2}^s & \cdots & \sum_{j \in \mathcal{N}_N} a_{Nj}^s \end{bmatrix}, \quad (5.11)$$

where $\mathbf{P}^s = -[a_{10}^s, \ldots, a_{N0}^s]^T$, $a_{i0}^s = c_i^s$, and $a_{i0}^s = 1$ if agent i in graph $\mathcal{G}(A^s)$ is pinned, $i = 1, \ldots, N$. According to the condition that, for each $s \in \mathcal{P}$, $\mathcal{G}(\tilde{A}^s)$ contains a directed spanning tree, then $\overline{\mathcal{L}}^s$ is a nonsingular M-matrix. Then, by using Lemma 2.15, we can get some positive definite matrices $\left(\overline{\Phi}^{\sigma(\bar{t}_\rho)} \overline{\mathcal{L}}^{\sigma(\bar{t}_\rho)} + \left(\overline{\mathcal{L}}^{\sigma(\bar{t}_\rho)} \right)^T \overline{\Phi}^{\sigma(\bar{t}_\rho)} \right)$ by letting $\overline{\Phi}^{\sigma(\bar{t}_\rho)} = \text{diag}\{\phi_1^{\sigma(\bar{t}_\rho)}, \ldots, \phi_N^{\sigma(\bar{t}_\rho)}\}$ with $\phi^{\sigma(\bar{t}_\rho)} = \left[\phi_1^{\sigma(\bar{t}_\rho)}, \ldots, \phi_N^{\sigma(\bar{t}_\rho)} \right]^T$ satisfies $\left(\overline{\mathcal{L}}^{\sigma(\bar{t}_\rho)} \right)^T \phi^{\sigma(\bar{t}_\rho)} = \mathbf{1}_N$.

For notational brevity, one may let

$$\widetilde{\lambda}_0^\rho = \min_{i \in \mathcal{Q}_{\text{sub}}^{t_{\min}^\rho}} \widetilde{\lambda}_0^i, \tag{5.12}$$

where $\mathcal{Q}_{\text{sub}}^{t_{\min}^\rho} = \{\sigma(t) : t \in [t_{\min}^\rho, \bar{t}_{\rho+1})\}$, $\widetilde{\lambda}_0^i$ is the smallest eigenvalue of $\overline{\mathcal{L}}^i + \left(\overline{\Phi}^{\sigma(\bar{t}_\rho)}\right)^{-1} \left(\overline{\mathcal{L}}^i\right)^T \overline{\Phi}^{\sigma(\bar{t}_\rho)}$. Furthermore, let

$$\mu = \max_{i \neq j, i, j \in \mathcal{P}} \frac{\phi_{\max}^i}{\phi_{\min}^j}, \tag{5.13}$$

where $\phi_{\min}^s = \min_{r=1,\dots,N} \phi_r^s$, $\phi_{\max}^s = \max_{r=1,\dots,N} \phi_r^s$, for each $s \in \mathcal{P}$.

Based on the above analysis, one may get the following theorem which summarizes the main results of this section.

Theorem 5.1 *Under Assumption 5.1, the consensus tracking in CNS (5.2) with target trajectory given in (5.3) could be achieved if there exists a positive scalar ε_0 such that, for each $\rho \in \mathbb{N}$, the following conditions hold:*

(1) $\alpha > \frac{2\lambda_{\max}(\Gamma)}{\overline{\lambda}_0^{\sigma(\bar{t}_\rho)}}$,

(2) $\overline{\gamma}_\rho > \frac{-\widetilde{\gamma}_\rho(\bar{t}_{\rho+1} - t_{\min}^\rho) + \ln \mu + \varepsilon_0}{(t_{\min}^\rho - \bar{t}_\rho)}$,

where $\overline{\gamma}_\rho = \alpha \overline{\lambda}_0^{\sigma(\bar{t}_\rho)} - 2\lambda_{\max}(\Gamma)$, $\widetilde{\gamma}_\rho = \alpha \widetilde{\lambda}_0^\rho - 2\lambda_{\max}(\Gamma)$, $\lambda_{\max}(\Gamma)$ is the largest eigenvalue of Γ with Γ given in (5.10), $\overline{\lambda}_0^{\sigma(\bar{t}_\rho)}$ is the smallest eigenvalue of $\overline{\mathcal{L}}^{\sigma(\bar{t}_\rho)} + \left(\overline{\Phi}^{\sigma(\bar{t}_\rho)}\right)^{-1} \left(\overline{\mathcal{L}}^{\sigma(\bar{t}_\rho)}\right)^T \overline{\Phi}^{\sigma(\bar{t}_\rho)}$, $\widetilde{\lambda}_0^\rho$ and μ are respectively defined in (5.12) and in (5.13).

Proof 5.1 *Note that consensus tracking in CNS (5.2) with target trajectory generated by (5.3) is achieved if and only if the zero equilibrium point of switched system (5.5) is globally attractive.*

For each $\rho \in \mathbb{N}$, one may construct the following MLFs for the switched systems (5.8) and (5.9):

$$V(t) = e^T(t) \left(\overline{\Phi}^{\sigma(\bar{t}_\rho)} \otimes I_n\right) e(t), \quad t \in [\bar{t}_\rho, \bar{t}_{\rho+1}), \tag{5.14}$$

for all $\rho \in \mathbb{N}$. For $t \in [\bar{t}_\rho, t_{\min}^\rho)$ and an arbitrarily given $\rho \in \mathbb{N}$, taking the time derivative of $V(t)$ along the trajectories of systems (5.8) gives

$$\dot{V}(t) = 2e^T(t) \left(\overline{\Phi}^{\sigma(\bar{t}_\rho)} \otimes I_n\right) F(e(t), t)$$

$$- \alpha e^T(t) \left[\left(\overline{\Phi}^{\sigma(\bar{t}_\rho)} \overline{\mathcal{L}}^{\sigma(\bar{t}_\rho)} + \left(\overline{\mathcal{L}}^{\sigma(\bar{t}_\rho)}\right)^T \overline{\Phi}^{\sigma(\bar{t}_\rho)}\right) \otimes I_n\right] e(t). \tag{5.15}$$

Based on the above analysis and by Assumption 5.1, one gets that

$$
\dot{V}(t) \leq 2e^T(t)\left(\overline{\Phi}^{\sigma(\bar{t}_\rho)} \otimes \Gamma\right)e(t)
$$
$$
- \alpha e^T(t)\left[\left(\overline{\Phi}^{\sigma(\bar{t}_\rho)}\overline{\mathcal{L}}^{\sigma(\bar{t}_\rho)} + \left(\overline{\mathcal{L}}^{\sigma(\bar{t}_\rho)}\right)^T\overline{\Phi}^{\sigma(\bar{t}_\rho)}\right) \otimes I_n\right]e(t) \tag{5.16}
$$

for $t \in [\bar{t}_\rho, t_{\min}^\rho)$. By using the properties of Kronecker product, it follows from (5.16) that

$$
\dot{V}(t) \leq \left(2\lambda_{\max}(\Gamma) - \alpha\overline{\lambda}_0^{\sigma(\bar{t}_\rho)}\right)V(t), \tag{5.17}
$$

where $\lambda_{\max}(\Gamma)$ is the largest eigenvalue of Γ, $\overline{\lambda}_0^{\sigma(\bar{t}_\rho)}$ is the smallest eigenvalue of $\overline{\mathcal{L}}^{\sigma(\bar{t}_\rho)} + \left(\overline{\Phi}^{\sigma(\bar{t}_\rho)}\right)^{-1}\left(\overline{\mathcal{L}}^{\sigma(\bar{t}_\rho)}\right)^T\overline{\Phi}^{\sigma(\bar{t}_\rho)}$. One may then get that

$$
\dot{V}(t) \leq -\overline{\gamma}_\rho V(t), \quad t \in [\bar{t}_\rho, t_{\min}^\rho), \tag{5.18}
$$

where $\overline{\gamma}_\rho = \alpha\overline{\lambda}_0^{\sigma(\bar{t}_\rho)} - 2\lambda_{\max}(\Gamma)$, $\rho \in \mathbb{N}$. It can be yielded from condition (1) that $\overline{\gamma}_\rho > 0$ for $\rho \in \mathbb{N}$.

For the case of $t_{\min}^\rho = \bar{t}_{\rho+1}$, it can be directly obtained from (5.18) and condition (2) that

$$
V(\bar{t}_{\rho+1}) \leq \mu V(\bar{t}_{\rho+1}^-) < \exp(-\varepsilon_0)V(\bar{t}_\rho), \tag{5.19}
$$

where μ is defined in (5.13).

For the case of $t_{\min}^\rho < \bar{t}_{\rho+1}$, taking the time derivative of $V(t)$ along the trajectories of systems (5.8) with $t \in [t_{\min}^\rho, \bar{t}_{\rho+1})$ yields

$$
\dot{V}(t) = 2e^T(t)\left(\overline{\Phi}^{\sigma(\bar{t}_\rho)} \otimes I_n\right)F(e(t), t)
$$
$$
- \alpha e^T(t)\left[\left(\overline{\Phi}^{\sigma(\bar{t}_\rho)}\overline{\mathcal{L}}^{\sigma(t)} + \left(\overline{\mathcal{L}}^{\sigma(t)}\right)^T\overline{\Phi}^{\sigma(\bar{t}_\rho)}\right) \otimes I_n\right]e(t)
$$
$$
\leq 2\lambda_{\max}(\Gamma)V(t) - \alpha e^T(t)\left[\left(\overline{\Phi}^{\sigma(\bar{t}_\rho)}\overline{\mathcal{L}}^{\sigma(t)} + \left(\overline{\mathcal{L}}^{\sigma(t)}\right)^T\overline{\Phi}^{\sigma(\bar{t}_\rho)}\right) \otimes I_n\right]e(t).
$$

Note that the matrix $\left(\overline{\Phi}^{\sigma(\bar{t}_\rho)}\overline{\mathcal{L}}^{\sigma(t)} + \left(\overline{\mathcal{L}}^{\sigma(t)}\right)^T\overline{\Phi}^{\sigma(\bar{t}_\rho)}\right) \otimes I_n$ may not be positive definite for $t \in [t_{\min}^\rho, \bar{t}_{\rho+1})$. However, it can be obtained from Lemma 2.7 that

$$
-\alpha e^T(t)\left[\left(\overline{\Phi}^{\sigma(\bar{t}_\rho)}\overline{\mathcal{L}}^{\sigma(t)} + \left(\overline{\mathcal{L}}^{\sigma(t)}\right)^T\overline{\Phi}^{\sigma(\bar{t}_\rho)}\right) \otimes I_n\right]e(t) \leq -\alpha\widetilde{\lambda}_0^{\sigma(t)}V(t)
$$

for $t \in [t_{\min}^\rho, \bar{t}_{\rho+1})$, where $\widetilde{\lambda}_0^{\sigma(t)}$ is the smallest eigenvalue of $\overline{\mathcal{L}}^{\sigma(t)} + \left(\overline{\Phi}^{\sigma(\bar{t}_\rho)}\right)^{-1}\left(\overline{\mathcal{L}}^{\sigma(t)}\right)^T\overline{\Phi}^{\sigma(\bar{t}_\rho)}$. Set $\mathcal{Q}_{\text{sub}}^{t_{\min}^\rho} = \{\sigma(t) : t \in [t_{\min}^\rho, \bar{t}_{\rho+1})\}$, one has that $\mathcal{Q}_{\text{sub}}^{t_{\min}^\rho}$

is a proper subset of $\mathcal{Q}$. Then, one may choose $\widetilde{\lambda}_0^\rho = \min\limits_{i \in \mathcal{Q}_{sub}^{t_{min}^\rho}} \widetilde{\lambda}_0^i$. Based on the above statements, one has

$$\dot{V}(t) \leq -\widetilde{\gamma}_\rho V(t), \ \forall \ t \in [t_{min}^\rho, \bar{t}_{\rho+1}), \tag{5.20}$$

where $\widetilde{\gamma}_\rho = \left(\alpha \widetilde{\lambda}_0^\rho - 2\lambda_{max}(\Gamma)\right)$. Then, for the case of $t_{min}^\rho < \bar{t}_{\rho+1}$, it can be obtained from (5.20) and condition (2) that

$$V(\bar{t}_{\rho+1}) \leq \mu V(\bar{t}_{\rho+1}^-) < \exp(-\varepsilon_0)V(\bar{t}_\rho).$$

It can be concluded from the above analysis that $V(\bar{t}_{\rho+1}) < \exp(-\varepsilon_0)V(\bar{t}_\rho)$ for an arbitrarily given $\rho \in \mathbb{N}$. For $\rho = 1$, one gets that $V(\bar{t}_2) < V(0) \cdot \exp(-\varepsilon_0)$. Furthermore, it can be obtained by recursion that

$$V(\bar{t}_{\rho+1}) \leq V(0) \cdot \exp(-\rho\varepsilon_0) \tag{5.21}$$

for any given positive integer ρ.

According to the fact that the dwell time $\tau_m > 0$, one knows that no Zeno behavior will be emerged as the considered CNS evolves with time [227]. Thus, for an arbitrarily given $t > 0$, there exits a positive integer z such that $\bar{t}_z < t \leq \bar{t}_{z+1}$.

When $t \in (0, t_{min}^1)$, one gets

$$V(t) < V(0) \cdot \exp(-\overline{\gamma}_1 t), \tag{5.22}$$

where $\overline{\gamma}_1 = \alpha \overline{\lambda}_0^{\sigma(\bar{t}_1)} - 2\lambda_{max}(\Gamma) > 0$. For the case of $t_{min}^1 = \bar{t}_2$, it can be obtained from (5.22) that

$$V(\bar{t}_2) < \mu V(0) \cdot \exp(-\overline{\gamma}_1 \bar{t}_2), \tag{5.23}$$

where μ is defined in (5.13). For the case of $t_{min}^1 < \bar{t}_2$ and $\widetilde{\gamma}_1 > 0$, some calculations give that

$$V(t) < \mu V(0) \cdot \exp(-\gamma_1 t), \ t \in \left[t_{min}^1, \bar{t}_2\right], \tag{5.24}$$

where $\gamma_1 = \min\{\overline{\gamma}_1, \widetilde{\gamma}_1\}$. For the case of $t_{min}^1 < \bar{t}_2$ and $\widetilde{\gamma}_1 \leq 0$, one has

$$\begin{aligned} V(t) &< \mu V(0) \cdot \exp\left(-\widetilde{\gamma}_1\left(t - t_{min}^1\right)\right) \cdot \exp(-[(\overline{\gamma}_1 t_{min}^1)/\bar{t}_2]t) \\ &< \mu V(0) \cdot \exp(-\widetilde{\gamma}_1 \overline{\tau}_M) \cdot \exp(-[(\overline{\gamma}_1 t_{min}^1)/\bar{t}_2]t), \ t \in \left[t_{min}^1, \bar{t}_2\right]. \end{aligned} \tag{5.25}$$

When $t \in (\bar{t}_z, t_{min}^z)$, $z \geq 2$, one has

$$\begin{aligned} V(t) &\leq V(\bar{t}_z) \cdot \exp(-\overline{\gamma}_z(t - \bar{t}_z)) \\ &< V(0) \cdot \exp(-[((z-1)\varepsilon_0)/(z\overline{\tau}_M)]t) \\ &< V(0) \cdot \exp(-[\varepsilon_0/(2\overline{\tau}_M)]t), \end{aligned} \tag{5.26}$$

where the last inequality is obtained since $z \geq 2$. For the case of $t^z_{\min} = \bar{t}_{z+1}$, it can be directly obtained from (5.26) that

$$V(\bar{t}_{z+1}) < \mu V(0) \cdot \exp(-[\varepsilon_0/(2\bar{\tau}_M)]\bar{t}_{z+1}). \tag{5.27}$$

For the case of $t^z_{\min} < \bar{t}_{z+1}$ and $\tilde{\gamma}_z > 0$, it can be obtained that

$$V(t) < \mu V(0) \cdot \exp(-\varrho_0 t), \quad t \in [t^z_{\min}, \bar{t}_{z+1}], \tag{5.28}$$

where $\varrho_0 = \min\{\varepsilon_0/(2\bar{\tau}_M), \tilde{\gamma}_z\}$. For the case of $t^z_{\min} < \bar{t}_{z+1}$ and $\tilde{\gamma}_z \leq 0$, some calculations give that

$$\begin{aligned}
V(t) &\leq \mu V(t_z) \cdot \exp(-\bar{\gamma}_z(t^z_{\min} - \bar{t}_z)) \cdot \exp(-\tilde{\gamma}_z(t - t^z_{\min})) \\
&< \mu V(0) \exp(-(z-1)\varepsilon_0) \cdot \exp(-\tilde{\gamma}_z \bar{\tau}_M) \\
&< \mu V(0) \exp(-\tilde{\gamma}_z \bar{\tau}_M) \cdot \exp\left(-\frac{\varepsilon_0}{2\bar{\tau}_M}t\right), \quad t \in [t^z_{\min}, \bar{t}_{z+1}]. \tag{5.29}
\end{aligned}$$

Then one gets from (5.22) to (5.29) that the zero equilibrium point of switched systems (5.8) and (5.9) is globally exponentially stable. One can thus conclude that the consensus tracking in CNSs (5.2) with target trajectory given in (5.3) could be achieved.

Suppose that Assumption 5.1 holds, it can be obtained from Theorem 5.1 that consensus tracking in the considered CNSs can be ensured if the conditions (1) and (2) given in Theorem 5.1 are simultaneously satisfied. Here, condition (1) means that the coupling strength among the neighboring agents is larger than a threshold value. With this condition, condition (2) can be equivalently rewritten as $(t^\rho_{\min} - \bar{t}_\rho) > [-\tilde{\gamma}_\rho(\bar{t}_{\rho+1} - t^\rho_{\min}) + \ln\mu + \varepsilon_0]/\bar{\gamma}_\rho$. Intuitively speaking, condition (2) implies that, over each time interval $[\bar{t}_\rho, \bar{t}_{\rho+1})$, $\rho \in \mathbb{N}$, the total activation time for the network topologies with a directed spanning tree is larger than a threshold quantity.

Remark 5.2 *By using tools from M-matrix theory and algebraic graph theory, a class of quadratic MLFs in the form of (5.14) has been succesfully constructed. It can be observed from the proof of Theorem 5.1 that the construction of such topology dependent quadratic Lyapunov functions provides an efficient tool for analyzing consensus behavior of the switched CNSs (5.2) with target system (5.3). Note that constructing a common quadratic Lyapunov function for analyzing the consensus behavior of CNSs under consideration is still challenging today. It is also worth noting that there may even have no common quadratic Lyapunov function for switched linear CNSs [119].*

Remark 5.3 *When applying Theorem 5.1 to solve the consensus tracking problem of practical CNSs, one needs to calculate $\lambda_{\max}(\Gamma)$, $\overline{\lambda}_0^{\sigma(\bar{t}_\rho)}$, and $\tilde{\lambda}_0^\rho$. The stability and precision of the numerical computation method adopted here should be fully considered since most of the typical CNSs are large-scale. One may use the traditional Jacobi eigenvalue algorithm or Givens eigenvalue algorithm [202] to calculate $\lambda_{\max}(\Gamma)$ and*

$\overline{\lambda}_0^{\sigma(\overline{t}_\rho)}$ since both Γ and $\left(\overline{\Phi}^{\sigma(\overline{t}_\rho)} \overline{\mathcal{L}}^{\sigma(\overline{t}_\rho)} + \left(\overline{\mathcal{L}}^{\sigma(\overline{t}_\rho)} \right)^T \overline{\Phi}^{\sigma(\overline{t}_\rho)} \right)$ are real and symmetric. Note that both the Jacobi eigenvalue algorithm and Givens eigenvalue algorithm have good numerical stability. Furthermore, to obtain $\widetilde{\lambda}_0^\rho$, one needs to calculate the eigenvalues of $\left(\overline{\Phi}^{\sigma(\overline{t}_\rho)} \right)^{-1} \left(\overline{\Phi}^{\sigma(\overline{t}_\rho)} \overline{\mathcal{L}}^i + \left(\overline{\mathcal{L}}^i \right)^T \overline{\Phi}^{\sigma(\overline{t}_\rho)} \right)$ which shares the same eigenvalues with symmetric matrix

$$M_{sy} = \sqrt{\left(\overline{\Phi}^{\sigma(\overline{t}_\rho)} \right)^{-1}} \left(\overline{\Phi}^{\sigma(\overline{t}_\rho)} \overline{\mathcal{L}}^i + \left(\overline{\mathcal{L}}^i \right)^T \overline{\Phi}^{\sigma(\overline{t}_\rho)} \right) \sqrt{\left(\overline{\Phi}^{\sigma(\overline{t}_\rho)} \right)^{-1}}, \quad (5.30)$$

where $\sqrt{\left(\overline{\Phi}^{\sigma(\overline{t}_\rho)} \right)^{-1}} = \text{diag}\left\{ \sqrt{\left(\phi_1^{\sigma(\overline{t}_\rho)} \right)^{-1}}, \sqrt{\left(\phi_2^{\sigma(\overline{t}_\rho)} \right)^{-1}}, \ldots, \sqrt{\left(\phi_N^{\sigma(\overline{t}_\rho)} \right)^{-1}} \right\}$. The above analysis indicates that, to calculate $\widetilde{\lambda}_0^\rho$, one may just need to use the traditional Jacobi eigenvalue algorithm or Givens eigenvalue algorithm to calculate the eigenvalues of M_{sy} given in (5.30).

Based on Theorem 5.1, one can obtain the following corollaries where the detailed proofs are omitted for brevity.

Corollary 5.1 *Under Assumption 5.1, the consensus tracking in CNS (5.2) with target trajectory given in (5.3) could be achieved if there exists a positive scalar ε_ρ for each $\rho \in \mathbb{N}$, such that the following conditions hold:*

(1) $\alpha > \frac{2\lambda_{\max}(\Gamma)}{\overline{\lambda}_0^{\sigma(\overline{t}_\rho)}}$,

(2) $\overline{\gamma}_\rho > \frac{-\widetilde{\gamma}_\rho \left(\overline{t}_{\rho+1} - t_{\min}^\rho \right) + \ln \mu_\rho + \varepsilon_\rho}{\left(t_{\min}^\rho - \overline{t}_\rho \right)}$,

where $\overline{\gamma}_\rho = \left(\alpha \overline{\lambda}_0^{\sigma(\overline{t}_\rho)} - 2\lambda_{\max}(\Gamma) \right)$, $\widetilde{\gamma}_\rho = \left(\alpha \widetilde{\lambda}_0^\rho - 2\lambda_{\max}(\Gamma) \right)$, $\lambda_{\max}(\Gamma)$ *is the largest eigenvalue of Γ with Γ given in (5.10), $\widetilde{\lambda}_0^\rho$ is defined in (5.12), and*

$$\mu_\rho = \frac{\phi_{\max}^{\sigma(\overline{t}_{\rho+1})}}{\phi_{\min}^{\sigma(\overline{t}_\rho)}}, \quad (5.31)$$

where $\phi_{\min}^{\sigma(\overline{t}_\rho)} = \min_{r=1,\ldots,N} \phi_r^{\sigma(\overline{t}_\rho)}$, $\phi_{\max}^{\sigma(\overline{t}_{\rho+1})} = \max_{r=1,\ldots,N} \phi_r^{\sigma(\overline{t}_{\rho+1})}$, $\phi^{\sigma(\overline{t}_\rho)} = [\phi_1^{\sigma(\overline{t}_\rho)}, \ldots, \phi_N^{\sigma(\overline{t}_\rho)}]^T = \left(\overline{\mathcal{L}}^{\sigma(\overline{t}_\rho)} \right)^{-T} \mathbf{1}_N$, and $\phi^{\sigma(\overline{t}_{\rho+1})} = \left[\phi_1^{\sigma(\overline{t}_{\rho+1})}, \ldots, \phi_N^{\sigma(\overline{t}_{\rho+1})} \right]^T = \left(\overline{\mathcal{L}}^{\sigma(\overline{t}_{\rho+1})} \right)^{-T} \mathbf{1}_N$.

Corollary 5.2 *Suppose that Assumption 5.1 holds and $\overline{\lambda}_0^{\sigma(\overline{t}_\rho)} (t_{\min}^\rho - \overline{t}_\rho) + \widetilde{\lambda}_0^\rho (\overline{t}_{\rho+1} - t_{\min}^\rho) > 0$, for $\rho \in \mathbb{N}$. Then, the consensus tracking in CNS (5.2) with target trajectory given in (5.3) could be achieved if there exists a positive constant ε_0 such that*

$$\alpha > \alpha_{th},$$

with $\alpha_{th} = \sup_{\rho \in \mathbb{N}} \alpha_\rho$, $\alpha_\rho = \frac{\varepsilon_0 + \ln \mu + 2\lambda_{\max}(\Gamma) \left(\overline{t}_{\rho+1} - \overline{t}_\rho \right)}{\overline{\lambda}_0^{\sigma(\overline{t}_\rho)} \left(t_{\min}^\rho - \overline{t}_\rho \right) + \widetilde{\lambda}_0^\rho \left(\overline{t}_{\rho+1} - t_{\min}^\rho \right)}$, *where $\widetilde{\lambda}_0^\rho$ is defined in (5.12),* μ *is defined in (5.13).*

Corollary 5.3 *Suppose that Assumption 5.1 holds and $\overline{\mathcal{G}} = \widehat{\mathcal{G}} = \{1\}$. Then, consensus tracking in CNS (5.2) with target trajectory given in (5.3) could be achieved if*

$$\alpha > \lambda_{th}, \tag{5.32}$$

where $\lambda_{th} = 2\lambda_{\max}(\Gamma)\lambda_0$, λ_0 is the largest eigenvalue of $\left(\overline{\Phi}^1 \overline{\mathcal{L}}^1 + \left(\overline{\mathcal{L}}^1 \right)^T \overline{\Phi}^1 \right)^{-1} \overline{\Phi}^1$.

Remark 5.4 *Compared with those given in Theorem 5.1, the conditions given in Corollary 5.1 are less conservative. However, the consensus conditions given in Theorem 5.1 are more convenient to use in practical applications since one does not need to calculate μ_ρ for all time intervals.*

Remark 5.5 *Corollary 5.3 indicates that consensus tracking can be ensured if the fixed network topology contains a directed spanning tree and the coupling strength is appropriately selected. We point out the results provided in Corollary 5.3 confirm those given in some existing literature on consensus tracking over CNSs with fixed topology, such as [74, 99, 138].*

Remark 5.6 *The coupling pattern among the agents in the considered network (5.1) is a full-state coupling. Also, it is required in the present section that the relative full-state information between the target system (5.3) and each pinned agent in network (5.1) is available for pinning feedback. However, it is challenging yet important to further study the consensus tracking problem for directed switching CNSs with partial-state coupling where only the relative output information between the target system and each pinned agent is available for pinning feedback. On the other hand, the present theoretical results are derived based on Assumption 5.1 which may be restricted in some applications. Therefore, it is still an unsolved problem about how to ensure consensus tracking in directed switching CNSs without Assumption 5.1.*

5.3 CONSENSUS TRACKING OF CNSS WITH LORENZ TYPE DYNAMICS

In this section, the consensus tracking in CNSs with Lorenz type dynamics which include the Chen and Lü systems as special cases will be addressed. As it is well known, the Lorenz system, as one of the paradigms of chaos, has been a focal subject in nonlinear control since 1963 [94]. As a dual Lorenz system, Chen system was found in 1999 [17]. The work in [17] is really a breakthrough in the study of Lorenz family systems. Then, the Lü system was introduced in [100] as a transition system between Lorenz and Chen systems. Since it is hard or impossible to verify whether the Lü system satisfies the Lipschitz condition, the consensus criteria derived in some existing works with Lipschitz condition may not be valid for coupled networks with Lorenz type agents [42]. Note that even for the nonlinear systems that satisfy the Lipschitz condition, some conservatism will be involved in calculating the Lipschitz constants. Thus, the consensus condition based on the derived Lipschitz constants will be conservative. Furthermore, another common assumption always made in existing literature on consensus tracking is that the communication topology is fixed. It is more interesting to study consensus tracking under switching communication topologies.

5.3.1 Model formulation

Consider a CNS with Lorenz type dynamics which are given by

$$\dot{x}_i(t) = Ax_i(t) + \beta x_i(t)Bx_i(t) + \alpha \sum_{j=1}^{N} a_{ij}(t)H(x_j(t) - x_i(t)), \tag{5.33}$$

where

$$A = \begin{bmatrix} -(25\gamma + 10) & (25\gamma + 10) & 0 \\ (28 - 35\gamma) & (29\gamma - 1) & 0 \\ 0 & 0 & -\frac{(\gamma+8)}{3} \end{bmatrix}, \quad B = \begin{bmatrix} 0 & 0 & 0 \\ 0 & 0 & -1 \\ 0 & 1 & 0 \end{bmatrix}, \tag{5.34}$$

$\beta = [1, 0, 0]$, $\gamma \in [0, 1]$ is a parameter, $\alpha > 0$ represents the coupling strength among the agents, $\mathcal{A}(t) = [a_{ij}(t)]_{N \times N}$ is the adjacency matrix of the communication topology at time t, and $H \in \mathbb{R}^{3 \times 3}$ is the positive definite inner linking matrix, $i = 1, \ldots, N$. Note that systems (5.33) will become the coupled Lorenz, Chen and Lü systems if $\gamma = 0$, 1, and 0.8, respectively. By the definition of the Laplacian matrix for a graph, it follows from (5.33) that

$$\dot{x}_i(t) = Ax_i(t) + \beta x_i(t)Bx_i(t) - \alpha \sum_{j=1}^{N} l_{ij}(t)Hx_j(t), \tag{5.35}$$

where $L(t) = [l_{ij}(t)]_{N \times N}$ is the Laplacian matrix of communication topology $\mathcal{G}(\mathcal{A}(t))$, $i = 1, \ldots, N$. It is assumed in this section that $t_0 = 0$.

The control goal here is to design some pinning controllers to some designed agents such that the states of all the agents in (5.33) to converge to a common target trajectory $s(t)$ in the sense of $\lim_{t \to \infty} \|x_i(t) - s(t)\| = 0$, for all $i = 1, \ldots, N$, with

$$\dot{s}(t) = As(t) + \beta s(t)Bs(t), \tag{5.36}$$

with arbitrarily given initial value $s(t_0) \in \mathbb{R}^3$. Motivated by the works in [74, 136, 205, 216], pinning CNS (5.33) by using some linear controllers $-\alpha c_i(t)H(x_i(t) - s(t))$ to agent i leads to

$$\dot{x}_i(t) = Ax_i(t) + \beta x_i(t)Bx_i(t)$$
$$- \alpha \sum_{j=1}^{N} l_{ij}(t)Hx_j(t) - \alpha c_i(t)H(x_i(t) - s(t)), \tag{5.37}$$

where $c_i(t) \in \{0, 1\}$ and $c_i(t) = 1$ if the agent i of (5.33) is pinned at time t.

Let $e_i(t) = x_i(t) - s(t)$, $i = 1, \ldots, N$, it thus follows from (5.37) that

$$\dot{e}_i(t) = Ae_i(t) + \beta x_i(t)Bx_i(t) - \beta s(t)Bs(t)$$
$$- \alpha \sum_{j=1}^{N} l_{ij}(t)He_j(t) - \alpha c_i(t)He_i(t), \tag{5.38}$$

where $i = 1, \ldots, N$. By the definition of B, it follows from (5.38) that

$$\dot{e}_i(t) = Ae_i(t) + \beta x_i(t) Be_i(t) + \beta e_i(t) Bs(t)$$

$$- \alpha \sum_{j=1}^{N} l_{ij}(t) He_j(t) - \alpha c_i(t) He_i(t), \ i = 1, \ldots, N. \tag{5.39}$$

Obviously, $e_i(t) = \mathbf{0}_n$ is a fixed point of system (5.39), for each $i = 1, \ldots, N$. Furthermore, it is not hard to verify that consensus tracking in CNS (5.35) with target trajectory given in (5.36) is achieved if and only if the zero equilibrium point of (5.39) is globally attractive.

5.3.2 Main results for directed fixed communication topology

In this subsection, consensus tracking of CNS (5.33) with target trajectory given in (5.36) under a fixed communication topology is studied. Without loss of generality, let $\mathcal{G}(\mathcal{A}(t)) = \mathcal{G}(\mathcal{A})$ for all $t \geq 0$. And we label the target as agent 0.

Assumption 5.2 *There exists at least one directed spanning tree rooted at agent 0 (i.e., the target) in the augmented communication topology $\mathcal{G}(\tilde{\mathcal{A}})$.*

It is clearly that Assumption 5.2 will hold if all the agents $1, \ldots, N$ are pinned, i.e., $c_i(t) = 1$, for all $i = 1, \ldots, N$ and $t \geq 0$. However, it is more interesting to study how to make Assumption 5.2 hold if only a small fraction of the agents in $\mathcal{G}(\mathcal{A})$ could be selected and pinned. To do this, the following algorithm is proposed to determine at least how many and what kinds of agents should be pinned such that Assumption 5.2 holds.

Algorithm 5.2 *Find the strongly connected components of $\mathcal{G}(\mathcal{A})$ by employing the Tarjan's algorithm [157]. Note that the time complexity of this operation is $O(N+E)$, where N and E are, respectively, the numbers of agents and links of $\mathcal{G}(\mathcal{A})$. Suppose that there are ω strongly connected components in $\mathcal{G}(\mathcal{A})$, labeled as W_1, W_2, $\ldots$, W_ω. Set $m_i = 0$, $i = 1, \ldots, \omega$, and $h = 1$. Then, execute the following steps*

(1) *Check whether there exists at least one agent n_k belonging to W_h which is reachable from an agent n_g belonging to W_j, $j = 1, \ldots, \omega, j \neq h$. If it holds, go to step (2); if it dose not hold, go to step (3).*

(2) *Check whether the following condition holds: $h < \omega$. If it holds, let $h = h + 1$ and re-perform step (1); else stop.*

(3) *Arbitrarily selected one agent in W_h and pinned, let $m_h = 1$; Check whether the following condition holds: $h < \omega$. If it holds, let $h = h + 1$ and re-perform step (1); else stop.*

Remark 5.7 *Note that one may use the* `graphconncomp` *function of Matlab to search the strongly connected components of $\mathcal{G}(\mathcal{A})$. Furthermore, it is not hard to verify that there exists a directed path from the target to each agent in $\mathcal{G}(\mathcal{A})$ if the $\delta = \sum_{j=1}^{\omega} m_i$*

selected agents in $\mathcal{G}(\mathcal{A})$ are pinned. And, δ is equal to the number of the strongly connected components without incoming links in $\mathcal{G}(\mathcal{A})$. Furthermore, the number δ and the set of the selected agents are respectively called the forest dimension and the agent basis of directed graph $\mathcal{G}(\mathcal{A})$ [2]. Note also that $\delta = 1$ if the network topology $\mathcal{G}(\mathcal{A})$ contains a directed spanning tree, i.e., there exists a directed path from the target, labeled as 0, to each agent in $\mathcal{G}(\mathcal{A})$ if the root agent of a directed spanning tree within this graph is selected and pinned. Alternatively, when the network topology $\mathcal{G}(\mathcal{A})$ consists of ς separate components with each of them having a directed spanning tree, it follows from Algorithm 5.2 that $\delta = \varsigma$ and the root agent of the spanning tree in each strongly connected component should be pinned. For an arbitrarily given $\mathcal{G}(\mathcal{A})$, it can be verified that δ determined in Algorithm 5.2 corresponds to the minimum number of agents should be pinned such that Assumption 5.2 holds. Obviously, if there are some additional agents beside the δ selected agents are also pinned, Assumption 5.2 still holds; however, Assumption 5.2 can never be ensured if there are only ϖ agents are selected and pinned, where $\varpi < \delta$.

Remark 5.8 *For the case of $\mathcal{G}(\mathcal{A})$ is an undirected graph, it can be seen from Algorithm 5.2 that δ is equal to the number of disjoint connected components in $\mathcal{G}(\mathcal{A})$.*

As the target is labelled as agent 0, the Laplacian matrix of the augmented graph $\mathcal{G}(\tilde{\mathcal{A}})$ can be written as

$$\tilde{\mathcal{L}} = \begin{bmatrix} 0 & \mathbf{0}_N^T \\ \mathbf{P} & \overline{\mathcal{L}} \end{bmatrix},$$

$$\overline{\mathcal{L}} = \begin{bmatrix} \sum_{j \in \mathcal{N}_1} a_{1j} & -a_{12} & \cdots & -a_{1N} \\ -a_{21} & \sum_{j \in \mathcal{N}_2} a_{2j} & \cdots & -a_{2N} \\ \vdots & \vdots & \ddots & \vdots \\ -a_{N1} & -a_{N2} & \cdots & \sum_{j \in \mathcal{N}_N} a_{Nj} \end{bmatrix},$$

where $\mathbf{P} = -[a_{10}, \cdots, a_{N0}]^T$ with $a_{i0} = c_i$, $i = 1, \ldots, N$. Under Assumption 5.2, it can be got from Lemma 2.15 that there exists a positive definite diagonal matrix $\Phi = \mathrm{diag}\{\phi_1, \ldots, \phi_N\}$ such that $\overline{\mathcal{L}}^T \Phi + \Phi \overline{\mathcal{L}} > 0$, where $\phi = [\phi_1, \ldots, \phi_N]^T$ can be obtained by solving the matrix equation $\overline{\mathcal{L}}^T \phi = \mathbf{1}_N$.

Based on the above analysis, one may get the following theorem which summarizes the main result of this section.

Theorem 5.2 *Suppose that Assumption 5.2 holds and the target trajectory $s(t)$ satisfies $\|s(t)\| \leq \iota$, for all $t \geq 0$ where ι is a given positive scalar. Then, consensus tracking in the CNS (5.37) with fixed communication topology $\mathcal{G}(\mathcal{A})$ can be achieved exponentially if there exists a positive scalar c_0 such that the following conditions hold: $\alpha > \left(\lambda_{\max}(A^T + A) + 2\iota + c_0\right) \Big/ \left(\lambda_{\min}(\overline{\mathcal{L}} + \Phi^{-1}\overline{\mathcal{L}}^T \Phi)\lambda_{\min}(H)\right)$.*

Proof 5.2 *Note that the global stability of the error systems (5.37) for its zero equilibrium point implies that the consensus tracking in CNS (5.39) will be achieved.*

Construct the following Lyapunov function

$$V(t) = \sum_{i=1}^{N} \phi_i e_i^T(t) e_i(t). \tag{5.40}$$

Taking the time derivative of $V(t)$ along the trajectories of (5.39) and using the facts $e_i^T(t)\beta x_i(t) Be_i(t) = 0$, for all $i = 1, \ldots, N$, yields

$$\dot{V}(t) = \sum_{i=1}^{N} \phi_i e_i^T(t) \left(A^T + A \right) e_i(t) + 2 \sum_{i=1}^{N} \phi_i e_i^T(t) \left(\beta e_i(t) Bs(t) \right)$$

$$- 2\alpha \sum_{i=1}^{N} \phi_i e_i^T(t) \sum_{j=1}^{N} l_{ij} H e_j(t) - 2\alpha \sum_{i=1}^{N} \phi_i c_i(t) e_i^T(t) H e_i(t). \tag{5.41}$$

Furthermore, according to (5.34), one has

$$e_i^T(t) \left(\beta e_i(t) Bs(t) \right) = e_{i1}(t) e_{i3}(t) s_2(t) - e_{i1}(t) e_{i2}(t) s_3(t)$$

$$\leq \iota \left(|e_{i1}(t) e_{i3}(t)| + |e_{i1}(t) e_{i2}(t)| \right) \leq \iota e_i^T(t) e_i(t), \tag{5.42}$$

where $e_i(t) = [e_{i1}(t), e_{i2}(t), e_{i3}(t)]^T$ and $\|s(t)\| \leq \iota$, for all $t \geq 0$. Then it follows from (5.41) and (5.42) that

$$\dot{V}(t) \leq e^T(t) \left\{ \Phi \otimes (A^T + A + 2\iota I_3) - \alpha \left[(\Phi \overline{\mathcal{L}} + \overline{\mathcal{L}}^T \Phi) \otimes H \right] \right\} e(t), \tag{5.43}$$

where $e(t) = [e_1^T(t), \ldots, e_N^T(t)]^T$. From the condition

$$\alpha \geq \frac{\lambda_{\max}(A^T + A) + 2\iota + c_0}{\lambda_{\min}(\overline{\mathcal{L}} + \Phi^{-1}\overline{\mathcal{L}}^T \Phi)\lambda_{\min}(H)}, \tag{5.44}$$

and (5.43), one obtains

$$\dot{V}(t) \leq -c_0 V(t), \ \forall \, t \geq 0. \tag{5.45}$$

Thus, one may get that the consensus tracking in CNS (5.37) with directed fixed communication topology can be achieved exponentially.

Remark 5.9 *Suppose that the consensus tracking in CNS (5.37) with communication topology $\mathcal{G}(\mathcal{A})$ can be achieved for some given coupling strength α. It can be seen from (5.45) that the convergence rate of consensus is characterized by the positive scalar c_0.*

From Theorem 5.2, one may directly get the following corollary where the proof is omitted for brevity.

Corollary 5.4 *Suppose that Assumption 5.2 holds and the target trajectory $s(t)$ satisfies $\|s(t)\| \leq \iota$ for all $t \geq 0$, where ι is a given positive scalar. Then, consensus tracking in CNS (5.37) with fixed communication topology $\mathcal{G}(\mathcal{A})$ can be achieved asymptotically if the coupling strength α satisfies the following condition:*

$$\alpha > \lambda_{\max}(\Psi),$$

where $\Psi = \left[\left(\Phi \overline{\mathcal{L}} + \overline{\mathcal{L}}^T \Phi \right) \otimes H \right]^{-1} [\Phi \otimes (A^T + A + 2\iota I_3)]$.

5.3.3 Main results for directed switching communication topologies

The underlying topology of the CNS considered in this subsection is modeled by directed switching graphs. Let $\overline{\mathcal{G}} = \{\mathcal{G}(\mathcal{A}^1), \ldots, \mathcal{G}(\mathcal{A}^\kappa)\}$, $\kappa \geq 2$, indicate the set of all possible directed communication topologies. Suppose that there exists an infinite sequence of uniformly bounded non-overlapping time intervals $[t_k, t_{k+1})$, $k \in \mathbb{N}$, with $t_0 = 0$, over which the interaction graph is fixed. The time sequence t_k, $k \in \mathbb{N}$ is then called the switching sequence, at which the interaction graph changes. Furthermore, introduce a switching signal $\sigma(t) : [0, +\infty) \mapsto \{1, \ldots, \kappa\}$. Then, let $\mathcal{G}(\mathcal{A}^{\sigma(t)})$ be the communication topology of the CNS at time t. Note that $\mathcal{G}(\mathcal{A}^{\sigma(t)}) \in \overline{\mathcal{G}}$, for all $t \geq 0$. The error dynamical system (5.39) can be rewritten as

$$\dot{e}_i(t) = Ae_i(t) + \beta x_i(t) Be_i(t) + \beta e_i(t) Bs(t) - \alpha \sum_{j=1}^{N} l_{ij}^{\sigma(t)} He_j(t)$$

$$- \alpha c_i(t) He_i(t), \ i = 1, \ldots, N, \tag{5.46}$$

where $\mathcal{L}^{\sigma(t)} = [l_{ij}^{\sigma(t)}]_{N \times N}$ is the Laplacian matrix of communication topology $\mathcal{G}(\mathcal{A}^{\sigma(t)})$. Throughout this section, the time derivatives of functions $e_i(t)$ and $x_i(t)$ at any switching instant represent its right derivative.

Assumption 5.3 *There exists at least one directed spanning tree rooted at agent 0 (i.e., the target) in the augmented communication topology $\mathcal{G}(\widetilde{\mathcal{A}}^{\sigma(t)})$.*

Remark 5.10 *Applying Algorithm 5.2 to each possible communication topology $\mathcal{G}(\mathcal{A}^i)$, $i = 1, \ldots, \kappa$, one gets that Assumption 5.3 will hold if the selected agents are pinned.*

Similar to the last subsection, the Laplacian matrix of $\mathcal{G}(\widetilde{\mathcal{A}}^i)$, $i = 1, \ldots, \kappa$, can be written as

$$\widetilde{\mathcal{L}}^i = \begin{bmatrix} 0 & \mathbf{0}_N^T \\ \mathbf{P}^i & \overline{\mathcal{L}}^i \end{bmatrix},$$

$$\overline{\mathcal{L}}^i = \begin{bmatrix} \sum_{j \in \mathcal{N}_1} a_{1j}^i & -a_{12}^i & \cdots & -a_{1N}^i \\ -a_{21}^i & \sum_{j \in \mathcal{N}_2} a_{2j}^i & \cdots & -a_{2N}^i \\ \vdots & \vdots & \ddots & \vdots \\ -a_{N1}^i & -a_{N2}^i & \cdots & \sum_{j \in \mathcal{N}_N} a_{Nj}^i \end{bmatrix},$$

where $\mathbf{P}^i = -[a_{10}^i, \cdots, a_{N0}^i]^T$. with $a_{j0}^i = c_j^i$, $j = 1, \ldots, N$. Under Assumption 5.3, it can be got from Lemma 2.15 that there exists a sequence of positive definite diagonal matrices $\Phi^i = \text{diag}\{\phi_1^i, \ldots, \phi_N^i\}$ such that $(\overline{\mathcal{L}}^i)^T \Phi^i + \Phi^i \overline{\mathcal{L}}^i > 0$, where $\phi^i = [\phi_1^i, \ldots, \phi_N^i]^T$ can be obtained by solving the matrix equation $(\overline{\mathcal{L}}^i)^T \phi^i = \mathbf{1}_N$, $i = 1, \ldots, \kappa$.

Based on the above analysis, one may get the following theorem which is the main result of this subsection.

Theorem 5.3 *Suppose that Assumption 5.3 holds and the target trajectory $s(t)$ satisfies $\|s(t)\| \leq \iota$ for all $t \geq 0$, where ι is a given positive scalar. Then, consensus tracking in CNSs (5.1) with directed switching topologies can be achieved if there exists a positive scalar ε_0 such that the following conditions hold:*

$$\alpha > \frac{\lambda_{\max}(A^T + A) + 2\iota + \varepsilon_0}{\left(\min_{i=1,\dots,\kappa} \lambda_{\min}(\overline{\mathcal{L}}^i + (\Phi^i)^{-1}(\overline{\mathcal{L}}^i)^T \Phi^i)\right) \lambda_{\min}(H)},$$

and

$$\tau_a > (\ln \mu)/c_0,$$

where $\mu = \phi_{\max}/\phi_{\min}$, $\phi_{\max} = \max_{i,j} \phi_j^i$, $\phi_{\min} = \min_{i,j} \phi_j^i$, $i \in \{1,\dots,\kappa\}$, and $j \in \{1,\dots,N\}$.

Proof 5.3 *Construct the following MLFs*

$$V(t) = \sum_{i=1}^{N} \phi_i^{\sigma(t)} e_i^T(t) e_i(t). \tag{5.47}$$

Taking the time derivative of $V(t)$ along the trajectories of (5.46) and using the facts $e_i^T(t)\beta x_i(t) B e_i(t) = 0$, for all $i = 1,\dots,N$, yields

$$\dot{V}(t) = \sum_{i=1}^{N} \phi_i^{\sigma(t)} e_i^T(t) \left(A^T + A\right) e_i(t) + 2 \sum_{i=1}^{N} \phi_i^{\sigma(t)} e_i^T(t) \left(\beta e_i(t) B s(t)\right)$$

$$- 2\alpha \sum_{i=1}^{N} \phi_i^{\sigma(t)} e_i^T(t) \sum_{j=1}^{N} l_{ij}^{\sigma(t)} H e_j(t) - 2\alpha \sum_{i=1}^{N} \phi_i^{\sigma(t)} c_i(t) e_i^T(t) H e_i(t). \tag{5.48}$$

Similar to the steps in the proof of Theorem 5.2, one has

$$\dot{V}(t) \leq e^T(t) \left\{ \Phi^{\sigma(t)} \otimes \left((A^T + A) + 2\iota I_3\right) \right.$$

$$\left. - \alpha \left[(\Phi^{\sigma(t)} \overline{\mathcal{L}}^{\sigma(t)} + (\overline{\mathcal{L}}^{\sigma(t)})^T \Phi^{\sigma(t)}) \otimes H\right] \right\} e(t), \tag{5.49}$$

where $e(t) = [e_1^T(t),\dots,e_N^T(t)]^T$. From the condition

$$\alpha > \frac{\lambda_{\max}(A^T + A) + 2\iota + \varepsilon_0}{\left(\min_{i=1,\dots,\kappa} \lambda_{\min}(\overline{\mathcal{L}}^i + (\Phi^i)^{-1}(\overline{\mathcal{L}}^i)^T \Phi^i)\right) \lambda_{\min}(H)},$$

and (5.49), one obtains

$$\dot{V}(t) \leq -\varepsilon_0 V(t), \ \forall \, t \geq 0. \tag{5.50}$$

Noticing that $V(t_k) \leq \mu V(t_k^-)$ with $\mu = \phi_{\max}/\phi_{\min}$, then it follows from (5.50) that

$$V(t_1) \leq \mu V(t_1^-) \leq \mu \exp(-\varepsilon_0(t_1 - t_0)) V(t_0).$$

By recursion, it is not difficult to deduce that

$$V(t_k) \leq \mu^k \exp(-\varepsilon_0(t_k - t_0))V(t_0), \ k = 1, 2, \ldots.$$

For arbitrarily given $t > t_0$, it can be got from the preceding inequality that

$$V(t) < \mu^{N_\sigma[t_0,t)} \exp(-\varepsilon_0(t - t_0))V(t_0)$$

$$\leq \mu^{N_0} \exp\left(-\left(\varepsilon_0 - \frac{\ln \mu}{\tau_a}\right)(t - t_0)\right) V(t_0), \tag{5.51}$$

where $N_\sigma[t_0, t)$ represents the number of switchings during $[t_0, t)$. Since $\tau_a > (\ln \mu)/c_0$, then $\lim_{t \to \infty} V(t) = 0$. So consensus tracking in CNS (5.37) with directed switching topologies can be achieved exponentially.

Under Assumption 5.3, Theorem 5.3 indicates that the consensus tracking in CNS (5.37) with directed switching topologies can be achieved if both the coupling strength between neighboring agents and the ADT for each possible communication topology are respectively larger than their thresholds. In some practical applications, one may only tune the coupling strength while can not design the dwell time for switching since the switching operation may be triggered by link breakdown, actuator failures and so forth. It is thus interesting to study under what conditions consensus can be ensured with an arbitrary switching. Before moving forward, the following assumption is given.

Assumption 5.4 *Each possible topology $\mathcal{G}(\mathcal{A}^i)$, $i = 1, \ldots, \kappa$, is undirected.*

Then, one may get the following corollary directly from Theorem 5.3.

Corollary 5.5 *Suppose that Assumptions 5.3, 5.4 hold, and the target trajectory $s(t)$ satisfies $\|s(t)\| \leq \iota$ for all $t \geq 0$, where ι is a given positive scalar. Then, consensus tracking in CNS (5.37) with arbitrarily switching topologies can be achieved if the coupling strength α satisfies the following condition:*

$$\alpha > \max_{i=1,\ldots,\kappa} \lambda_{\max}(\Psi_i), \tag{5.52}$$

where $\Psi_i = \left\{\left[\overline{\mathcal{L}}^i + \left(\overline{\mathcal{L}}^i\right)^T\right] \otimes H\right\}^{-1} [I \otimes (A^T + A + 2\iota I_3)], \ i = 1, \ldots, \kappa.$

Proof 5.4 *From Assumptions 5.3 and 5.4, one gets that the matrix $\overline{\mathcal{L}}^{\sigma(t)}$ is positive-definite. Thus, one may construct the following common Lyapunov function for the error dynamical system (5.46):*

$$V(t) = \sum_{i=1}^{N} e_i^T(t)e_i(t). \tag{5.53}$$

Following the steps given in the proof of Theorem 5.3, this corollary can be proved.

Remark 5.11 *Note that whether it is possible to construct a common Lyapunov function for error dynamical system (5.46) without Assumption 5.4 is a challenging issue up to date.*

Remark 5.12 *Since the threshold value of coupling strength for achieving consensus in fixed or switching networks of Lorenz type agents is derived by using a Lyapunov-based approach, it is thus sometimes conservative. Note also that it is very hard or even impossible to obtain some necessary and sufficient conditions on selecting the coupling strength for synchronization of the considered networks as the agent dynamics are indeed nonlinear. Nevertheless, the topic of how to find a less conservative coupling strength for synchronizing the states of coupled Lorenz-type agents is interesting and derives future research.*

5.4   NUMERICAL SIMULATIONS

In this section, we will give two examples to validate the theoretical results given in the last two sections, respectively.

Example 1: Consider the following linearly coupled neural networks:

$$\dot{x}_i(t) = -Ax_i(t) + g(Wx_i(t)) + J(t) + \alpha \sum_{j=0}^{10} a_{ij}(t)(x_j(t) - x_i(t)), \qquad (5.54)$$

where $x_i(t) = [x_{i1}(t), \ldots, x_{in}(t)]^T \in \mathbb{R}^n$ is the neuron state, $i = 1, \ldots, 10$, $W = \left[\hat{W}_1^T, \ldots, \hat{W}_n^T\right]$ stands for the synaptic connection weights, $J = [J_1(t), \ldots, J_n(t)]^T$ represents the external inputs, $g(Wx_i(t)) = \left[g_1(\hat{W}_1x_i(t)), \ldots, g_n(\hat{W}_nx_i(t))\right]^T$ is the activation of neurons. Particularly, choose $A = \begin{bmatrix} 0.125 & 0 \\ 0 & 0.125 \end{bmatrix}$, $W = \begin{bmatrix} -0.15 & 0.45 \\ 0.6 & 0.65 \end{bmatrix}$, $g(y(t)) = [\tanh(y_1(t)), \tanh(y_2(t))]^T$, for all $y(t) = [y_1(t), y_2(t)]^T \in \mathbb{R}^2$, $J(t) = [30\sin(30t), 30\cos(30t)]^T$. In view of Assumption 5.1, it can be obtained that $\Gamma = 0.7234I_2$. Let $\hat{\mathcal{G}} = \left\{\mathcal{G}(\tilde{A}^1), \mathcal{G}(\tilde{A}^2), \mathcal{G}(\tilde{A}^3), \mathcal{G}(\tilde{A}^4)\right\}$ and $\overline{\mathcal{G}} = \{\mathcal{G}(\tilde{A}^1), \mathcal{G}(\tilde{A}^3)\}$. Topologies $\mathcal{G}(\tilde{A}^i)$, $i = 1, \ldots, 4$, are respectively given in Figs. 5.2–5.5, where the weight on each edge is assumed to be 1. In Figs. 5.2–5.5, the neighboring relationships between the agents in CNS (5.2) and the target (labeled as agent 0) are depicted by solid lines with arrows. Let ι_0^i be the minimum number of the agents that should be pinned such that the augmented communication topology $\mathcal{G}(\tilde{A}^i)$ contains at least one directed spanning tree, $i = 1, \ldots, 4$. According to Algorithm 5.1, one gets that $\iota_0^1 = \iota_0^2 = \iota_0^3 = 4$, $\iota_0^4 = 6$. Choose

$$\sigma(t) = \begin{cases} 1 & t \in [k,\, k+0.4), \\ 2, & t \in [k+0.4,\, k+0.5), \\ 3, & t \in [k+0.5,\, k+0.9), \\ 4, & t \in [k+0.9,\, k+1), \end{cases} \qquad (5.55)$$

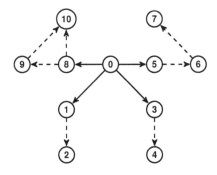

Figure 5.2 The communication graph $\mathcal{G}(\widetilde{\mathcal{A}}^1)$.

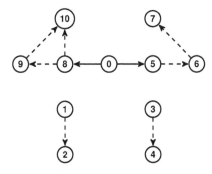

Figure 5.3 The communication graph $\mathcal{G}(\widetilde{\mathcal{A}}^2)$.

where $k = 0, 1, 2, \ldots$. Furthermore, one has that $\bar{t}_{2k-1} = k - 1$, $t_{\min}^{2k-1} = k - 0.6$, $\bar{t}_{2k} = k - 0.5$, $t_{\min}^{2k} = k - 0.1$, where $k \in \mathbb{N}^+$. It can be thus obtained from (5.12) that $\widetilde{\lambda}_0^{2k-1} = -0.7321$ and $\widetilde{\lambda}_0^{2k} = -0.4495$. The switching signal is depicted in Fig. 5.6. Also, one may get from (5.13) that $\mu = 6$. In the simulations, the target system is given as

$$\dot{x}_0(t) = -Ax_0(t) + g(Wx_0(t)) + J(t), \tag{5.56}$$

where $x_{11}(0) = [40, -20]^T$, the other parameters are defined the same as those in (5.54). Choose $\alpha = 8.8$, it can be obtained from Theorem 5.1 that consensus tracking can be ensured in the switched CNS (5.54) with the target given in (5.56). The state trajectories of the CNSs are respectively shown in Figs. 5.7–5.8. Use $\text{Error}(t) = (1/10)\sqrt{\sum_{j=1}^{10} \|x_j(t) - x_0(t)\|^2}$ to denote the consensus error of the considered CNSs. Figure 5.9 indicates that the consensus tracking problem is indeed solved.

Example 2: Consider the switched CNS (5.33) that consists of seven identical Chen systems, described as follows:

$$\dot{x}_i(t) = Ax_i(t) + \beta x_i(t)Bx_i(t) + \alpha \sum_{j=1}^{7} a_{ij} H(x_j(t) - x_i(t)),$$

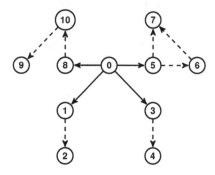

Figure 5.4 The communication graph $\mathcal{G}(\widetilde{\mathcal{A}}^3)$.

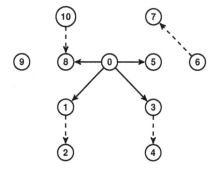

Figure 5.5 The communication graph $\mathcal{G}(\widetilde{\mathcal{A}}^4)$.

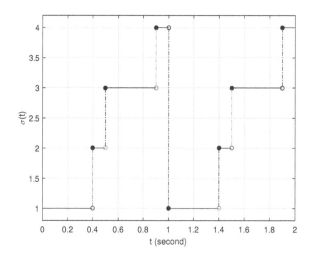

Figure 5.6 Switching signal $\sigma(t)$ in Example 1.

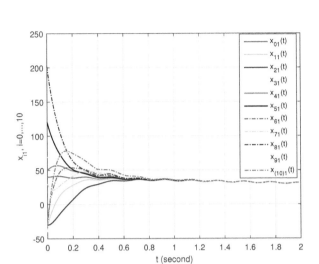

Figure 5.7 The agents' state trajectories $x_{i1}(t)$, $i = 0, \ldots, 10$, in Example 1.

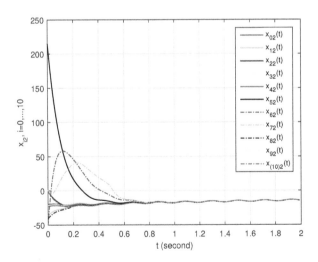

Figure 5.8 The agents' state trajectories $x_{i2}(t)$, $i = 0, \ldots, 10$, in Example 1.

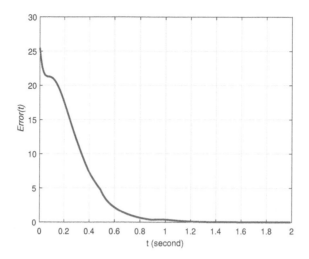

Figure 5.9 Trajectories of the consensus error Error(t) in Example 1.

where

$$A=\begin{bmatrix} -35 & 35 & 0 \\ -7 & 28 & 0 \\ 0 & 0 & -3 \end{bmatrix}, \quad B=\begin{bmatrix} 0 & 0 & 0 \\ 0 & 0 & -1 \\ 0 & 1 & 0 \end{bmatrix}, \quad \beta=[1, 0, 0],$$

$H = \mathrm{diag}\{30, 30, 30\}$, $i = 1,\ldots,7$. The network topologies $\mathcal{G}(\mathcal{A}^5)$ and $\mathcal{G}(\mathcal{A}^6)$ are given in Figs. 5.10 and 5.11, respectively, where the weights on each link are assumed to be 1. By using Algorithm 5.2, one gets that $\mathcal{G}(\mathcal{A}^5)$ contains two disjoint connected components, and $\mathcal{G}(\mathcal{A}^6)$ contains three disjoint connected components. To make each possible augmented communication topology contains a directed spanning tree, the nodes labeled as 1, 4 in $\mathcal{G}(\mathcal{A}^5)$ are pinned (see Figure 5.12), and the nodes labeled as 1, 3, and 5 in $\mathcal{G}(\mathcal{A}^6)$ are pinned (see Figure 5.13). In this example, the node labeled as 0 is the leader whose dynamics are described by (5.36) with the initial states $s(0) = [0.75, 0.4, -0.4]^T$. It is theoretically shown in [100] that all the solutions of system (5.36) are globally bounded. Let $D_1 = 31$, $D_2 = 36.5$, and $D_3 = 61$, one gets that the solution $s(t) = [s_1(t), s_2(t), s_3(t)]^T$ of Chen system (5.36) satisfies $s_1(t) \leq D_1$, $s_2(t) \leq D_2$, and $s_3(t) \leq D_3$. Thus, the parameter ι can be set as $\iota = 77.56$. The state trajectory of system (5.36) is given in Figure 5.14 by using the fourth order Runge-Kutta method. According to Theorem 5.3, consensus tracking of the CNS can be realized if the coupling strength $\alpha \geq 12.2781$ and the ADT $\tau_a > 0.0347$. Set $N_0 = 4$ and suppose the topologies switched at the time instants $t = 0.01, 0.03, 0.05, 0.08\,\mathrm{s}$, and switched periodically every 0.4 s after $t = 0.08\,\mathrm{s}$, then $\tau_a = 0.04\,\mathrm{s}$. Choose $\alpha = 13$, the state trajectories of the agents are respectively given in Figs. 5.15–5.17 from which we can see the consensus tracking is indeed achieved.

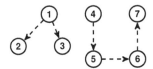

Figure 5.10 The communication graph $\mathcal{G}(\mathcal{A}^5)$.

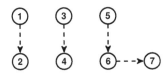

Figure 5.11 The communication graph $\mathcal{G}(\mathcal{A}^6)$.

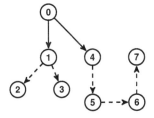

Figure 5.12 The augmented graph $\mathcal{G}(\widetilde{\mathcal{A}}^5)$.

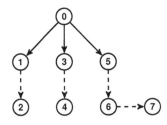

Figure 5.13 The augmented graph $\mathcal{G}(\widetilde{\mathcal{A}}^6)$.

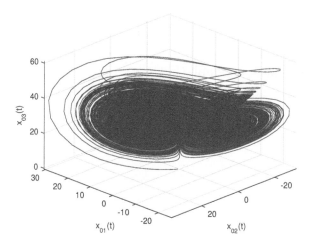

Figure 5.14 Trajectory of the Chen system in Example 2.

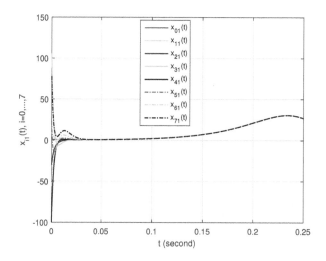

Figure 5.15 The agents' state trajectories $x_{i1}(t)$, $i = 0, 1, \ldots, 7$.

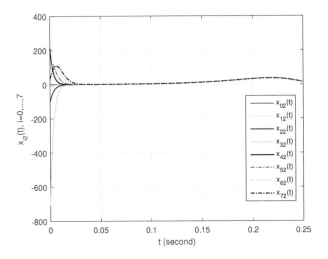

Figure 5.16 The agents' state trajectories $x_{i2}(t)$, $i = 0, 1, \ldots, 7$.

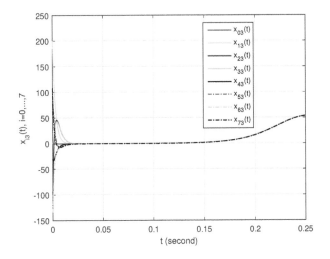

Figure 5.17 The agents' state trajectories $x_{i3}(t)$, $i = 0, 1, \ldots, 7$.

5.5 CONCLUSIONS

This chapter has solved the consensus tracking problem for switched CNSs with Lipschitz nonlinear dynamics as well as Lorenz nonlinear dynamics. First, to make each possible augmented communication topology contains a directed spanning tree, several algorithms were proposed to select the pinning agents. Furthermore, by using the stability theory of switched systems and developing some appropriate MLFs, some sufficient consensus criteria were given. Finally, two numerical simulations were performed to validate the obtained theoretical results.

CHAPTER 6

Consensus tracking of CNSs with higher-order dynamics and directed switching topologies

This chapter studies the consensus tracking of CNSs with higher-order dynamics and directed switching topologies. This chapter begins by overviewing some previous works and by indicating our motivations. Section 6.2 firstly studies the case with Lipschitz nonlinear dynamics and directed fixed topology. Then we extend the results to directed switching topologies with each topology contains a directed spanning tree. This section finally studies the case with directed switching topologies that frequently contain a directed spanning tree. Section 6.3 studies the case with general linear dynamics and occasionally missing control inputs. This section presents some sufficient criteria for achieving consensus tracking. Moreover, the convergence rate is discussed. Finally, some simulations are given to validate the theoretical results.

6.1 INTRODUCTION

In contrast to CNSs with first-order nonlinear dynamics, CNSs with second-order non-linear dynamics are more interesting as it can describe a large class of real networked systems, including coupled pendulums [5] and coupled point-mass systems with or without nonlinear disturbances [165]. Leaderless consensus problem for CNSs with second-order nonlinear dynamics and a fixed weakly connected topology was investigated in [217]. In [137], the consensus tracking problem for CNSs with second-order nonlinear dynamics in the presence of a leader under an arbitrarily given directed topology was studied from pinning control approach. Furthermore, consensus tracking problem for CNSs with higher-order Lipschitz type agent dynamics and a fixed topology was studied in [79].

In the existing literature on the consensus tracking problem for CNSs with nonlinear dynamics, it is commonly assumed that the communication topology is fixed.

However, this may not be the case in reality due to technological limitations of sensors or external disturbances on communication channels. Motivated by this observation and based on the aforementioned works, this chapter makes further endeavors to consider the consensus tracking problem for CNSs with higher-order nonlinear dynamics and directed switching topologies, even for the case where the time-varying topology only frequently but not always contains a directed spanning tree.

6.2 CONSENSUS TRACKING OF CNSs WITH HIGHER-ORDER NONLINEAR DYNAMICS

6.2.1 Problem formulation

Consider a CNS consisting of a leader and N followers, where the leader is labelled as agent 0 and the followers are labelled as agents $1, \ldots, N$. The dynamics of agent i, $i = 0, 1, \ldots, N$, are given by

$$\dot{x}_i(t) = Ax_i(t) + Cf(x_i(t), t) + Bu_i(t), \tag{6.1}$$

where $x_i(t) \in \mathbb{R}^n$ represent the states of agent i, $f(\cdot, \cdot) : \mathbb{R}^n \times [0, +\infty) \mapsto \mathbb{R}^p$ is a continuously differentiable vector-valued function representing the intrinsic nonlinear dynamics, $u_i(t) \in \mathbb{R}^m$ is the control input to be designed, $A \in \mathbb{R}^{n \times n}$, $B \in \mathbb{R}^{n \times m}$, and $C \in \mathbb{R}^{n \times p}$ are constant real matrices. It is assumed that the matrix pair (A, B) is stabilizable. In this section, it is assumed that the leader will not being affected by any followers, i.e. $u_0(t) \equiv \mathbf{0}_m$ in CNS (6.1). Before moving on, the following assumption is made.

Assumption 6.1 *There exists a nonnegative constant ϱ, such that*

$$\|f(y, t) - f(z, t)\| \le \varrho \|y - z\|, \ \forall y, z \in \mathbb{R}^n, \ t \ge 0.$$

To achieve consensus tracking, the following distributed consensus tracking protocol is proposed for each follower i:

$$u_i(t) = \alpha F \sum_{j=0}^{N} a_{ij}(t) \left(x_j(t) - x_i(t) \right), \quad i = 1, \ldots, N, \tag{6.2}$$

where $\alpha > 0$ represents the coupling strength, $F \in \mathbb{R}^{m \times n}$ is the feedback gain matrix to be designed, and $\mathcal{A}(t) = \left[a_{ij}(t) \right]_{(N+1) \times (N+1)}$ is the adjacency matrix of graph $\mathcal{G}(t)$. Here, $\mathcal{G}(t)$ describes the underlying communication topology among the $N+1$ agents at time t.

Remark 6.1 *System (6.1) is quite general since it covers the CNSs with integrator-type dynamics. For example, system (6.1) becomes a second-order CNS with nonlinear dynamics, studied in [217], if*

$$A = \begin{bmatrix} 0 & I_r \\ 0 & 0 \end{bmatrix}, \quad B = \begin{bmatrix} 0 \\ I_r \end{bmatrix}, \quad C = \begin{bmatrix} 0 & 0 \\ 0 & I_r \end{bmatrix}, \tag{6.3}$$

where r is the dimension of the position state vector of an agent. Furthermore, it is easy to check that matrix pair (A, B) with A and B given in (6.3) is controllable, therefore also stabilizable. Note also that system (6.1) reduces to the commonly studied linear CNSs [78] if C is a zero matrix.

6.2.2 Main results for fixed topology containing a directed spanning tree

In this section, distributed consensus tracking is addressed for CNS (6.1) with a fixed communication topology containing a directed spanning tree.

Without loss of generality, let $\mathcal{G}(t) = \mathcal{G}$ for all $t \geq 0$ since the communication topology is assumed to be fixed in this subsection. To derive the main results, the following assumption is needed.

Assumption 6.2 *The communication topology $\mathcal{G}$ contains a directed spanning tree with agent 0 (i.e. the leader) as the root.*

Under Assumption 6.2, , the Laplacian matrix of directed graph $\mathcal{G}$ can be written as

$$\mathcal{L} = \begin{bmatrix} 0 & \mathbf{0}_N^T \\ \mathbf{P} & \overline{\mathcal{L}} \end{bmatrix}, \quad \overline{\mathcal{L}} = \begin{bmatrix} \sum_{j \in \mathcal{N}_1} a_{1j} & -a_{12} & \cdots & -a_{1N} \\ -a_{21} & \sum_{j \in \mathcal{N}_2} a_{2j} & \cdots & -a_{2N} \\ \vdots & \vdots & \ddots & \vdots \\ -a_{N1} & -a_{N2} & \cdots & \sum_{j \in \mathcal{N}_N} a_{Nj} \end{bmatrix}, \quad (6.4)$$

where $\mathbf{P} = -[a_{10}, \ldots, a_{N0}]^T$. It can be thus obtained from Lemma 2.15 that there exists a positive definite diagonal matrix $\Phi = \mathrm{diag}\{\phi_1, \ldots, \phi_N\}$ such that $\overline{\mathcal{L}}^T \Phi + \Phi \overline{\mathcal{L}} > 0$. One such $\phi = [\phi_1, \ldots, \phi_N]^T$ can be obtained by solving the matrix equation $\overline{\mathcal{L}}^T \phi = \mathbf{1}_N$.

Since $u_0(t) \equiv \mathbf{0}_m$, one has

$$\dot{x}_0(t) = Ax_0(t) + Cf(x_0(t), t).$$

Furthermore, substituting (6.2) into (6.1) gives a closed-loop system:

$$\dot{x}_i(t) = Ax_i(t) + Cf(x_i(t), t) + \alpha BF \sum_{j=0}^N a_{ij} (x_j(t) - x_i(t)), \ i = 1, \ldots, N,$$

where $\mathcal{A} = [a_{ij}]_{(N+1) \times (N+1)}$ is the adjacency matrix of graph $\mathcal{G}$.

Define $e_i(t) = x_i(t) - x_0(t)$, $i = 1, \ldots, N$, and $e(t) = [e_1^T(t), \ldots, e_N^T(t)]^T$. Based on the above analysis, one has the following error dynamical system:

$$\dot{e}_i(t) = Ae_i(t) + C\left(f(x_i(t), t) - f(x_0(t), t)\right) - \alpha BF \sum_{j=1}^N \bar{l}_{ij}(t)e_j(t). \quad (6.5)$$

Rewriting (6.5) into a compact form, one has

$$\dot{e}(t) = (I_N \otimes A) e(t) + (I_N \otimes C)\tilde{f}(x(t), t) - \alpha \left(\overline{\mathcal{L}} \otimes BF\right) e(t), \quad (6.6)$$

where $\tilde{f}(x(t), t) = \left[(f(x_1(t), t) - f(x_0(t), t))^T, \ldots, (f(x_N(t), t) - f(x_0(t), t))^T \right]^T$ and $x(t) = [x_0^T(t), x_1^T(t), \ldots, x_N^T(t)]^T$.

Before moving on, a multi-step design procedure is given for selecting the control parameters in protocol (6.2) under a fixed topology $\mathcal{G}$.

Algorithm 6.1 *Under Assumptions 6.1 and 6.2, the consensus protocol (6.2) can be designed as follows:*

(1) Select two scalars $c > 0$ and $\beta > 0$. Solve the LMI

$$\begin{bmatrix} AS + SA^T - cBB^T + \varrho^2 CC^T + \beta S & S \\ S & -I \end{bmatrix} < 0 \qquad (6.7)$$

to get a matrix $S > 0$. Then, take $F = \frac{1}{2} B^T S^{-1}$.

(2) Choose the coupling strength $\alpha > 2c/\lambda_0$, where $\lambda_0 = \lambda_{\min}(\overline{\mathcal{L}} + \Phi^{-1}\overline{\mathcal{L}}^T\Phi)$, $\overline{\mathcal{L}}$ is given in (6.4).

Then, one can establish the following theorem.

Theorem 6.1 *Suppose that Assumptions 6.1 and 6.2 hold, and the LMI (6.7) has a feasible solution. Then, the consensus tracking problem for the CNS (6.1) can be solved by protocol (6.2) with control parameters constructed in Algorithm 6.1.*

Proof 6.1 *Construct the following Lyapunov function for the error dynamical system (6.6):*

$$V(t) = e^T(t) \left(\Phi \otimes S^{-1} \right) e(t), \qquad (6.8)$$

where $S > 0$ is a feasible solution of (6.7).

Taking the time derivative of $V(t)$ along the trajectories of system (6.6) gives

$$\dot{V}(t) = e^T(t) \left(\Phi \otimes (S^{-1}A + A^T S^{-1}) \right) e(t)$$

$$+ 2 \sum_{i=1}^N \phi_i e_i^T(t) S^{-1} C(f(x_i(t), t) - f(x_0(t), t))$$

$$- 2\alpha e^T(t) \left(\Phi \overline{\mathcal{L}} \otimes S^{-1} BF \right) e(t). \qquad (6.9)$$

Substituting $F = \frac{1}{2} B^T S^{-1}$ into (6.9) yields

$$\dot{V}(t) = e^T(t) \left[\Phi \otimes (S^{-1}A + A^T S^{-1}) \right] e(t)$$

$$+ 2 \sum_{i=1}^N \phi_i e_i^T(t) S^{-1} C(f(x_i(t), t) - f(x_0(t), t))$$

$$- \alpha e^T(t) \left(\Phi \overline{\mathcal{L}} \otimes S^{-1} BB^T S^{-1} \right) e(t). \qquad (6.10)$$

Based on Assumption 6.1, it follows from (6.10) that

$$\dot{V}(t) \leq e^T(t) \left(\Phi \otimes (S^{-1}A + A^T S^{-1}) \right) e(t)$$

$$+ \sum_{i=1}^{N} \phi_i e_i^T(t) \left(\varrho^2 S^{-1} C C^T S^{-1} + I \right) e_i(t)$$

$$- \alpha e^T(t) \left(\Phi \overline{\mathcal{L}} \otimes S^{-1} B B^T S^{-1} \right) e(t)$$

$$= e^T(t) [\Phi \otimes (S^{-1} A + A^T S^{-1} + \varrho^2 S^{-1} C C^T S^{-1} + I)] e(t)$$

$$- \frac{\alpha}{2} e^T(t) \left[\left(\Phi \overline{\mathcal{L}} + \overline{\mathcal{L}}^T \Phi \right) \otimes S^{-1} B B^T S^{-1} \right] e(t). \tag{6.11}$$

Let $\varepsilon(t) = [\varepsilon_1^T(t), \ldots, \varepsilon_N^T(t)]^T$, where $\varepsilon_i(t) = S^{-1} e_i(t)$, $i = 1, \ldots, N$. Obviously, $e(t) = (I_N \otimes S) \varepsilon(t)$. It thus follows from (6.11) that

$$\dot{V}(t) \leq \varepsilon^T(t) \left[\Phi \otimes \left(AS + SA^T + \varrho^2 C C^T + S^T S \right) \right] \varepsilon(t)$$

$$- \frac{\alpha}{2} \varepsilon^T(t) \left\{ \left[\Phi \overline{\mathcal{L}} + \overline{\mathcal{L}}^T \Phi \right] \otimes B B^T \right\} \varepsilon(t). \tag{6.12}$$

Then, one has

$$\dot{V}(t) \leq \varepsilon^T(t) \left[\Phi \otimes \left(AS + SA^T + \varrho^2 C C^T + S^T S \right) \right] \varepsilon(t)$$

$$- \frac{\alpha \lambda_0}{2} \varepsilon^T(t) \left(\Phi \otimes B B^T \right) \varepsilon(t), \tag{6.13}$$

where $\lambda_0 = \lambda_{\min} \left(\overline{\mathcal{L}} + \Phi^{-1} \overline{\mathcal{L}}^T \Phi \right)$. Since $\alpha > 2c/\lambda_0$, it follows from (6.13) that

$$\dot{V}(t) \leq \varepsilon^T(t) \left[\Phi \otimes \left(AS + SA^T + \varrho^2 C C^T + S^T S \right) \right] \varepsilon(t)$$

$$- c \varepsilon^T(t) \left(\Phi \otimes B B^T \right) \varepsilon(t). \tag{6.14}$$

Using (6.7) and the Schur complement lemma, it follows from (6.14) that

$$\dot{V}(t) \leq -\beta \varepsilon^T(t) \left(\Phi \otimes S \right) \varepsilon(t)$$

$$= -\beta e^T(t) \left(\Phi \otimes S^{-1} \right) e(t).$$

Thus, one gets

$$V(t) < \exp(-\beta t) V(t_0) \tag{6.15}$$

for all $t > t_0$. Then, one concludes that $\|e(t)\| \to 0$ as $t \to +\infty$. Thus, the consensus tracking problem in CNS (6.1) is solved by the distributed consensus tracking protocol (6.2) with control parameters constructed in Algorithm 6.1.

Remark 6.2 According to Algorithm 6.1, one gets that the existence of protocol (6.2) depends on the solvability of the LMI (6.7). It can be seen from (6.15) that consensus tracking can be achieved in the closed-loop system (6.1) with protocol (6.2) constructed in Algorithm 6.1 if LMI (6.7) is solvable for some given positive scalars β and c. More specifically, the selections of β and c do not influence the qualitative results given in Theorem 6.1. The above analysis indicates that both β and c are free positive scalars *in (6.7). Thus, the solvability of (6.7) is equivalent to the following feasible problem: there exist a scalar $\iota > 0$ and a matrix $P > 0$ such that*

$$\begin{bmatrix} AP + PA^T - \iota BB^T + \varrho^2 CC^T & P \\ P & -I \end{bmatrix} < 0. \qquad (6.16)$$

By using Finsler's Lemma and the Schur complement lemma, one gets that there exist a scalar $\iota > 0$ and a matrix $P > 0$ such that (6.16) holds if and only if there exist matrices $P > 0$ and $E \in \mathbb{R}^{m \times n}$ such that the following algebraic Riccati inequality holds:

$$(A - BE)P + P(A - BE)^T + \varrho^2 CC^T + P^2 < 0. \qquad (6.17)$$

According to the bounded real lemma, one gets that (6.17) holds if and only if there exists a matrix $E \in \mathbb{R}^{m \times n}$ such that

$$\left\| \varrho C^T \left(sI - (A - BE)^T \right)^{-1} \right\|_\infty < 1,$$

i.e.,

$$\left\| C^T \left(sI - (A - BE)^T \right)^{-1} \right\|_\infty < 1/\varrho. \qquad (6.18)$$

Thus, LMI (6.7) is solvable if and only if there exists a matrix $E \in \mathbb{R}^{m \times n}$ such that (6.18) holds.

Remark 6.3 *Note that a necessary and sufficient condition for the solvability of LMI (6.7) was provided in Remark 6.2. Noticeably, it can be seen from Remark 6.2 that (A, B) is stabilizable is a necessary condition for the solvability of LMI (6.7).*

6.2.3 Main results for switching topologies with each topology containing a directed spanning tree

Based on the results given in subsection 6.2.2, consensus tracking is considered in this subsection for CNS (6.1) with directed switching topologies where each possible topology contains a directed spanning tree rooted at the leader node.

Suppose that the communication topologies switch at time instants t_1, t_2, ... And let $t_0 = 0$ be the initial time instant. It is assumed that the switching sequence satisfies $0 < \tau_m \leq t_{k+1} - t_k \leq \tau_M < +\infty$, $k = 0, 1, \ldots$. And it is assumed that $\mathcal{G}(t) \in \hat{\mathcal{G}} = \{\mathcal{G}^1, \ldots, \mathcal{G}^\kappa\}$ with $\kappa > 1$ and $\kappa \in \mathbb{N}$.

From the above statements, it can be seen that the communication topology $\mathcal{G}(t) = \mathcal{G}^{\sigma(t)}$ is fixed for $t \in [t_k, t_{k+1})$, $k = 0, 1, \ldots$.

Assumption 6.3 *Each possible communication topology $\mathcal{G}^i$, $i \in \{1, \ldots, \kappa\}$, contains a directed spanning tree with agent 0 (i.e. the leader) as the root.*

Based on the analysis, the Laplacian matrix of directed graph $\mathcal{G}^{\sigma(t)}$ can be written as

$$\mathcal{L}^{\sigma(t)} = \begin{bmatrix} 0 & \mathbf{0}_N^T \\ \mathbf{P}^{\sigma(t)} & \overline{\mathcal{L}}^{\sigma(t)} \end{bmatrix},$$

$$
\overline{\mathcal{L}}^{\sigma(t)} = \begin{bmatrix} \sum_{j\in\mathcal{N}_1} a_{1j}^{\sigma(t)} & -a_{12}^{\sigma(t)} & \cdots & -a_{1N}^{\sigma(t)} \\ -a_{21}^{\sigma(t)} & \sum_{j\in\mathcal{N}_2} a_{2j}^{\sigma(t)} & \cdots & -a_{2N}^{\sigma(t)} \\ \vdots & \vdots & \ddots & \vdots \\ -a_{N1}^{\sigma(t)} & -a_{N2}^{\sigma(t)} & \cdots & \sum_{j\in\mathcal{N}_N} a_{Nj}^{\sigma(t)} \end{bmatrix}, \tag{6.19}
$$

where $\mathbf{P}^{\sigma(t)} = -[a_{10}^{\sigma(t)},\dots,a_{N0}^{\sigma(t)}]^T$ and $\mathcal{A}^{\sigma(t)} = [a_{ij}^{\sigma(t)}]_{(N+1)\times(N+1)}$ is the adjacency matrix of graph $\mathcal{G}^{\sigma(t)}$. Under Assumption 6.3, it can be thus obtained from Lemma 2.15 that there exists a positive definite diagonal matrix $\Phi^j = \text{diag}\{\phi_1^j,\dots,\phi_N^j\}$ such that $(\overline{\mathcal{L}}^j)^T\Phi^j + \Phi^j\overline{\mathcal{L}}^j > 0$, $j = 1,\dots,\kappa$, where Φ^j satisfies $(\overline{\mathcal{L}}^j)^T\phi^j = \mathbf{1}_N$ with $\phi^j = [\phi_1^j,\dots,\phi_N^j]^T$.

Since $\mathcal{G}(t) = \mathcal{G}^{\sigma(t)}$ for all $t \geq 0$, the consensus tracking protocol (6.2) can be rewritten as

$$
u_i(t) = \alpha F \sum_{j=0}^N a_{ij}^{\sigma(t)}(x_j(t) - x_i(t)), \quad i = 1,\dots,N, \tag{6.20}
$$

where $\alpha > 0$ represents the coupling strength, $F \in \mathbb{R}^{m\times n}$ is the feedback gain matrix to be designed.

Define $e_i(t) = x_i(t) - x_0(t)$, $i = 1,\dots,N$, and $e(t) = [e_1^T(t),\dots,e_N^T(t)]^T$. Similar to the analysis given in subsection 6.2.2, one has the following error dynamical system:

$$
\dot{e}_i(t) = Ae_i(t) + C\left(f(x_i(t),t) - f(x_0(t),t)\right) - \alpha BF\sum_{j=0}^N \bar{l}_{ij}^{\sigma(t)}(t)e_j(t), \tag{6.21}
$$

where $\overline{\mathcal{L}} = [\bar{l}_{ij}^{\sigma(t)}]_{N\times N}$ is defined in (6.19). Rewriting (6.21) into a compact form, one has

$$
\dot{e}(t) = (I_N \otimes A)e(t) + (I_N \otimes C)\tilde{f}(x(t),t) - \alpha\left(\overline{\mathcal{L}}^{\sigma(t)} \otimes BF\right)e(t), \tag{6.22}
$$

where $\tilde{f}(x(t),t) = [(f(x_1(t),t) - f(x_0(t),t))^T,\dots,(f(x_N(t),t) - f(x_0(t),t))^T]^T$ and $x(t) = [x_0^T(t), x_1^T(t),\dots,x_N^T(t)]^T$.

Before moving on, a multi-step design procedure is given for selecting the control parameters in protocol (6.20).

Algorithm 6.2 *Under Assumptions 6.1 and 6.3, the consensus tracking protocol (6.20) can be designed as follows:*

(1) Select two scalars $c > 0$ and $\beta > 0$. Solve the LMI

$$
\begin{bmatrix} AS + SA^T - cBB^T + \varrho^2 CC^T + \beta S & S \\ S & -I \end{bmatrix} < 0 \tag{6.23}
$$

to get a matrix $S > 0$. Then, take $F = \frac{1}{2}B^T S^{-1}$.

(2) Choose the coupling strength $\alpha > 2c/\lambda_0$, where $\lambda_0 = \min_{i=1,\dots,\kappa}\lambda_{\min}(\overline{\mathcal{L}}^i + (\Phi^i)^{-1}(\overline{\mathcal{L}}^i)^T\Phi^i)$, and $\Phi^i = \text{diag}\{\phi_1^i,\dots,\phi_N^i\} > 0$, $i = 1,\dots,\kappa$.

Then, one can establish the following theorem.

Theorem 6.2 *Suppose that Assumptions 6.1 and 6.3 hold, and the LMI (6.23) has a feasible solution. Then, the consensus tracking problem for the CNS (6.1) can be solved by protocol (6.20) with control parameters constructed in Algorithm 6.2, if the dwell time $\tau_m > (\ln\mu)/\beta$, where $\mu = \bar{\phi}/\underline{\phi}$, $\bar{\phi} = \max_{i,j}\phi_j^i$, $\underline{\phi} = \min_{i,j}\phi_j^i$, $i \in \{1,\ldots,\kappa\}$, and $j = 1,\ldots,N$.*

Proof 6.2 *Construct the following MLFs for the error dynamical system (6.22):*

$$V(t) = e^T(t)\left(\Phi^{\sigma(t)} \otimes S^{-1}\right)e(t), \tag{6.24}$$

where $\Phi^{\sigma(t)} \in \{\Phi^1,\ldots,\Phi^\kappa\}$ and $S > 0$ is a feasible solution of (6.23).

Note that the communication topology $\mathcal{G}^{\sigma(t)}$ is fixed for $t \in [t_0, t_1)$. Then, similar to the proof of Theorem 6.1, one gets that

$$\dot{V}(t) \leq \varepsilon^T(t)\left[\Phi^{\sigma(t)} \otimes \left(AS + SA^T + \varrho^2 CC^T + S^T S\right)\right]\varepsilon(t)$$
$$- \frac{\alpha\lambda_0}{2}\varepsilon^T(t)\left(\Phi^{\sigma(t)} \otimes BB^T\right)\varepsilon(t), \quad t \in [t_0, t_1), \tag{6.25}$$

where $\varepsilon(t) = [\varepsilon_1^T(t),\ldots,\varepsilon_N^T(t)]^T$, $\varepsilon_i(t) = S^{-1}e_i(t)$, $i = 1, \ldots, N$, and $\lambda_0 = \min_{i=1,\ldots,\kappa}\lambda_{\min}\left(\overline{\mathcal{L}}^i + (\Phi^i)^{-1}(\overline{\mathcal{L}}^i)^T\Phi^i\right)$. Since $\alpha > 2c/\lambda_0$, it follows from (6.25) that

$$\dot{V}(t) < \varepsilon^T(t)\left[\Phi^{\sigma(t)} \otimes \left(AS + SA^T + \varrho^2 CC^T + S^T S - cBB^T\right)\right]\varepsilon(t). \tag{6.26}$$

Using (6.23) and the Schur complement lemma, it follows from (6.26) that

$$\dot{V}(t) < -\beta\varepsilon^T(t)\left(\Phi^{\sigma(t)} \otimes S\right)\varepsilon(t)$$
$$= -\beta e^T(t)\left(\Phi^{\sigma(t)} \otimes S^{-1}\right)e(t). \tag{6.27}$$

Note that the closed-loop CNS (6.1) with protocol (6.20) switches at $t = t_1$. It thus follows from the above analysis that

$$V(t_1^-) < V(t_0)\exp(-\beta(t_1 - t_0))$$
$$\leq \exp(-\beta\tau_m)V(t_0).$$

According to (6.24), one gets that

$$V(t_1) \leq \mu V(t_1^-),$$

where $\mu = \bar{\phi}/\underline{\phi}$, $\bar{\phi} = \max_{i,j}\phi_j^i$, $\underline{\phi} = \min_{i,j}\phi_j^i$, $i \in \{1,\ldots,\kappa\}$, and $j \in \{1,\ldots,N\}$.
Thus, one gets

$$V(t_1) < \mu\exp(-\beta\tau_m)V(t_0),$$

i.e.,

$$V(t_1) < \exp(-\beta \tau_m + \ln \mu)V(t_0). \tag{6.28}$$

According to fact that $\tau_m > (\ln \mu)/\beta$, one gets that $\beta - (\ln \mu)/\tau_m > 0$. Based on the above analysis, it follows from (6.28) that

$$V(t_1) < \exp(-\upsilon \tau_m)V(t_0), \tag{6.29}$$

where $\upsilon = \beta - \frac{\ln \mu}{\tau_m} > 0$. For an arbitrarily given $t > t_1$, there exists a positive integer $z \geq 1$ such that $t_z < t \leq t_{z+1}$. Furthermore, for an arbitrarily given $h \in \mathbb{N}$, one gets the following inequality by recursion:

$$\begin{aligned} V(t_{h+1}) &< \exp(-\upsilon \tau_m)V(t_h) \\ &< \exp(-(h+1)\upsilon \tau_m)V(t_0). \end{aligned} \tag{6.30}$$

When $t \in (t_z, t_{z+1})$, based on the above analysis, one gets

$$\begin{aligned} V(t) &< \exp(-\beta(t - t_z))V(t_z) \\ &< \exp(-[\beta(t - t_z) + z\upsilon \tau_m])V(t_0) \\ &< \exp(-z\upsilon \tau_m)V(t_0) \\ &< \exp\left(-\frac{z\tau_m}{(z+1)\tau_M}\upsilon t\right)V(t_0). \end{aligned} \tag{6.31}$$

Since $z \geq 1$, it follows from (6.31) that

$$V(t) < \exp\left(-\frac{\tau_m \upsilon}{2\tau_M}t\right)V(t_0), \ t \in (t_z, t_{z+1}). \tag{6.32}$$

For the case of $t = t_{z+1}$, one gets from (6.30) that

$$V(t) < \exp\left(-\frac{\tau_m \upsilon}{\tau_M}t\right)V(t_0). \tag{6.33}$$

According to (6.32) and (6.33), one concludes that $\|e(t)\| \to 0$ as $t \to +\infty$. Thus, the consensus tracking problem in CNS (6.1) is solved by protocol (6.20) with control parameters constructed in Algorithm 6.2.

Remark 6.4 *The existence of protocol (6.20) depends on the solvability of the LMI (6.23). It is not hard to check that the LMI (6.23) is feasible if and only if the LMI (6.16) is feasible. Thus, the solvability conditions for the LMI (6.16) provided in Remark 6.2 are applicable for the LMI (6.23).*

6.2.4 Main results for switching topologies frequently containing a directed spanning tree

In this subsection, distributed consensus tracking is investigated for CNS (6.1) with directed switching topologies that only frequently but not always contains a directed spanning tree.

In the context of CNSs, the initial communication topology of the multiple agents always meets some connection conditions [143, 182]. However, as the system evolves with time, agents might not always be able to well sense their neighbors due to sensor failures or the distances between them are larger than the effective sensing range.

Motivated by this observation and for convenience of analysis, it is now assumed that the initial communication topology of the CNS (6.1) has a directed spanning tree with the leader agent being the root and all the other possible topologies do not contain any directed spanning tree. To achieve consensus tracking in such a scenario, a communication restoration mechanism is employed here, i.e. the sensing devices may be able to recover from failures through some backup or repairing efforts, though it may take a short period of time. Without loss of generality, it is assumed that $\mathcal{G}^{\sigma(t_0)} = \mathcal{G}^1$ which contains a directed spanning tree rooted at the leader agent. Then, based on the above analysis, one may suppose that there exists an infinite sequence of uniformly bounded non-overlapping time intervals $[t'_k, t'_{k+1})$ with $t'_0 = t_0 = 0$, $\inf_{k \in \mathbb{N}} (t'_{k+1} - t'_k) \geq \tilde{\tau}_m > \tau_m > 0$, $\sup_{k \in \mathbb{N}} (t'_{k+1} - t'_k) < \tilde{\tau}_M$, such that $\mathcal{G}^{\sigma(t)} = \mathcal{G}^1$, $t \in [t'_k, t'_k + \delta_k)$, where $\tau_m < \delta_k < t'_{k+1} - t'_k$, $k \in \mathbb{N}$. The objective in this subsection is to construct a distributed consensus tracking algorithm to realize consensus tracking in the CNS (6.1) from an intermittent control approach. Specifically, the multiple agents only share their information with their neighbors when $t \in [t'_k, t'_k + \delta_k)$, $k \in \mathbb{N}$. In this case, the protocol (6.2) can be specified as

$$
u_i(t) = \begin{cases} \alpha F \sum_{j=0}^{N} a^1_{ij} (x_j(t) - x_i(t)), & t \in [t'_k, t'_k + \delta_k), \\ \mathbf{0}_m, & t \in [t'_k + \delta_k, t'_{k+1}), \end{cases} \quad k \in \mathbb{N}, \tag{6.34}
$$

where $\alpha > 0$ represents the coupling strength, $F \in \mathbb{R}^{m \times n}$ is the feedback gain matrix to be designed, and $\mathcal{A}^1 = \left[a^1_{ij}\right]_{(N+1) \times (N+1)}$ is the adjacency matrix of $\mathcal{G}^1$.

Remark 6.5 *Generally speaking, it is more difficult to solve the consensus tracking problem for nonlinear CNSs with communication topology frequently having a directed spanning tree than the case where each possible topology containing a directed spanning tree. To deal with this challenging case, a communication restoration mechanism is employed to restore the topology to its initial form. Furthermore, it is assumed that the multiple agents have the ability to discard communications when the network topology does not contain any directed spanning tree. Specifically, each agent will evolve according to its own intrinsic dynamics by discarding the information from its neighbors when the communication topology does not contain any spanning tree. Obviously, the condition that the topologies frequently contain a directed spanning tree is stronger than that the topologies jointly have a directed spanning tree. Additionally, it is unknown whether it is possible to solve the consensus tracking problem of the CNS (6.1) with topologies only jointly have a directed spanning tree.*

Since $\mathcal{G}^1$ contains a directed spanning tree with the leader agent being the root, it can be obtained from Lemma 2.15 that there exists a positive definite diagonal matrix $\Phi^1 = \text{diag}\{\phi^1_1, \ldots, \phi^1_N\}$ such that $(\overline{\mathcal{L}}^1)^T \Phi^1 + \Phi^1 \overline{\mathcal{L}}^1 > 0$, where $\phi^1 = [\phi^1_1, \ldots, \phi^1_N]^T$ satisfies $(\overline{\mathcal{L}}^1)^T \phi^1 = \mathbf{1}_N$.

Furthermore, substituting (6.34) into (6.1) yields

$$\dot{x}_i(t) = Ax_i(t) + Cf(x_i(t),t) + \alpha BF \sum_{j=0}^{N} a_{ij}^1 \left(x_j(t) - x_i(t) \right), \ t \in [t'_k, t'_k + \delta_k),$$

$$\dot{x}_i(t) = Ax_i(t) + Cf(x_i(t),t), \ t \in [t'_k + \delta_k, t'_{k+1}), \tag{6.35}$$

where $k \in \mathbb{N}$ and $i = 1, \ldots, N$. Let $e_i(t) = x_i(t) - x_0(t)$, $i = 1, \ldots, N$, and $e(t) = [e_1^T(t), \ldots, e_N^T(t)]^T$. Then, one has the following error dynamic system

$$\dot{e}(t) = (I_N \otimes A) e(t) + (I_N \otimes C)\widetilde{f}(x(t),t) - \alpha \left(\overline{\mathcal{L}}^1 \otimes BF \right) e(t), \ t \in [t'_k, t'_k + \delta_k),$$

$$\dot{e}(t) = (I_N \otimes A) e(t) + (I_N \otimes C)\widetilde{f}(x(t),t), \ t \in [t'_k + \delta_k, t'_{k+1}), \tag{6.36}$$

where $\widetilde{f}(x(t),t) = \left[\left((f(x_1(t),t) - f(x_0(t),t))^T, \ldots, (f(x_N(t),t) - f(x_0(t),t))^T \right]^T \right.$ and $x(t) = [x_0^T(t), x_1^T(t), \ldots, x_N^T(t)]^T$.

In the sequel, a multi-step design procedure is given to select the control parameters in protocol (6.34) for achieving consensus tracking.

Algorithm 6.3 *The consensus protocol (6.34) can be designed as follows:*

(1) Select two scalars $c > 0$ and $\beta > 0$. Solve the following LMI

$$\begin{bmatrix} AS + SA^T - cBB^T + \varrho^2 CC^T + \beta S & S \\ S & -I \end{bmatrix} < 0 \tag{6.37}$$

to get a matrix $S > 0$. Then, take $F = \frac{1}{2}B^T S^{-1}$.

(2) Solve the LMI

$$\begin{bmatrix} A^T Q + QA + I_n - \gamma Q & \varrho QC \\ \varrho C^T Q & -I \end{bmatrix} < 0 \tag{6.38}$$

to get a matrix $Q > 0$ and a scalar $\gamma > 0$.

(3) Choose the coupling strength $\alpha > 2c/\nu_0$, where c is defined in (6.37), $\nu_0 = \lambda_{\min} \left(\overline{\mathcal{L}}^1 + (\Phi^1)^{-1}(\overline{\mathcal{L}}^1)^T \Phi^1 \right)$, $\Phi^1 = \text{diag}\{\phi_1^1, \ldots, \phi_N^1\} > 0$.

Define $r_k = \delta_k/(t'_{k+1} - t'_k)$, which indicates the communication rate on the kth time interval $[t'_k, t'_{k+1})$, $k \in \mathbb{N}$. Then, one can establish the following theorem.

Theorem 6.3 *Suppose that Assumption 6.1 holds, the graph $\mathcal{G}^1$ has a directed spanning tree root at the leader, and the LMIs (6.37) and (6.38) have feasible solutions. Then, the consensus tracking problem of the CNS (6.1) can be solved by the protocol (6.34) with control parameters constructed in Algorithm 6.3, if the communication rate $r_k > \frac{\gamma}{\beta+\gamma} + \frac{2\ln\mu}{(\beta+\gamma)(t'_{k+1}-t'_k)}$, where $\mu = \max\left\{ \frac{\lambda_{\max}(S^{-1})}{\lambda_{\min}(Q)}, \frac{\lambda_{\max}(Q)}{\lambda_{\min}(S^{-1})} \right\}$, in which $k \in \mathbb{N}$, S and Q are positive definite solutions of (6.37) and (6.38), respectively.*

Proof 6.3 *Construct the following MLFs for the error dynamical system (6.36):*

$$
V(t) = \begin{cases} e^T(t)\left(\Phi^1 \otimes S^{-1}\right)e(t), & t \in [t'_k, t'_k + \delta_k), \\ e^T(t)\left(\Phi^1 \otimes Q\right)e(t), & t \in [t'_k + \delta_k, t'_{k+1}), \end{cases} \quad k \in \mathbb{N}, \tag{6.39}
$$

where $\Phi^1 = \text{diag}\{\phi_1^1, \ldots, \phi_N^1\}$, matrices S and Q are the positive definite solutions of (6.37) and (6.38), respectively.

For $t \in [t'_k, t'_k + \delta_k)$, $k \in \mathbb{N}$, let $\varepsilon(t) = [\varepsilon_1^T(t), \ldots, \varepsilon_N^T(t)]^T$, where $\varepsilon_i(t) = S^{-1}e_i(t)$, $i = 1, \ldots, N$. Obviously, $e(t) = (I_N \otimes S)\varepsilon(t)$. It thus follows from (6.36) that

$$
\dot{V}(t) \leq \varepsilon^T(t)\left[\Phi^1 \otimes \left(AS + SA^T + \varrho^2 CC^T + S^T S\right)\right]\varepsilon(t)
$$
$$
- \frac{\alpha}{2}\varepsilon^T(t)\left[\left((\overline{\mathcal{L}}^1)^T\Phi^1 + \Phi^1\overline{\mathcal{L}}^1\right) \otimes BB^T\right]\varepsilon(t).
$$

Based on the above analysis and according to step (1) of Algorithm 6.3, one has

$$
\dot{V}(t) \leq \varepsilon^T(t)\left[\Phi^1 \otimes \left(AS + SA^T + \varrho^2 CC^T + S^T S\right)\right]\varepsilon(t)
$$
$$
- \frac{\alpha\nu_0}{2}\varepsilon^T(t)\left(\Phi^1 \otimes BB^T\right)\varepsilon(t)
$$
$$
\leq \varepsilon^T(t)\left[\Phi^1 \otimes \left(AS + SA^T + \varrho^2 CC^T + S^T S\right)\right]\varepsilon(t)
$$
$$
- c\varepsilon^T(t)\left(\Phi^1 \otimes BB^T\right)\varepsilon(t), \tag{6.40}
$$

where $\nu_0 = \lambda_{\min}\left(\overline{\mathcal{L}}^1 + (\Phi^1)^{-1}(\overline{\mathcal{L}}^1)^T\Phi^1\right)$. Using (6.37) and the Schur complement lemma, it follows from (6.40) that

$$
\dot{V}(t) < -\beta\varepsilon^T(t)\left(\Phi^1 \otimes S\right)\varepsilon(t)
$$
$$
= -\beta e^T(t)\left(\Phi^1 \otimes S^{-1}\right)e(t). \tag{6.41}
$$

For $t \in [t'_k + \delta_k, t'_{k+1})$, $k \in \mathbb{N}$, taking the time derivative of $V(t)$ along the trajectories of system (6.36) gives

$$
\dot{V}(t) = e^T(t)\left(\Phi^1 \otimes (QA + A^T Q)\right)e(t)
$$
$$
+ 2\sum_{i=1}^{N}\phi_i^1 e_i^T(t)QC\left(f(x_i(t), t) - f(x_0(t), t)\right).
$$

Based on the above analysis, one gets

$$
\dot{V}(t) \leq e^T(t)\left(\Phi^1 \otimes (QA + A^T Q)\right)e(t)
$$
$$
+ \sum_{i=1}^{N}\phi_i^1 e_i^T(t)\left(\varrho^2 QCC^T Q + I\right)e_i(t)
$$
$$
= e^T(t)\left[\Phi^1 \otimes \left(QA + A^T Q + \varrho^2 QCC^T Q + I\right)\right]e(t)
$$
$$
< \gamma e^T(t)\left(\Phi^1 \otimes Q\right)e(t), \tag{6.42}
$$

where the last inequality follows from (6.38) and the Schur complement lemma.

Note that systems (6.36) switch at $t = t'_k$ *and* $t = t'_k + \delta_k$, $k \in \mathbb{N}$. *Therefore, based on (6.41) and (6.42), one obtains*

$$
\begin{aligned}
V(t'_1) &< \mu \exp(\gamma(t'_1 - t'_0 - \delta_0))V(t'_0 + \delta_0) \\
&< \mu^2 \exp(-\beta\delta_0 + \gamma(t'_1 - t'_0 - \delta_0))V(t'_0) \\
&= \exp(-\tilde{v}_0)V(t_0),
\end{aligned}
\tag{6.43}
$$

with $\tilde{v}_0 = \beta\delta_0 - \gamma(t'_1 - \delta_0) - 2\ln\mu$, *where the last equation in (6.43) is derived by using the fact of* $t'_0 = t_0 = 0$. *According to the condition* $r_0 = \frac{\delta_0}{t'_1 - t'_0} > \left(\frac{\gamma}{\beta+\gamma} + \frac{2\ln\mu}{(\beta+\gamma)(t'_1 - t'_0)} \right)$, *one has* $\tilde{v}_0 > 0$. *By recursion, for any* $k > 1$ *and* $k \in \mathbb{N}$, *one has*

$$
V(t'_k) < V(t_0) \exp\left(-\sum_{i=0}^{k-1} \tilde{v}_i \right),
\tag{6.44}
$$

where $\tilde{v}_i = \beta\delta_i - \gamma(t'_{i+1} - t'_i - \delta_i) - 2\ln\mu > 0$, $i = 0, \ldots, k-1$.

For any $t > t'_1$, *there exists a positive integer* $z \geq 1$ *such that* $t'_z < t \leq t'_{z+1}$. *Let* $\bar{v} = \inf_{h \in \mathbb{N}} \tilde{v}_h > 0$. *When* $t \in (t'_z, t'_z + \delta_z)$, $z \in \mathbb{N}$, *based on the above analysis and the fact that* $\sup_{k \in \mathbb{N}}(t'_{k+1} - t'_k) < \tilde{\tau}_M$, *one gets*

$$
\begin{aligned}
V(t) &< V(t'_z) \exp(-\beta\delta_z) < V(t_0) \exp\left(-\sum_{j=0}^{z-1} \tilde{v}_j \right) \\
&\leq V(t_0) \exp(-z\bar{v}) < V(t_0) \exp\left(-\frac{z\bar{v}}{(z+1)\tilde{\tau}_M}t \right) \\
&< V(t_0) \exp\left(-\frac{\bar{v}}{2\tilde{\tau}_M}t \right),
\end{aligned}
\tag{6.45}
$$

where the last inequality is obtained since $z \geq 1$. *When* $t = t'_z + \delta_z$, $z \in \mathbb{N}$, *the above analysis indicates that*

$$
V(t) < \mu V(t_0) \exp\left(-\frac{\bar{v}}{2\tilde{\tau}_M}t \right).
\tag{6.46}
$$

For the case of $t \in (t'_z + \delta_z, t'_{z+1})$, *some simple calculations give that*

$$
\begin{aligned}
V(t) &< \exp(\gamma\tilde{\tau}_M)V(t'_z + \delta_z) \\
&< \mu \exp(\gamma\tilde{\tau}_M)V(t_0) \exp\left(-\sum_{j=0}^{z-1} \tilde{v}_j \right) \\
&< \mu \exp(\gamma\tilde{\tau}_M)V(t_0) \exp(-z\bar{v}) \\
&< \mu \exp(\gamma\tilde{\tau}_M)V(t_0) \exp\left(-\frac{\bar{v}}{2\tilde{\tau}_M}t \right).
\end{aligned}
\tag{6.47}
$$

When $t = t'_{z+1}$, $z \in \mathbb{N}$, *it follows from (6.47) that*

$$
V(t) < \mu^2 \exp(\gamma\tilde{\tau}_M)V(t_0) \exp\left(-\frac{\bar{v}}{2\tilde{\tau}_M}t \right).
\tag{6.48}
$$

From the above analysis, one gets that $\|e(t)\| \to 0$ *as* $t \to +\infty$. *This indicates that the consensus tracking problem in CNS (6.1) is indeed solved by protocol (6.34) with control parameters constructed by Algorithm 6.3.*

Remark 6.6 *The condition that the communication topology frequently has a directed spanning tree is stronger than that of the topology jointly having a directed spanning tree. By using the MLFs based approach, it has been shown in Theorem 6.3 that the consensus tracking problem in the CNS (6.1) can be solved by protocol (6.34) with control parameters appropriately designed. It is also worth mentioning that how to construct a distributed protocol to guarantee consensus tracking in the CNS (6.1) with topology jointly containing a directed spanning tree remains a challenging issue today.*

Remark 6.7 *It is not hard to see that the solvability conditions for the LMI (6.7) provided in Remark 6.2 are applicable for the LMI (6.37). By using Schur complement lemma, one gets that the LMI (6.38) holds if and only if there exist a positive scalar* $\gamma > 0$ *and* $Q > 0$ *such that*

$$A^T Q + QA + I_n + \varrho^2 QCC^T Q - \gamma Q < 0. \tag{6.49}$$

Obviously, the LMI (6.49) is solvable if $\gamma > \overline{\lambda}$, *where* $\overline{\lambda}$ *is the maximum eigenvalue of* $A + A^T + \varrho^2 CC^T + I_n$.

Remark 6.8 *Under Assumption 6.1 and the condition that* $\mathcal{G}^1$ *contains a directed spanning tree rooted at the leader, it follows from Theorem 6.3 that the consensus tracking for system (6.1) with protocol (6.34) designed by Algorithm 6.3 can be achieved if the LMIs (6.37), (6.38) have feasible solutions and the communication rate* r_k *is larger than a threshold value. It can be observed that, for given* β, γ, *and topology* $\mathcal{G}^1$, *the minimum admissible communication rate depends only on the eigenvalue ratio of* S^{-1} *and* Q. *However, the LMIs (6.37) and (6.38) in Algorithm 6.3 are solved independently, which may introduce conservatism in seeking an admissible communication rate to satisfy the consensus tracking conditions. Thus, it is important to further study, for given parameters* β, γ, *and a topology* $\mathcal{G}^1$ *containing a directed spanning tree, how large the minimum admissible communication rate is needed to achieve the intended consensus tracking. For this purpose, further investigation is needed.*

Remark 6.9 *It should be noted that distributed consensus tracking for CNSs with homogeneous Lipschitz-type nonlinear dynamics has been studied in the present section. For CNSs with general heterogeneous nonlinear dynamics, developing a distributed consensus tracking protocol becomes more involved. First, for general nonlinear CNSs, it is a challenge to design distributed tracking protocols based only on the relative states of neighboring agents over directed networks to eliminate the effects of the nonlinear term. Second, within the context of CNSs, it is unclear how to deal with the heterogeneous dynamics since the coupling terms will vanish owing to the diffusive property of the Laplacian matrix when consensus tracking is achieved; from this viewpoint, the states of neighboring agents will diverge from each other again when the relative states of them are very small.*

Remark 6.10 *It should be noted that the design of the coupling strength α of the protocols provided here relies on the minimum eigenvalue of some positive-definite matrices depending on the Laplacian matrices associated with the communication topologies, which indeed is a piece of global information within the context of CNSs. However, in practice, one could calculate the lower bound of this positive quantity off-line, since the numbers of the agents and the topological structures are finite. Nevertheless, it is more interesting to construct some fully distributed protocols such that consensus tracking can be ensured without using any global information. However, whether it is possible to construct such a fully distributed tracking protocol for CNSs over fixed or directed switching topologies is still an open problem.*

6.3 CONSENSUS TRACKING OF CNSS WITH OCCASIONALLY MISSING CONTROL INPUTS

6.3.1 Model formulation

In this section, the challenging issue of distributed consensus tracking for higher-order linear CNSs with directed switching topologies is considered. For this problem, the existing approaches are inapplicable. Specifically, most of the approaches on consensus tracking of higher-order linear CNSs with switching topologies were based on two common assumptions: The system matrix of each agent has no unstable eigenvalue, and each possible topology is undirected. In this section, these two assumptions are removed. It is only assumed that the possible topologies are directed graphs containing a directed spanning tree rooted at the leader. Compared with the existing literature, another distinctive feature of this section is to solve the consensus tracking problem in the presence of aperiodic control input loss which might be caused by temporal actuator failures or network-induced packet loss [150, 181, 183, 228].

Consider a CNS consisting of a leader and N followers, where the leader is labelled as agent 0 and the followers are labelled as agents $1, \ldots, N$. The dynamics of the leader are described by

$$\dot{x}_0(t) = Ax_0(t), \tag{6.50}$$

where $x_0(t) \in \mathbb{R}^n$ is the state of the leader and matrix $A \in \mathbb{R}^{n \times n}$. Furthermore, the dynamics of agent i, $i = 1, \ldots, N$, are described by

$$\dot{x}_i(t) = Ax_i(t) + Bu_i(t), \tag{6.51}$$

where $x_i(t) \in \mathbb{R}^n$ is the state of agent i, $u_i(t) \in \mathbb{R}^m$ is the control input to be designed, and $B \in \mathbb{R}^{n \times m}$ is the input matrix. Throughout this section, the matrix pair (A, B) is assumed to be stabilizable. Note that the control input acting on each follower i, $i = 1, \ldots, N$, is designed based only on the relative information of neighboring agents rather than the absolute information of agents in the context of CNSs. Furthermore, to make model (6.51) be able to characterize the CNSs with occasionally missing control inputs due to temporary actuator failures, network-induced packet loss or purposeful suspension of actuators for saving power and prolonging the life of device [150, 181, 183, 228], it is further assumed that the control inputs act only over some disconnected time intervals. Note that the agents will evolve according to their own intrinsic linear dynamics when the control inputs are absent over some time intervals.

Suppose that there exists an infinite sequence of uniformly bounded non-overlapping time intervals $[t_k, t_{k+1})$, $k \in \mathbb{N}$, with $t_0 = 0$, $0 < \tau_m \leq t_{k+1} - t_k \leq \tau_M < +\infty$, such that for each $k \in \mathbb{N}$, there exits a positive integer h_k and a finite sequence of time points $t_k = t_k^1 < t_k^2 < \ldots < t_k^{h_k-1} < t_k^{h_k} = t_{k+1}$, for which the underlying topology is time-invariant for all $t \in [t_k^i, t_k^{i+1})$, $i = 1, \ldots, h_k - 2$. Furthermore, suppose that the control inputs are missed for the agents when $t \in \bigcup_{k \in \mathbb{N}} [t_k^{h_k-1}, t_k^{h_k})$. It is further assumed that $2 < h_k < h_{\max}$ for each $k \in \mathbb{N}$ and some given positive integer $h_{\max}$. Within this context, the state of each agent evolves according to its own intrinsic linear dynamics when $t \in [t_k^{h_k-1}, t_{k+1})$ for each $k \in \mathbb{N}$. This indicates that each agent may only share its state information with its neighbors when $t \in \bigcup_{k \in \mathbb{N}} [t_k, t_k^{h_k-1})$. To achieve consensus tracking, the following distributed consensus tracking protocol is proposed for each follower i

$$u_i(t) = \begin{cases} \alpha F \sum_{j=0}^{N} a_{ij}^{\sigma(t)} (x_j(t) - x_i(t)), & t \in [t_k, t_k^{h_k-1}), \\ \mathbf{0}_m, & t \in [t_k^{h_k-1}, t_{k+1}), \end{cases} \quad k \in \mathbb{N}, \quad (6.52)$$

where $i = 1, \ldots, N$, $\alpha > 0$ represents the coupling strength, $F \in \mathbb{R}^{m \times n}$ is the feedback gain matrix to be designed, $\mathcal{A}^{\sigma(t)} = [a_{ij}^{\sigma(t)}]_{(N+1) \times (N+1)}$ is the adjacency matrix of graph $\mathcal{G}^{\sigma(t)}$. It is assumed that $\mathcal{G}^{\sigma(t)} \in \{\mathcal{G}^1, \ldots, \mathcal{G}^\kappa\}$.

Assumption 6.4 *For each $j = 1, \ldots, \kappa$, the graph $\mathcal{G}^j$ contains a directed spanning tree with agent 0 (i.e. the leader) being the root.*

Under Assumption 6.4, the Laplacian matrix of directed graph $\mathcal{G}^{\sigma(t)}$ can be written as

$$\mathcal{L}^{\sigma(t)} = \begin{bmatrix} 0 & \mathbf{0}_N^T \\ \mathbf{P}^{\sigma(t)} & \overline{\mathcal{L}}^{\sigma(t)} \end{bmatrix},$$

$$\overline{\mathcal{L}}^{\sigma(t)} = \begin{bmatrix} \sum_{j \in \mathcal{N}_1} a_{1j}^{\sigma(t)} & -a_{12}^{\sigma(t)} & \cdots & -a_{1N}^{\sigma(t)} \\ -a_{21}^{\sigma(t)} & \sum_{j \in \mathcal{N}_2} a_{2j}^{\sigma(t)} & \cdots & -a_{2N}^{\sigma(t)} \\ \vdots & \vdots & \ddots & \vdots \\ -a_{N1}^{\sigma(t)} & -a_{N2}^{\sigma(t)} & \cdots & \sum_{j \in \mathcal{N}_N} a_{Nj}^{\sigma(t)} \end{bmatrix}, \quad (6.53)$$

where $\mathbf{P}^{\sigma(t)} = -[a_{10}^{\sigma(t)}, \ldots, a_{N0}^{\sigma(t)}]^T$ and $\mathcal{A}^{\sigma(t)} = [a_{ij}^{\sigma(t)}]_{(N+1) \times (N+1)}$ is the adjacency matrix of graph $\mathcal{G}^{\sigma(t)}$. It can thus be obtained from Lemma 2.15 that there exists a positive definite diagonal matrix $\Phi^j = \text{diag}\{\phi_1^j, \ldots, \phi_N^j\}$ such that $(\overline{\mathcal{L}}^j)^T \Phi^j + \Phi^j \overline{\mathcal{L}}^j > 0$, where $\Phi^j = \text{diag}\{\phi_1^j, \ldots, \phi_N^j\}$ with $[\phi_1^j, \ldots, \phi_N^j]^T = (\overline{\mathcal{L}}^j)^{-T} \cdot \mathbf{1}_N$, $j = 1, \ldots, \kappa$.

6.3.2 Main results

In this section, the main theoretical results are presented and discussed.

Let $e_i(t) = x_i(t) - x_0(t)$, $i = 1, \ldots, N$, and $e(t) = [e_1^T(t), \ldots, e_N^T(t)]^T$. Then, one has the following consensus error systems:

$$\dot{e}_i(t) = \begin{cases} Ae_i(t) - \alpha BF \sum_{j=1}^N \bar{l}_{ij}^{\sigma(t)} e_j(t), & t \in [t_k, t_k^{h_k-1}), \\ Ae_i(t), & t \in [t_k^{h_k-1}, t_{k+1}), \end{cases} \quad k \in \mathbb{N}, \qquad (6.54)$$

where $\overline{\mathcal{L}} = [\bar{l}_{ij}^{\sigma(t)}]_{N \times N}$ is defined in (6.53). Rewriting (6.54) into a compact form, one has

$$\dot{e}(t) = \begin{cases} (I_N \otimes A) e(t) - \alpha \left(\overline{\mathcal{L}}^{\sigma(t)} \otimes BF \right) e(t), & t \in [t_k, t_k^{h_k-1}), \\ (I_N \otimes A) e(t), & t \in [t_k^{h_k-1}, t_{k+1}), \end{cases} \quad k \in \mathbb{N}. \quad (6.55)$$

Before moving forward, a two-step design procedure is given to select the control parameters of protocol (6.52) for achieving consensus tracking in the closed-loop system (6.51) with a leader described by (6.50).

Algorithm 6.4 *Suppose that the matrix pair (A, B) is stabilizable and Assumption 6.4 holds. Then, the consensus protocol (6.52) can be designed as follows:*

(1) Select two scalars $c > 0$ and $\beta > 0$. Solve the LMI

$$AP + PA^T - cBB^T + \beta P < 0 \qquad (6.56)$$

to get a matrix $P > 0$. Then, take $F = \frac{1}{2}B^T P^{-1}$.

(2) Choose the coupling strength $\alpha > 2c/\lambda_0$, where $\lambda_0 = \min_{j=1,\ldots,\kappa} \lambda_{\min}(\overline{\mathcal{L}}^j + (\Phi^j)^{-1}(\overline{\mathcal{L}}^j)^T \Phi^j)$.

Remark 6.11 *According to Algorithm 6.4, the existence of protocol (6.52) depends on the solvability of LMI (6.56). On the other hand, since β is a free parameter, it is not hard to verify that LMI (6.56) is feasible if and only if there exists a scalar $\tilde{c} > 0$ such that the following LMI has a feasible solution $R > 0$:*

$$AR + RA^T - \tilde{c}BB^T < 0. \qquad (6.57)$$

According to Finsler's lemma, LMI (6.57) is feasible if and only if there exist $R > 0$ and $\tilde{K} \in \mathbb{R}^{m \times n}$ such that

$$AR + RA^T + KB^T + BK^T < 0. \qquad (6.58)$$

Pre- and post-multiplying (6.58) by R^{-1} and its transpose, respectively, gives that

$$R^{-1}A + A^T R^{-1} + R^{-1}KB^T R^{-1} + R^{-1}BK^T R^{-1} < 0. \qquad (6.59)$$

Furthermore, under the assumption that (A, B) is stabilizable, one gets that there exist $Z > 0$ and $\tilde{K} \in \mathbb{R}^{m \times n}$ such that

$$A^T Z + ZA + \tilde{K}^T B^T Z + ZB\tilde{K} < 0. \qquad (6.60)$$

Then, from (6.59) and (6.60), one can conclude that LMI (6.56) is feasible if and only if (A, B) is stabilizable. It is also worth noting that the assumption that (A, B) is stabilizable is very mild since most practical linear systems satisfy this condition.

For notational convenience, let

$$\delta_k = t_k^{h_k-1} - t_k^1, \tag{6.61}$$

$$\rho_k = t_k^{h_k} - t_k^{h_k-1}, \tag{6.62}$$

respectively, $k \in \mathbb{N}$. Now, one can establish the following theorem.

Theorem 6.4 *Suppose that (A, B) is stabilizable and Assumption 6.4 holds. Then, the distributed consensus tracking problem of CNS with followers given by (6.51) and a leader given by (6.50) can be solved by the protocol (6.52) with control parameters constructed by Algorithm 6.4, if the following condition holds for an arbitrarily given $\epsilon_0 > 0$:*

$$\beta\delta_k > \gamma\rho_k + (h_k - 1)\ln\mu + \epsilon_0, \tag{6.63}$$

where β is defined in (6.56), δ_k and ρ_k are respectively defined in (6.61) and (6.62), $\gamma > 0$ such that $AP + PA^T < \gamma P$ with P defined in (6.56), $\mu = \bar{\phi}/\underline{\phi}$, $\bar{\phi} = \max_{i,j}\phi_j^i$, $\underline{\phi} = \min_{i,j}\phi_j^i$, $i \in \{1, \ldots, \kappa\}$, and $j \in \{1, \ldots, N\}$.

Proof 6.4 *Construct the following MLFs for the error system (6.55):*

$$V(t) = \begin{cases} e^T(t)\left(\Phi^{\sigma(t)} \otimes P^{-1}\right)e(t), & t \in [t_k, \, t_k^{h_k-1}), \\ e^T(t)\left(I_N \otimes P^{-1}\right)e(t), & t \in [t_k^{h_k-1}, \, t_{k+1}), \end{cases} \tag{6.64}$$

where P is the solution of (6.56), $k \in \mathbb{N}$.

For $t \in [t_0^1, t_0^2)$, taking the time derivative of $V(t)$ along the trajectories of system (6.55) gives

$$\dot{V}(t) = e^T(t)\left[\Phi^{\sigma(t)} \otimes \left(P^{-1}A + A^T P^{-1}\right)\right]e(t)$$
$$- 2\alpha e^T(t)\left[\left(\Phi^{\sigma(t)}\overline{\mathcal{L}}^{\sigma(t)}\right) \otimes \left(P^{-1}BF\right)\right]e(t). \tag{6.65}$$

Substituting $F = \frac{1}{2}B^T P^{-1}$ into (6.65) yields

$$\dot{V}(t) = e^T(t)\left[\Phi^{\sigma(t)} \otimes \left(P^{-1}A + A^T P^{-1}\right)\right]e(t)$$
$$- \alpha e^T(t)\left[\left(\Phi^{\sigma(t)}\overline{\mathcal{L}}^{\sigma(t)}\right) \otimes \left(P^{-1}BB^T P^{-1}\right)\right]e(t)$$
$$= e^T(t)\left[\Phi^{\sigma(t)} \otimes \left(P^{-1}A + A^T P^{-1}\right)\right]e(t)$$
$$- \frac{\alpha}{2}e^T(t)\left[\left(\Phi^{\sigma(t)}\overline{\mathcal{L}}^{\sigma(t)} + \left(\overline{\mathcal{L}}^{\sigma(t)}\right)^T \Phi^{\sigma(t)}\right) \otimes \left(P^{-1}BB^T P^{-1}\right)\right]e(t). \tag{6.66}$$

Let $\varepsilon(t) = [\varepsilon_1^T(t), \ldots, \varepsilon_N^T(t)]^T$, where $\varepsilon_i(t) = P^{-1}e_i(t)$, $i = 1, \ldots, N$. Obviously, $e(t) = (I_N \otimes P)\varepsilon(t)$. It thus follows from (6.66) that

$$\dot{V}(t) \leq \varepsilon^T(t)\left[\Phi^{\sigma(t)} \otimes \left(AP + PA^T\right)\right]\varepsilon(t)$$

$$-\frac{\alpha}{2}\varepsilon^T(t)\left[\left(\Phi^{\sigma(t)}\overline{\mathcal{L}}^{\sigma(t)}+\left(\overline{\mathcal{L}}^{\sigma(t)}\right)^T\Phi^{\sigma(t)}\right)\otimes BB^T\right]\varepsilon(t). \qquad (6.67)$$

From (6.67), one has

$$\dot{V}(t)\leq\varepsilon^T(t)\left[\Phi^{\sigma(t)}\otimes\left(AP+PA^T\right)\right]\varepsilon(t)-\frac{\alpha\lambda_0}{2}\varepsilon^T(t)\left(\Phi^{\sigma(t)}\otimes BB^T\right)\varepsilon(t), \quad (6.68)$$

where $\lambda_0=\min_{i=1,\dots,\kappa}\lambda_{\min}\left(\overline{\mathcal{L}}^i+(\Phi^i)^{-1}\left(\overline{\mathcal{L}}^i\right)^T\Phi^i\right)>0$. *Since* $\alpha>2c/\lambda_0$, *it follows from (6.68) that*

$$\dot{V}(t)\leq\varepsilon^T(t)\left[\Phi^{\sigma(t)}\otimes\left(AP+PA^T-cBB^T\right)\right]\varepsilon(t). \qquad (6.69)$$

Using (6.56), it follows from (6.69) that

$$\begin{aligned}\dot{V}(t)&\leq-\beta\varepsilon^T(t)\left(\Phi^{\sigma(t)}\otimes P\right)\varepsilon(t)\\&=-\beta e^T(t)\left(\Phi^{\sigma(t)}\otimes P^{-1}\right)e(t).\end{aligned} \qquad (6.70)$$

Note that the closed-loop CNS (6.51) with protocol (6.52) switches when $t=t_0^2$. *It thus follows from the above analysis that*

$$V(t_0^{2-})<V(t_0)\exp(-\beta(t_0^2-t_0^1)). \qquad (6.71)$$

On the other hand, it follows from (6.64) that

$$V(t_0^2)\leq\mu V(t_0^{2-}), \qquad (6.72)$$

with $\mu=\bar{\phi}/\underline{\phi}$, $\bar{\phi}=\max_{i,j}\phi_j^i$, $\underline{\phi}=\min_{i,j}\phi_j^i$, $i\in\{1,\dots,\kappa\}$, *and* $j\in\{1,\dots,N\}$.
   Thus, one gets

$$V(t_0^2)<\mu\exp(-\beta(t_0^2-t_0^1))V(t_0), \qquad (6.73)$$

i.e.,

$$V(t_0^2)<\exp([-\beta(t_0^2-t_0^1)+\ln\mu])V(t_0). \qquad (6.74)$$

By recursion, and according to (6.64), one gets that

$$V(t_0^{h_0-1})<\frac{1}{\underline{\phi}}\exp([-\beta(t_0^{h_0-1}-t_0)+(h_0-2)\ln\mu])V(t_0), \qquad (6.75)$$

where $V(t_0^{h_0-1})$ *is given in (6.75) and* $h_0\geq3$. *Then, since* $AP+PA^T\leq\gamma P$, *one gets that*

$$V(t_1^-)\leq\exp\left(\gamma(t_1-t_0^{h_0-1})\right)V(t_0^{h_0-1}). \qquad (6.76)$$

Since the closed-loop CNS (6.51) with protocol (6.52) switches when $t=t_1$, *it follows from (6.76) that*

$$V(t_1)\leq\bar{\phi}V(t_1^-)$$

$$< (\bar{\phi}/\underline{\phi}) \exp([-\beta\delta_0 + \gamma\rho_0 + (h_0 - 2)\ln\mu])V(t_0)$$
$$= \exp(-\kappa_0)V(t_0), \tag{6.77}$$

where $\kappa_0 = [\beta\delta_0 - \gamma\rho_0 - (h_0 - 1)\ln\mu]$. From the condition that $\beta\delta_0 > \gamma\rho_0 + (h_0 - 1)\ln\mu$, for $k \in \mathbb{N}$, one knows that $\kappa_0 > 0$. Similar to the above analysis, one can conclude that, for any given $k \in \mathbb{N}$,

$$V(t_{k+1}) \le \exp\left(-\sum_{j=0}^{k} \kappa_j\right) V(t_0)$$
$$\le \exp\left(-(k+1)\epsilon_0\right) V(t_0), \tag{6.78}$$

where $\kappa_j = \beta\delta_j - \gamma\rho_j - (h_j - 1)\ln\mu > 0$ for all $j = 0, 1, \ldots, k$, and $\epsilon_0 = \inf_{j=0,1,\ldots} \kappa_j$.

For any given $t > t_1$, there exists a positive integer z ($z \ge 1$) such that $t_z < t \le t_{z+1}$. For the case that $t \in (t_z, t_z^{h_z - 1})$, based on the above analysis, one gets

$$V(t) < \exp(-\beta(t - t_z) + (h_z - 2)\ln\mu)V(t_z)$$
$$\le \varsigma_0 \exp\left(-\frac{z\epsilon_0}{(z+1)\tau_M}t\right) V(t_0), \tag{6.79}$$

where $\varsigma_0 = \exp((h^{\sup} - 2)\ln\mu)$, $h^{\sup} = \sup_{z \in \mathbb{N}} h_z$, and τ_M is a positive scalar such that $(t_{k+1} - t_k) \le \tau_M$, $k \in \mathbb{N}$. Since $z \ge 1$, it follows from (6.79) that

$$V(t) < \varsigma_0 \exp\left(-\frac{\epsilon_0}{2\tau_M}t\right) V(t_0), \quad t \in (t_z, t_z^{h_z - 1}). \tag{6.80}$$

For the case of $t = t_z^{h_z - 1}$, one gets that

$$V(t) < \mu\varsigma_0 \exp\left(-\frac{z\epsilon_0}{(z+1)\tau_M}t\right) V(t_0)$$
$$< \mu\varsigma_0 \exp\left(-\frac{\epsilon_0}{2\tau_M}t\right) V(t_0). \tag{6.81}$$

For the case of $t \in (t_z^{h_z - 1}, t_{z+1}]$, it follows from the fact $V(t) \le V(t_z)$ and (6.81) that

$$V(t) \le \exp\left(-\sum_{j=0}^{z} \kappa_j\right) V(t_0)$$
$$\le \exp\left(-\frac{\epsilon_0}{2\tau_M}t\right) V(t_0). \tag{6.82}$$

According to (6.80)–(6.82), one can conclude that $\|e(t)\| \to 0$ as $t \to +\infty$. Thus, the consensus tracking problem for CNS with followers given in (6.51) and a leader given in (6.50) is solved by protocol (6.52) with control parameters constructed by Algorithm 6.4.

Remark 6.12 *Some sufficient conditions for achieving consensus tracking in the closed-loop CNSs with followers given in (6.51) and a leader given in (6.50) have been provided in Theorem 6.4. Note that one prerequisite in Theorem 6.4 is that there exists* a scalar $\gamma > 0$ such that $AP+PA^T < \gamma P$, where P is a feasible solution of LMI (6.56). It is not hard to verify that this condition can be ensured if $\gamma > c\lambda_{\max}(P^{-\frac{1}{2}}BB^TP^{-\frac{1}{2}})$, where $P^{-\frac{1}{2}}$ is the inverse matrix of $P^{\frac{1}{2}}$, $\lambda_{\max}(P^{-\frac{1}{2}}BB^TP^{-\frac{1}{2}})$ represents the largest eigenvalue of $P^{-\frac{1}{2}}BB^TP^{-\frac{1}{2}}$. The above analysis indicates that under the condition that (A, B) is stabilizable, there always exist some positive scalars c, β, γ such that the following two LMIs share a common positive solution P: $AP+PA^T - cBB^T + \beta P < 0$ and $AP + PA^T < \gamma P$.

In the case that there are no occasionally missing control inputs, i.e., $t_k^{h_k-1} = t_{k+1}$, for all $k \in \mathbb{N}$, the distributed consensus tracking protocol $u_i(t)$ reduces to

$$u_i(t) = \alpha F \sum_{j=0}^{N} a_{ij}^{\sigma(t)} (x_j(t) - x_i(t)), \quad k \in \mathbb{N}, \tag{6.83}$$

where $\alpha > 0$ is the coupling strength, $F \in \mathbb{R}^{m \times n}$ is the feedback gain matrix to be designed, and $\mathcal{A}^{\sigma(t)} = [a_{ij}^{\sigma(t)}]_{(N+1) \times (N+1)}$ is the adjacency matrix of graph $\mathcal{G}^{\sigma(t)}$. Then, one can directly get the following corollary from Theorem 6.4, for which the proof is omitted for brevity.

Corollary 6.1 *Suppose that (A, B) is stabilizable and Assumption 6.4 holds. Then, the distributed consensus tracking problem of CNS with followers given by (6.51) and a leader given by (6.50) can be solved by the protocol (6.83) with control parameters constructed by Algorithm 6.4 if there are no occasionally missing control inputs and the following condition holds for an arbitrarily given $\epsilon_0 > 0$:*

$$\beta\delta_k > (h_k - 1)\ln\mu + \epsilon_0,$$

where β is defined in (6.56), δ_k is defined in (6.61), $\mu = \bar{\phi}/\underline{\phi}$, $\bar{\phi} = \max_{i,j}\phi_j^i$, $\underline{\phi} = \min_{i,j}\phi_j^i$, $i \in \{1, \ldots, \kappa\}$, and $j \in \{1, \ldots, N\}$.

Remark 6.13 *It can be observed from Algorithm 6.4 that the design of the feedback gain matrix in (6.56) is separated from the topology, i.e., the first step of Algorithm 6.4 deals with the agents' dynamics and determines the feedback gain matrix of the distributed control protocol, while the effect of the communication topologies on consensus is handled in the second step by designing the coupling strength. This separation helps us to successfully construct a class of MLFs with partial common structures to analyze the evolution behavior of the consensus error system (6.55).*

6.3.3 Discussions on the convergence rate

Suppose that the distributed consensus tracking problem of CNS with followers given by (6.51) and a leader given by (6.50) can be solved by the protocol (6.52) with control parameters constructed by Algorithm 6.4. It can be seen from (6.80) to (6.82) that the convergence rate of consensus tracking in the closed-loop CNSs is characterized by $\epsilon_0 = \inf_{j \in \mathbb{N}} \kappa_j$ with $\kappa_j = \beta\delta_k - \gamma\rho_j - (h_j - 1)\ln\mu$. Specifically, the larger ϵ_0, the faster distributed consensus tracking. For a given CNS, the convergence rate of the distributed consensus tracking can be increased by maximizing β and minimizing γ. It can be seen from LMI (6.56) that the parameter β can be chosen as arbitrarily large if (A, B) is controllable; however, for the case that (A, B) is stabilizable but not controllable, β should not be larger than -2χ, where χ is the largest real part of the uncontrollable stable eigenvalues of A. Also, γ can be chosen as arbitrarily small if A has no unstable eigenvalue; but if A contains some unstable eigenvalues, the parameter γ should be larger than 2ϖ, where ϖ is the largest real part of the unstable eigenvalues of A. However, it is still unclear how to select β and γ such that the LMIs (6.56) and $AP + PA^T < \gamma P$ share a common solution P while the above-mentioned ϵ_0 attains its maximum value. Nevertheless, this optimal design can be solved after β is fixed. Specifically, let $\gamma = \gamma_0$ be fixed, and $\beta_{\max}$ and $\beta_{\min}$ be, respectively, the maximal and minimal allowable values of β such that the LMIs (6.56) and $AP + PA^T < \gamma P$ share a common solution $P > 0$ and the condition (6.63) holds. Then, the CNS with followers given by (6.51) and a leader given by (6.50) equipped with the protocol (6.52) constructed by Algorithm 6.4 with $\beta = \beta_{\max}$ yields a fast convergence rate.

6.4 NUMERICAL SIMULATIONS

In this section, we will give two examples to validate some theoretical results given in sections 6.2 and 6.3, respectively.

Example 1: In this example, we consider a CNS of five single-link manipulators with switching topologies $\mathcal{G}^1$ and $\mathcal{G}^2$ as shown in Figure 6.1, where node 0 represents the leader and each edge weight is assumed to be 1. Direct calculation gives $\lambda_0 = 0.7839$ and $\mu = 8$. It is clearly that both $\mathcal{G}^1$ and $\mathcal{G}^2$ contain a directed spanning tree with node 0 as the root, hence Assumption 6.3 holds. The dynamics of agent i are described by (6.1) with $x_i(t) = [x_{i1}(t), x_{i2}(t), x_{i3}(t), x_{i4}(t)]^T$,

$$
A = \begin{bmatrix} 0 & 1 & 0 & 0 \\ -48.6 & -1.26 & 48.6 & 0 \\ 0 & 0 & 0 & 10 \\ 1.95 & 0 & -1.95 & 0 \end{bmatrix}, \quad B = \begin{bmatrix} 0 \\ 21.6 \\ 0 \\ 0 \end{bmatrix}, \quad C = \begin{bmatrix} 0 & 0 & 0 & 0 \\ 0 & 0 & 0 & 0 \\ 0 & 0 & 0 & 0 \\ 0 & 0 & 0 & 1 \end{bmatrix},
$$

$f(x_i(t), t) = [0, 0, 0, 0.22\sin(x_{i3}(t))]^T$, $i = 0, 1, \ldots, 4$. It is easy to check that (A, B) is controllable, and thus stabilizable. In view of Assumption 6.1, one gets $\varrho = 0.22$. Choosing $c = 2$, $\beta = 3$, one gets from Algorithm 6.2 that $F = [24.0464, 1.1012, -8.7391, 70.1804]$ and $\alpha = 5.2 > 5.1024$. According to Theorem 6.2, consensus tracking can be achieved in the considered CNS if $\tau_m > 0.6931$. Suppose that the communication topology switches between $\mathcal{G}^1$ and $\mathcal{G}^2$ in every 0.7 s. The state trajectories of the closed loop CNS (6.1) under protocol (6.20) are shown in Figs. 6.2–6.5 which show that consensus tracking is indeed achieved.

Example 2: In this example, we consider the CNS with followers given in (6.51) and a leader given in (6.50), where the topology $\mathcal{G}^{\sigma(t)} = \mathcal{G}^3$ for $t \in [k, k + 0.40)s$, $\mathcal{G}^{\sigma(t)} = \mathcal{G}^4$ for $t \in [k + 0.40, k + 0.90)s$, and the control inputs are missing for $t \in [k + 0.90, k + 1)s$, $k \in \mathbb{N}$. Topologies $\mathcal{G}^3$ and $\mathcal{G}^4$ are shown in Figure 6.6, where the weights on each edge is assumed to be 1. Hence Assumption 6.4

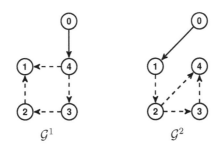

Figure 6.1 The communication graphs $\mathcal{G}^1$ and $\mathcal{G}^2$.

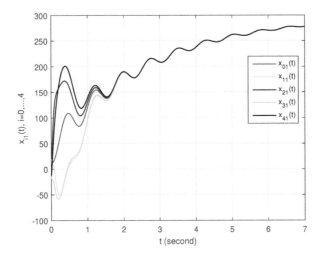

Figure 6.2 The agents' state trajectories $x_{i1}(t)$ in Example 1, $i = 0, 1, \ldots, 4$.

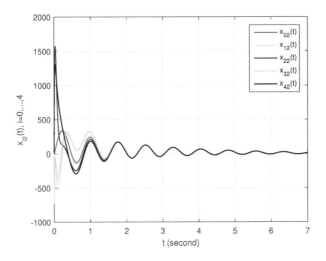

Figure 6.3 The agents' state trajectories $x_{i2}(t)$ in Example 1, $i = 0, 1, \ldots, 4$.

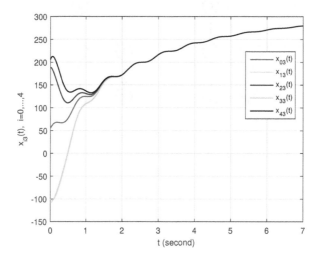

Figure 6.4 The agents' state trajectories $x_{i3}(t)$ in Example 1, $i = 0, 1, \ldots, 4$.

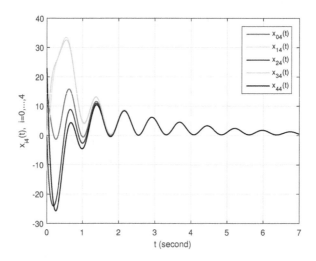

Figure 6.5 The agents' state trajectories $x_{i4}(t)$ in Example 1, $i = 0, 1, \ldots, 4$.

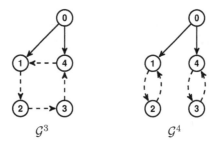

Figure 6.6 The communication graphs $\mathcal{G}^3$ and $\mathcal{G}^4$.

holds. Direct calculation gives $\lambda_0 = 0.6419$ and $\mu = 2.2$. The agents in this example are selected as the Caltech multi-vehicle wireless testbed vehicles [52], with $x_i(t) = [x_{i1}(t), x_{i2}(t), x_{i3}(t), x_{i4}(t), x_{i5}(t), x_{i6}(t)]^T \in \mathbb{R}^6$,

$$A = \begin{bmatrix} 0 & 0 & 0 & 1 & 0 & 0 \\ 0 & 0 & 0 & 0 & 1 & 0 \\ 0 & 0 & 0 & 0 & 0 & 1 \\ 0 & 0 & -0.2003 & -0.2003 & 0 & 0 \\ 0 & 0 & 0.2003 & 0 & -0.2003 & 0 \\ 0 & 0 & 0 & 0 & 0 & -1.6129 \end{bmatrix},$$

$$B = \begin{bmatrix} 0 & 0 \\ 0 & 0 \\ 0 & 0 \\ 0.9441 & 0.9441 \\ 0.9441 & 0.9441 \\ -28.7097 & 28.7097 \end{bmatrix},$$

where $x_{i1}(t)$ and $x_{i2}(t)$ are respectively the positions of i-th vehicle along the x and y coordinates, and $x_{i3}(t)$ is the orientation of the i-th vehicle, $\dot{x}_{i4}(t) = x_{i1}(t)$, $\dot{x}_{i5}(t) = x_{i2}(t)$, $\dot{x}_{i6}(t) = x_{i3}(t)$, $i = 0, 1, \dots, 4$. It is easy to check that (A, B) is controllable. Constructing the distributed controller (6.52) according to Algorithm 6.4 with parameters $c = 1$, $\beta = 2$, and $\gamma = 4$ yields

$$F = \begin{bmatrix} 1.4651 & 0.1895 & -0.3903 & 1.8563 & -0.4348 & -0.1200 \\ 0.1895 & 1.4651 & 0.3903 & -0.4348 & 1.8563 & 0.1200 \end{bmatrix}.$$

Set $\alpha = 3.2 > 2c/\lambda_0 = 3.1159$. According to Theorem 6.4, consensus tracking in the closed-loop CNSs is achieved since $0.9\beta - 0.1\gamma - \ln\mu = 0.6115 > 0$. The state trajectories of the closed loop CNS (6.51) under protocol (6.52) are shown in Figs. 6.7–6.12 which show that consensus tracking is indeed achieved.

The followings two cases are discussed:

(1) $\beta = 2$, $\gamma = 4$;

(2) $\beta = 1.6$, $\gamma = 4$.

According to Algorithm 6.4, the feedback gain matrix under case (2) can be selected as

$$F = \begin{bmatrix} 1.1839 & -0.0506 & -0.5096 & 1.8610 & -0.6680 & -0.1875 \\ -0.0506 & 1.1839 & 0.5096 & -0.6680 & 1.8610 & 0.1875 \end{bmatrix}.$$

It follows from Theorem 6.4 that the consensus tracking problem can be solved by protocol (6.52) constructed by Algorithm 6.4 for both cases (1) and (2). Use $\text{Error}(t) = \frac{1}{4}\sqrt{\sum_{i=1}^{4}(x_{i1}(t) - x_{01}(t))^2}$ to denote the consensus tracking errors of the closed-loop CNSs. The trajectories of $\text{Error}(t)$ in these two cases are shown in Figure 6.13 which verifies the analysis given in subsection 6.3.3.

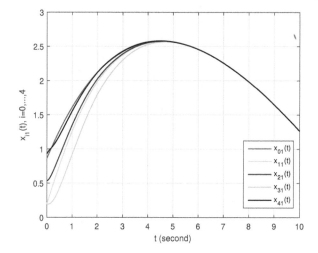

Figure 6.7 The agents' state trajectories $x_{i1}(t)$ in Example 2, $i = 0, 1, \ldots, 4$.

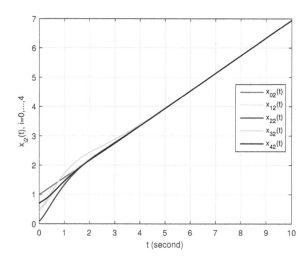

Figure 6.8 The agents' state trajectories $x_{i2}(t)$ in Example 2, $i = 0, 1, \ldots, 4$.

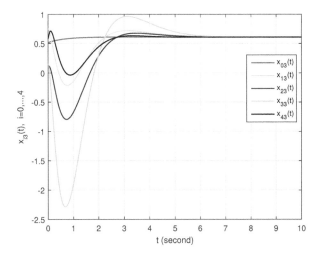

Figure 6.9 The agents' state trajectories $x_{i3}(t)$ in Example 2, $i = 0, 1, \ldots, 4$.

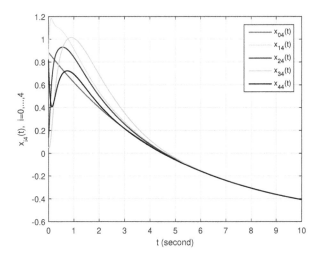

Figure 6.10 The agents' state trajectories $x_{i4}(t)$ in Example 2, $i = 0, 1, \ldots, 4$.

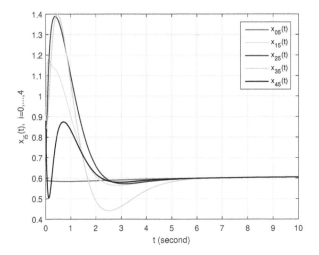

Figure 6.11 The agents' state trajectories $x_{i5}(t)$ in Example 2, $i = 0, 1, \ldots, 4$.

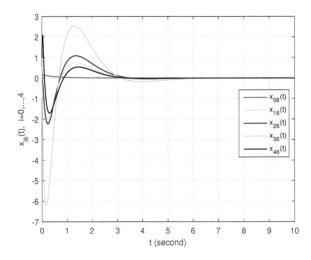

Figure 6.12 The agents' state trajectories $x_{i6}(t)$ in Example 2, $i = 0, 1, \ldots, 4$.

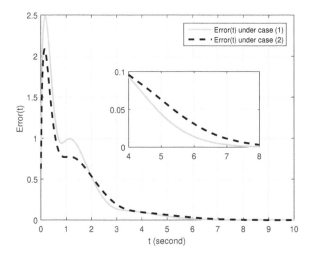

Figure 6.13 Trajectories of the consensus error $\mathrm{Error}(t)$ versus different β.

6.5 CONCLUSIONS

This chapter has solved the consensus tracking problem for CNSs with higher-order dynamics and directed switching topologies. First, for CNSs with Lipschitz nonlinear dynamics and topologies always (or frequently) contain a directed spanning tree rooted at the leader, this chapter has shown that consensus tracking can be achieved by choosing appropriate control parameters provided that the ADT is larger than a given threshold. Furthermore, some consensus criteria were proposed such that consensus tracking can be achieved in the CNSs with linear dynamics and occasionally missing control inputs.

CHAPTER 7

H-infinity consensus of CNSs with directed switching topologies

It is known that a practical system is unavoidably affected by various disturbances. This chapter studies the $\mathcal{H}_\infty$ consensus of CNSs with directed switching topologies. Section 7.1 briefly reviews some previous results. Section 7.2 studies the general linear dynamical models. This section begins by presenting the definition of $\mathcal{H}_\infty$ consensus. Then the main theoretical result is given and the convergence rate is discussed. Section 7.3 studies the Lipschitz nonlinear dynamical models with aperiodic sampled data communications. This section firstly studies the case with directed fixed topology. Then the result is extended to the case with directed switching topologies and sampled-data communications. Section 7.4 presents some simulations to validate the theoretical results.

7.1 INTRODUCTION

In most of the practical applications, the evolutions of CNSs are unavoidably affected by various disturbances such as sensor noises. Partly motivated by this observation, distributed $\mathcal{H}_\infty$ consensus for higher-order CNSs subject to external disturbances was investigated in [77, 229]. Note that most of the above-mentioned works mainly focus on solving $\mathcal{H}_\infty$ consensus under a fixed communication topology. In reality, the underlying topology of the mobile agents may switch among some possible topologies due to, for instance, limited sensing radius, temporary sonar equipment failures or the presence of communication obstacles. It is thus meaningful to further investigate the distributed $\mathcal{H}_\infty$ consensus over switching networks. Research along this line could not only yield some profound theoretical results but also help researchers and engineers implement distributed coordination control strategies in real CNSs.

Based on the aforementioned results, distributed $\mathcal{H}_\infty$ consensus for CNSs with directed switching topologies is considered in this chapter. Compared with the existing works in this field, the underlying topology among the multiple agents is assumed to be switching in the present framework. More precisely, the topology switches over some given directed graphs according to a piecewise constant switching signal as time evolves. This chapter addresses the following two issues. First, the case of CNSs with higher-order linear dynamics is studied. Furthermore, the case of CNSs with Lipschitz nonlinear dynamics and aperiodic sampled-data communications is studied.

7.2 $\mathcal{H}_\infty$ CONSENSUS OF LINEAR CNSS WITH DISTURBANCES

7.2.1 Model formulation

Consider a CNS consisting of N agents with general linear dynamics, described by

$$\dot{x}_i(t) = Ax_i(t) + Bu_i(t) + Dw_i(t), \tag{7.1}$$

where $x_i(t) \in \mathbb{R}^n$ is the state, $u_i(t) \in \mathbb{R}^m$ is the control input, $w_i(t) \in \mathbb{L}_2[0, +\infty)$ is the external disturbance, A, B, and D are constant real matrices. It is assumed that matrix pair (A, B) is stabilizable.

The communication topology among the N agents switches at the time instants $t_1, t_2, \ldots$. It is assumed that $t_0 = 0$ and $t_{k+1} - t_k \geq \tau_m > 0$, $k \in \mathbb{N}$. And it is assumed that $\mathcal{G}^{\sigma(t)} \in \{\mathcal{G}^1, \ldots, \mathcal{G}^\kappa\}$ with $\kappa \geq 1$ and $\kappa \in \mathbb{N}$.

To achieve consensus, the following consensus protocol based only on the relative information between agent i and its neighbors is proposed:

$$u_i(t) = \alpha K \sum_{j=1}^{N} a_{ij}^{\sigma(t)}[x_j(t) - x_i(t)], \ i = 1, \ldots, N, \tag{7.2}$$

where $\alpha > 0$ is the coupling strength to be selected, $K \in \mathbb{R}^{m \times n}$ is the feedback gain matrix to be designed, and $\mathcal{A}^{\sigma(t)} = [a_{ij}^{\sigma(t)}]_{N \times N}$ is the adjacency matrix of the graph $\mathcal{G}^{\sigma(t)}$.

Substituting (7.2) into (7.1) gives that

$$\dot{x}_i(t) = Ax_i(t) + \alpha BK \sum_{j=1}^{N} a_{ij}^{\sigma(t)}[x_j(t) - x_i(t)] + Dw_i(t), \ i = 1, \ldots, N. \tag{7.3}$$

Let $x(t) = [x_1^T(t), \ldots, x_N^T(t)]^T$ and $\omega(t) = [\omega_1^T(t), \ldots, \omega_N^T(t)]^T$, one gets

$$\dot{x}(t) = [(I_N \otimes A) - \alpha(\mathcal{L}^{\sigma(t)} \otimes BK)]x(t) + (I_N \otimes D)\omega(t), \tag{7.4}$$

where $\mathcal{L}^{\sigma(t)}$ is the Laplacian matrix of the graph $\mathcal{G}^{\sigma(t)}$. Before moving forward, the following assumption is made.

Assumption 7.1 *There exists a common positive vector $\theta = [\theta_1, \ldots, \theta_N]^T \in \mathbb{R}^N$ such that $\theta^T \mathcal{L}^i = 0_N^T$ and $1_N^T \theta = 1$, where $\mathcal{L}^i$ is the Laplacian matrix of the strongly connected graph $\mathcal{G}^i$, $i = 1, \ldots, \kappa$.*

Under Assumption 7.1, the generalized algebraic connectivity [217] of $\mathcal{G}^i$, $i = 1, \ldots, \kappa$, is defined as

$$a(\mathcal{L}^i) = \min_{y^T \theta = 0, \, y \neq \mathbf{0}_N} \frac{y^T \frac{\Theta \mathcal{L}^i + (\mathcal{L}^i)^T \Theta}{2} y}{y^T \Theta y}, \tag{7.5}$$

where $\Theta = \text{diag}\{\theta_1, \ldots, \theta_N\}$ with $\theta = [\theta_1, \ldots, \theta_N]^T$ given in Assumption 7.1.

Remark 7.1 *Assumption 7.1 holds if the directed (undirected) graph $\mathcal{G}^i$, $i = 1, \ldots, \kappa$, is strongly connected and balanced (connected) [217]. Additionally, note that $a(\mathcal{L}^i) > 0$ if $\mathcal{G}^i$ is strongly connected.*

Let $e(t) = (\widehat{\Theta} \otimes I_n)x(t)$, where $\widehat{\Theta} = (I_N - \mathbf{1}_N \theta^T) \in \mathbb{R}^{N \times N}$. Furthermore, taking the following similarity transformation $\Gamma = T^{-1}\widehat{\Theta}T$ with

$$T = \begin{bmatrix} 1 & \mathbf{0}_{N-1}^T \\ \mathbf{1}_{N-1} & I_{N-1} \end{bmatrix} \in \mathbb{R}^{N \times N}, \tag{7.6}$$

yields

$$\Gamma = \begin{bmatrix} 0 & \widetilde{\xi}^T \\ \mathbf{0}_{N-1} & I_{N-1} \end{bmatrix} \in \mathbb{R}^{N \times N}, \tag{7.7}$$

where $\widetilde{\xi} = [-\theta_2, -\theta_3, \ldots, -\theta_N]^T \in \mathbb{R}^{N-1}$. It can be verified that 0 is a simple eigenvalue of $\widehat{\Theta}$. In addition, it is easy to check that $\mathbf{1}_N$ is a right eigenvector associated with the eigenvalue 0 of $\widehat{\Theta}$. Thus, one can conclude that $e(t) = \mathbf{0}_{Nn}$ if and only if $x_1(t) = x_2(t) = \ldots = x_N(t)$, for $t \geq 0$. Here, $e(t)$ is called the *disagreement vector*.

Definition 7.1 *The distributed $\mathcal{H}_\infty$ consensus problem of CNS (7.1) is said to be solved by protocol (7.2) if the following requirements are satisfied:*

(1) The states of the closed-loop system (7.4) with $\omega(t) \equiv \mathbf{0}_{Nq}$ can achieve consensus in the sense of $\lim_{t \to \infty} \|x_i(t) - x_j(t)\| = 0$, for all $i, j = 1, 2, \ldots, N$;

(2) For a given weighting positive definite matrix $R \in \mathbb{R}^{n \times n}$ and a scalar $\gamma > 0$, the worst-case norm of consensus error vector $e(t)$ over all admissible exogenous disturbances $\omega(t)$ and initial disagreement vector $e(0)$, defined by $\gamma_\omega = \sup_{\|z\|^2 \neq 0} \frac{\|e(t)\|_{\mathbb{L}_2}}{z}$, is less than γ, i.e., $\gamma_\omega < \gamma$, where $z = [\|\omega(t)\|_{\mathbb{L}_2}^2 + e^T(0)(I_N \otimes R)e(0)]^{1/2}$.

Remark 7.2 *In most existing literature on $\mathcal{H}_\infty$ consensus of CNSs (e.g., [77, 172, 229]), the design goal is to construct some distributed protocols such that the $\mathcal{H}_\infty$ norm of the transfer matrix from exogenous signals to the disagreement vector is less than a prescribed positive number. In order to use tools from frequency domain analysis, it is usually assumed that the initial states of all agents are zero. However, there exist some practical situations where the initial states of agents under consideration are nonzero and possibly unknown, especially in large-scale CNSs, since it is very hard or even impossible to set the initial states of all agents to be a common vector. It is thus interesting and important to synthesize consensus protocols to provide attenuating over both exogenous and initial disturbances. Note that the idea for defining such a performance index is borrowed from the theory of $\mathcal{H}_\infty$ control with transients [43, 68].*

7.2.2 Main results

According to the definition of $e(t)$, it follows from (7.4) that

$$\dot{e}(t) = \left[(I_N \otimes A) - \alpha \left(\mathcal{L}^{\sigma(t)} \otimes BK\right)\right] e(t)$$
$$+ \left[(I_N - \mathbf{1}_N \theta^T) \otimes D\right] \omega(t). \tag{7.8}$$

Before moving forward, the following algorithm is provided for selecting the feedback gain matrix K and the coupling strength α of protocol (7.2) to achieve consensus.

Algorithm 7.1 *Suppose that (A, B) is stabilizable and Assumption 7.1 holds. The consensus protocol (7.2) can be designed as follows:*

(1) For a given positive scalar γ. Solve the LMI

$$\begin{bmatrix} AP + PA^T - BB^T + \varsigma DD^T & P \\ P & -(1/\widetilde{\theta})I_n \end{bmatrix} < 0 \tag{7.9}$$

to get $P > 0$, where $\varsigma = 1/(\widehat{\theta}\gamma^2)$, $\widehat{\theta} = 1/\max_{i=1,\ldots,N}\theta_i$, and $\widetilde{\theta} = 1/\min_{i=1,\ldots,N}\theta_i$. Then, take $K = B^T P^{-1}$.

(2) Choose the coupling strength $\alpha \geq \alpha_{\text{th}}$, where $\alpha_{\text{th}} = \frac{1}{2\kappa_0}$, with $\kappa_0 = \min_{i=1,\ldots,\kappa}a(\mathcal{L}^i)$, and $a(\mathcal{L}^i)$ is the generalized algebraic connectivity of graph $\mathcal{G}^i$ that is defined in (7.5), $i \in \{1, \ldots, \kappa\}$.

Based on the above scheme, one can get the following theorem.

Theorem 7.1 *Suppose that (A, B) is stabilizable and Assumption 7.1 holds. Then, the distributed $\mathcal{H}_\infty$ consensus with performance index γ and weighting matrix $R > 0$ for CNS (7.1) can be solved by protocol (7.2) with control parameters constructed by Algorithm 7.1 for any given dwell time $\tau_m > 0$, if $P > \varsigma R^{-1}$, where $\varsigma = 1/(\widehat{\theta}\gamma^2)$, $\widehat{\theta} = 1/\max_{i=1,\ldots,N}\theta_i$.*

Proof 7.1 *Construct the following Lyapunov function for the switch system (7.8):*

$$V(t) = e^T(t) \left(\Theta \otimes P^{-1}\right) e(t), \tag{7.10}$$

where $\Theta = \text{diag}\{\theta_1, \ldots, \theta_N\}$, with $\theta = [\theta_1, \ldots, \theta_N]^T$ being the common positive left eigenvector of Laplacian matrices $\mathcal{L}^i$, $i \in \{1, \ldots, \kappa\}$, associated with the zero eigenvalue, satisfying $\mathbf{1}_N^T \theta = 1$, and P is a positive definite solution of LMI (7.9).

Taking the time derivative of $V(t)$ along the trajectories of system (7.8) yields

$$\dot{V}(t) = e^T(t)\left\{\Theta \otimes \left(A^T P^{-1} + P^{-1}A\right)\right.$$
$$- \alpha[(\mathcal{L}^{\sigma(t)})^T \Theta \otimes (K^T B^T P^{-1}) + \Theta\mathcal{L}^{\sigma(t)} \otimes (P^{-1}BK)]\Big\} e(t)$$
$$+ 2e^T(t)\left\{\left[\Theta(I_N - \mathbf{1}_N\theta^T)\right] \otimes \left(P^{-1}D\right)\right\}\omega(t). \tag{7.11}$$

Substituting $K = B^T P^{-1}$ *into (7.11) gives*

$$
\begin{aligned}
\dot{V}(t) = e^T(t)\{\Theta \otimes (A^T P^{-1} + P^{-1} A) \\
- \alpha[((\mathcal{L}^{\sigma(t)})^T \Theta + \Theta \mathcal{L}^{\sigma(t)}) \otimes (P^{-1} BB^T P^{-1})]\} e(t) \\
+ 2e^T(t)\Big[[\Theta(I_N - \mathbf{1}_N \theta^T)] \otimes (P^{-1} D)\Big]\omega(t).
\end{aligned} \tag{7.12}
$$

Let $\varepsilon(t) = [\varepsilon_1^T(t), \ldots, \varepsilon_N^T(t)]^T$, *with* $\varepsilon_i(t) = P^{-1} e_i(t)$, $i = 1, \ldots, N$. *It is easy to check that* $e(t) = (I_N \otimes P)\varepsilon(t)$. *It then follows from (7.12) that*

$$
\begin{aligned}
\dot{V}(t) = \varepsilon^T(t)\Big\{\Theta \otimes \big(PA^T + AP\big) - \alpha[((\mathcal{L}^{\sigma(t)})^T \Theta + \Theta \mathcal{L}^{\sigma(t)}) \otimes (BB^T)]\Big\}\varepsilon(t) \\
+ 2\varepsilon^T(t)\Big\{\big[\Theta(I_N - \mathbf{1}_N \theta^T)\big] \otimes D\Big\}\omega(t).
\end{aligned} \tag{7.13}
$$

For the case of $\omega(t) \equiv \mathbf{0}_{Nq}$, *it follows from (7.13) that*

$$
\dot{V}(t) \leq \varepsilon(t)^T[\Theta \otimes (PA^T + AP - 2\alpha \kappa_0 BB^T)]\varepsilon(t), \tag{7.14}
$$

where $\kappa_0 = \min_{i=1,\ldots,\kappa} a(\mathcal{L}^i)$, *and* $a(\mathcal{L}^i)$ *is the generalized algebraic connectivity of graph* $\mathcal{G}^i$, $i \in \{1, \ldots, \kappa\}$. *It follows from the fact of* $\alpha > 1/(2\kappa_0)$ *and LMI (7.9) that* $\dot{V}(t) < 0$, *for all* $t \geq 0$. *Thus, the consensus problem of CNS (7.1) is solved by protocol (7.2) with control parameters constructed in Algorithm 7.1.*

For the case that $\omega(t)$ *is not a constant zero vector, the* $\mathcal{H}_\infty$ *performance, i.e., the worst-norm of the consensus error vector* $e(t)$ *over all admissible exogenous disturbances* $\omega(t)$ *and initial states* $x(0)$, *is now analyzed. First, one has*

$$
\begin{aligned}
\dot{V}(t) &\leq e^T(t)\Big\{\Theta \otimes \big[A^T P^{-1} + P^{-1} A - 2\alpha \kappa_0 P^{-1} BB^T P^{-1}\big]\Big\}e(t) \\
&\quad + 2e^T(t)\Big\{\big[\Theta(I_N - \mathbf{1}_N \theta^T)\big] \otimes \big(P^{-1} D\big)\Big\}\omega(t) \\
&\leq e^T(t)\Big\{\Theta \otimes [A^T P^{-1} + P^{-1} A - 2\alpha \kappa_0 P^{-1} BB^T P^{-1} + \tilde{\theta} I_n] - I_{Nn}\Big\}e(t) \\
&\quad - \widehat{\theta}\gamma^2 \omega^T(t)\,(\Theta \otimes I_n)\,\omega(t) + \gamma^2 \omega^T(t)\omega(t) \\
&\quad + 2e^T(t)\Big\{\big[\Theta(I_N - \mathbf{1}_N \theta^T)\big] \otimes \big(P^{-1} D\big)\Big\}\omega(t),
\end{aligned} \tag{7.15}
$$

where $\tilde{\theta} = 1/\min_{i=1,2,\ldots,N}\theta_i$, $\widehat{\theta} = 1/\max_{i=1,2,\ldots,N}\theta_i$ *and* γ *is a given positive scalar. By using the Schur complement lemma, one has that* $\dot{V}(t) + e^T(t)e(t) - \gamma^2 \omega^T(t)\omega(t) < 0$ *if and only if* $\Delta < 0$, *where*

$$
\begin{aligned}
\Delta = &\Theta \otimes [A^T P^{-1} + P^{-1} A - 2\alpha \kappa_0 P^{-1} BB^T P^{-1} + \tilde{\theta} I_n] \\
&+ \Big[\Theta(I_N - \mathbf{1}_N \theta^T)\Theta^{-1}(I_N - \mathbf{1}_N \theta^T)^T \Theta^T\Big] \otimes (\varsigma P^{-1} DD^T P^{-1}),
\end{aligned}
$$

with $\varsigma = 1/(\widehat{\theta}\gamma^2)$. *Furthermore, it follows from the facts that* $(I_N - \mathbf{1}_N \theta^T)^T \Theta^T = \Theta(I_N - \mathbf{1}_N \theta^T)$ *and* $(I_N - \mathbf{1}_N \theta^T)^2 = I_N - \mathbf{1}_N \theta^T$ *that* $[\Theta(I_N - \mathbf{1}_N \theta^T)\Theta^{-1}(I_N - \mathbf{1}_N \theta^T)^T \Theta^T] \otimes (\varsigma P^{-1} DD^T P^{-1}) = [\Theta(I_N - \mathbf{1}_N \theta^T)] \otimes (\varsigma P^{-1} DD^T P^{-1})$. *Noticing that* $\Delta < 0$ *if* $\tilde{\Delta} < 0$, *where*

$$
\tilde{\Delta} = \Theta \otimes (A^T P^{-1} + P^{-1} A - 2\alpha \kappa_0 P^{-1} BB^T P^{-1} + \tilde{\theta} I_n
$$

$$+\lambda_{\max}(I_N - \mathbf{1}_N\theta^T)\varsigma P^{-1}DD^TP^{-1}),$$

where $\lambda_{\max}(I_N-\mathbf{1}_N\theta^T)$ is the largest eigenvalue of $I_N-\mathbf{1}_N\theta^T$. According to (7.7), one has that $\Delta < 0$ if there exists a $P > 0$ such that $A^TP^{-1}+P^{-1}A-2\alpha\kappa_0 P^{-1}BB^TP^{-1}+\tilde{\theta}I_n + \varsigma P^{-1}DD^TP^{-1} < 0$, which is equivalent to

$$\begin{bmatrix} AP + PA^T - 2\alpha\kappa_0 BB^T + \varsigma DD^T & P \\ P & -(1/\tilde{\theta})I_n \end{bmatrix} < 0.$$

Based on the above analysis and by Algorithm 7.1, one has

$$\dot{V}(t) + e^T(t)e(t) - \gamma^2\omega^T(t)\omega(t) < 0. \tag{7.16}$$

Next, integrating both sides of inequality (7.16) over the infinite horizon gives

$$\|e(t)\|^2_{\mathbb{L}_2} < \gamma^2\|\omega(t)\|^2_{\mathbb{L}_2} + e^T(0)\left(\Theta \otimes P^{-1}\right)e(0). \tag{7.17}$$

Noting that $P > \varsigma R^{-1}$ with $\varsigma = 1/(\hat{\theta}\gamma^2)$ and $\hat{\theta} = 1/\max_{i=1,2,\dots,N}\theta_i$, it follows from (7.17) that

$$\|e(t)\|^2_{\mathbb{L}_2} < \gamma^2\left[\|\omega(t)\|^2_{\mathbb{L}_2} + e^T(0)\left(I_N \otimes R\right)e(0)\right], \tag{7.18}$$

which indicates that the distributed $\mathcal{H}_\infty$ consensus problem with performance γ and weighting matrix $R > 0$ for CNS (7.1) is solved.

Remark 7.3 It can be observed from Algorithm 7.1 that the existence of protocol (7.2) depends on the solvability of LMI (7.9). It is easy to check that LMI (7.9) is feasible if and only if there exists a positive definite matrix P such that

$$PA^T+AP-BB^T+\tilde{\theta}P^2+\varsigma DD^T<0, \tag{7.19}$$

where $\tilde{\theta} = 1/\min_{i=1,2,\dots,N}\theta_i$. By using Finsler's lemma, one has that there exists a $P > 0$ such that (7.19) holds if and only if there exist matrices $P > 0$ and $F \in \mathbb{R}^{m\times n}$ such that the following algebraic Riccati inequality holds:

$$P(A - BF)^T+(A - BF)P + \tilde{\theta}P^2+\varsigma DD^T < 0. \tag{7.20}$$

Based on the above analysis, one has that LMI (7.9) is feasible if and only if there exist matrices $Q > 0$ and $F \in \mathbb{R}^{m\times n}$ such that

$$(A - BF)^TQ+Q(A - BF) + \tilde{\theta}I_n+\varsigma QDD^TQ<0. \tag{7.21}$$

By using the bounded real lemma, one gets that (7.9) holds if and only if there exists a matrix $F \in \mathbb{R}^{m\times n}$ such that

$$\left\|[sI - (A - BF)]^{-1}D\right\|_\infty < (\tilde{\theta}\varsigma)^{-1/2}, \tag{7.22}$$

where $\tilde{\theta} = 1/\min_{i=1,\dots,N}\theta_i$, $\varsigma = 1/(\hat{\theta}\gamma^2)$, and $\hat{\theta} = 1/\max_{i=1,\dots,N}\theta_i$. Thus, LMI (7.9) is solvable if and only if there exists a matrix $F \in \mathbb{R}^{m\times n}$ such that (7.22) holds. Also, it can be seen from the above analysis that (A, B) is stabilizable is a necessary condition for the solvability of LMI (7.9).

Remark 7.4 *Since a common Lyapunov function (7.10) is constructed and employed for the error system (7.8), consensus in the closed-loop CNSs can be achieved for an arbitrarily given dwell time $\tau_m > 0$. It is also worth noting that the present $\mathcal{H}_\infty$ consensus problem reduces to that addressed in [229] if the initial states of all agents could be synchronously set as zero vectors and the topology is a fixed connected undirected graph.*

7.2.3 Discussions on the convergence rate

In the case of $\omega(t) \equiv \mathbf{0}_{Nq}$, it follows from Theorem 7.1 that consensus in CNS (7.1) with protocol (7.2) constructed by Algorithm 7.1 can be achieved exponentially. However, the convergence rate is not explicitly given in Algorithm 7.1, i.e., it is still unclear how fast consensus in nominal CNS (7.1) can be achieved. In this subsection, a modified algorithm is proposed to redesign protocol (7.2) such that consensus in the closed-loop nominal CNSs can be achieved with a given exponential convergence rate $c_0 > 0$.

Algorithm 7.2 *Suppose that (A, B) is stabilizable and Assumption 7.1 holds. The consensus protocol (7.2) with an exponential convergence rate c_0 can be designed as follows:*

(1) Solve the LMI

$$AP + PA^T - BB^T + 2c_0 P < 0 \tag{7.23}$$

to get one feasible solution: matrix $P > 0$ and scalar $c_0 > 0$. Then, take $K = B^T P^{-1}$.

(2) Choose the coupling strength $\alpha \geq \alpha_{\mathrm{th}}$, where α_{th} is defined in Step (2) of Algorithm 7.1.

Theorem 7.2 *Suppose that Assumption 7.1 holds and LMI (7.23) is feasible. Then, consensus in the closed-loop nominal CNS (7.1) with protocol (7.2) constructed by Algorithm 7.2 can be achieved with an exponential rate c_0 for any given dwell time $\tau_m > 0$.*

Proof 7.2 *Construct the same Lyapunov function $V(t)$ as that used in the proof of Theorem 7.1. One has that*

$$V(t) \leq V(t_0) \exp(-2c_0(t - t_0)), \ \forall \, t \geq t_0. \tag{7.24}$$

It thus follows from (7.24) that $\|e(t)\| \leq \exp(-c_0(t - t_0)) \sqrt{V(t_0)/\lambda_{\min}(P^{-1})}$, where $\lambda_{\min}(P^{-1})$ denotes the smallest eigenvalue of P^{-1}. Then the proof is completed.

Remark 7.5 *In the case that (A, B) is controllable, there is a matrix K_1 and a positive definite matrix Q such that $(A - BK_1)^T Q + Q(A - BK_1) + 2c_0 Q < 0$ for any given $c_0 > 0$. By using Finsler's lemma, it is not hard to verify that LMI (7.23) is always feasible under the condition that (A, B) is controllable, i.e., consensus in the closed-loop CNS (7.1) with protocol (7.2) can be achieved with an arbitrarily given* *convergence rate c_0 by appropriately choosing the control parameters. Suppose that (A, B) is stabilizable but not controllable. Let $-\rho_0 < 0$ be the largest real part of the uncontrollable mode. Then, one can verify that LMI (7.23) is feasible for any given $c_0 \in [0, \rho_0]$.*

7.3 $\mathcal{H}_\infty$ CONSENSUS OF CNSS WITH LIPSCHITZ NONLINEAR DYNAMICS AND APERIODIC SAMPLED DATA COMMUNICATIONS

7.3.1 Model formulation

This section considers the following CNS consisting of N agents with aperiodic sampled-data-based diffusive couplings:

$$\dot{x}_i(t) = f(x_i(t), t) + \alpha \sum_{j=1}^{N} a_{ij} \left(x_j(t_k) - x_i(t_k) \right) + \omega_i(t), \ t \in [t_k, t_{k+1}), \qquad (7.25)$$

where $x_i(t) \in \mathbb{R}^n$ is the state of agent i, $\alpha > 0$ is the coupling strength, $\mathcal{A} = [a_{ij}]_{N \times N}$ is the adjacency matrix of the communication topology $\mathcal{G}$, $\omega_i(t) \in \mathbb{L}_2[0, +\infty)$ are external disturbances, $i = 1, \ldots, N$, $f(\cdot, \cdot) : \mathbb{R}^n \times [0, +\infty) \mapsto \mathbb{R}^n$ is a nonlinear function satisfying the following global Lipschitz condition:

$$\|f(y, t) - f(\tilde{y}, t)\| \le \varrho \|y - \tilde{y}\|, \ \forall y, \tilde{y} \in \mathbb{R}^n, \qquad (7.26)$$

for some given $\varrho > 0$.

For the case of $\omega_i(t) \equiv \mathbf{0}_n$, $i = 1, \ldots, N$, the consensus tracking is said to be achieved if, under pinning control, the states of all agents in the CNS (7.25) converge to a prescribed trajectory $s(t)$ in the sense of $\lim_{t \to \infty} \|x_i(t) - s(t)\| = 0$ for any given initial conditions, where $s(t)$ is generated by

$$\dot{s}(t) = f(s(t), t), \qquad (7.27)$$

for arbitrary $s(t_0) \in \mathbb{R}^n$. For notational brevity, let $e(t) = [e_1^T(t), \ldots, e_N^T(t)]^T$ with $e_i(t) = x_i(t) - s(t)$, $i = 1, \ldots, N$. The $\mathcal{H}_\infty$ consensus tracking with performance index $\gamma > 0$ is said to be achieved if, under pinning control, the following two conditions are satisfied:

(1) For $\omega(t) \equiv \mathbf{0}_{Nn}$, consensus tracking in the CNS (7.25) with the target given by (7.27) can be achieved, where $\omega(t) = [\omega_1^T(t), \ldots, \omega_N^T(t)]^T$.

(2) For a given scalar $\gamma > 0$ and initial condition $e(t_0) = \mathbf{0}_{Nn}$, the worst-case norm of consensus error vector $e(t)$ over all admissible exogenous disturbances $\omega(t)$, defined by

$$\gamma_\omega = \sup_{\omega(t) \in \mathbb{L}_2[0, +\infty)} \frac{\|e(t)\|_{\mathbb{L}_2}}{\|\omega(t)\|_{\mathbb{L}_2}} \qquad (7.28)$$

is less than γ, i.e., $\gamma_\omega < \gamma$.

Remark 7.6 *The notion of $\mathcal{H}_\infty$ consensus is borrowed from the idea of $\mathcal{H}_\infty$ control for dynamical systems in the context of modern control theory [231]. The classical $\mathcal{H}_\infty$ control problem can be described as how to construct some stabilizing controllers, such that the closed-loop systems are internal stable and the $\mathcal{H}_\infty$ norm of the transfer matrix from exogenous signals to the performance variable is less than a prescribed positive number. Within the context of consensus of CNSs, the consensus error is always taken as the performance variable [77, 172]. However, for the convenience of analysis, it is always assumed in the literature on $\mathcal{H}_\infty$ consensus of CNSs that the information can be transmitted continuously [77, 172]. This indicates that each agent needs to share its state information with its neighbors continuously. However, there exist some practical situations where such information exchange only happens at some discrete time instants. It is thus practically important to study how to realize $\mathcal{H}_\infty$ consensus with sampled-data communications.*

For notational brevity, we label the virtual target (7.27) as agent 0. Then the Laplacian matrix of the augmented graph $\mathcal{G}(\tilde{\mathcal{A}})$ can be written as

$$\tilde{\mathcal{L}} = \begin{bmatrix} 0 & \mathbf{0}_N^T \\ \mathbf{P} & \overline{\mathcal{L}} \end{bmatrix},$$

$$\overline{\mathcal{L}} = \begin{bmatrix} \sum_{j\in\mathcal{N}_1} a_{1j} & -a_{12} & \cdots & -a_{1N} \\ -a_{21} & \sum_{j\in\mathcal{N}_2} a_{2j} & \cdots & -a_{2N} \\ \vdots & \vdots & \ddots & \vdots \\ -a_{N1} & -a_{N2} & \cdots & \sum_{j\in\mathcal{N}_N} a_{Nj} \end{bmatrix},$$

where $\mathbf{P} = -[a_{10}, \ldots, a_{N0}]^T$ with $a_{i0} > 0$ if agent i is pinned, and $a_{i0} = 0$ otherwise, $i = 1, \ldots, N$.

In this section, some sampled-data-based negative feedback injections will be employed to the CNS (7.25). The closed-loop CNS under pinning control is described by:

$$\dot{x}_i(t) = f(x_i(t), t) + \alpha \sum_{j=1}^N a_{ij} \left(x_j(t_k) - x_i(t_k) \right) + \omega_i(t)$$

$$- \alpha a_{i0}(x_i(t_k) - s(t_k)), \ t \in [t_k, \ t_{k+1}), \ k \in \mathbb{N}. \tag{7.29}$$

Let $e(t) = [e_1^T(t), \ldots, e_N^T(t)]^T$ with $e_i(t) = x_i(t) - s(t)$, $i = 1, \ldots, N$, one has

$$\dot{e}(t) = f(x(t); s(t)) - \alpha(\overline{\mathcal{L}} \otimes I_n)e(t_k) + \omega(t), \tag{7.30}$$

where $f(x(t); s(t)) = [f^T(x_1(t); s(t)), \ldots, f^T(x_N(t); s(t))]^T$, $f(x_i(t); s(t)) = f(x_i(t), t) - f(s(t), t)$, $i = 1, \ldots, N$, $\omega(t) = [\omega_1^T(t), \ldots, \omega_N^T(t)]^T$, $t \in [t_k, t_{k+1})$.

Let $d_k(t) = t - t_k$, for $t \in [t_k, t_{k+1})$, $k \in \mathbb{N}$. One then gets that $t_k = t - d_k(t)$ with $0 \leq d_k(t) < h$, for $t \in [t_k, t_{k+1})$, $k \in \mathbb{N}$. Then, the error system (7.30) can be rewritten as the following retarded functional differential equation:

$$\dot{e}(t) = f(x(t); s(t)) - \alpha(\overline{\mathcal{L}} \otimes I_n)e(t - d_k(t)) + \omega(t), \tag{7.31}$$

where $t \in [t_k, t_{k+1})$, $k \in \mathbb{N}$. The initial condition of (7.31) is set as $e(\theta) \equiv e(t_0)$ for all $\theta = [-h, 0]$.

Before ending this subsection, two assumptions are made.

Assumption 7.2 *There is a constant $h > 0$ such that $t_{k+1} - t_k \leq h$ for $k \in \mathbb{N}$.*

Assumption 7.3 *The augmented graph $\mathcal{G}(\tilde{\mathcal{A}})$ contains a directed spanning tree with agent 0 as the root.*

7.3.2 Selective pinning strategy

In this subsection, a graph search algorithm with linear time complexity is provided for choosing the agents to be pinned in the CNS (7.25). The important issues of at least how many and which agents should be pinned such that Assumption 7.3 holds will be addressed.

Algorithm 7.3 *Let $\mathcal{G}(\mathcal{A})$ be the communication topology of the CNS (7.25). Then, Assumption 7.3 will hold if the r_0 agents searched by the following procedures are selected and pinned.*

(1) Use Tarjan's algorithm [157] to find all the agents with zero in-degree and strongly connected components of $\mathcal{G}(\mathcal{A})$. Suppose that there are ι_1 ($\iota_1 \geq 0$) agents with zero in-degree, labeled as $v_1, \ldots, v_{\iota_1}$, and ι_2 ($\iota_2 \geq 0$) strongly connected components, represented by $\mathcal{G}(\mathcal{V}_1, \mathcal{E}_1, \mathcal{A}_1), \ldots, \mathcal{G}(\mathcal{V}_{\iota_2}, \mathcal{E}_{\iota_2}, \mathcal{A}_{\iota_2})$ in $\mathcal{G}(\mathcal{A})$. Set $r_i = 0$, for $i = 0, 1, \ldots, \iota_2$, and $g = 1$.

(2) All the ι_1 agents with zero in-degree should be selected and pinned. Then, update the value of r_0 by $r_0 = r_0 + \iota_1$.

(3) Check the condition $\iota_2 \neq 0$? If it does not hold, stop; else go to step (4).

(4) Check whether there exists at least one node in $\mathcal{V}_g$ which is reachable from a node belonging to the node set $\mathcal{V} \backslash \mathcal{V}_g$. If it holds, go to step (5); otherwise, go to step (6).

(5) Check the following condition: $g < \iota_2$? If it holds, let $g = g + 1$ and re-perform step (4); else stop.

(6) Arbitrarily select one agent in $\mathcal{V}_g$ to be pinned, update the value of r_0 by $r_0 = r_0 + 1$. Check the following condition: $g < \iota_2$? If it holds, let $g = g + 1$ and go to step (4); else stop.

It can be checked that there exists at least one agent in $\mathcal{G}(\mathcal{A})$ which is not reachable from agent 0 if there are less than r_0 agents in $\mathcal{G}(\mathcal{A})$ that are selected and pinned. Furthermore, it is worth noting that the complexity of Algorithm 7.3 is $\mathcal{O}(N + |\mathcal{E}|)$, where $|\mathcal{E}|$ is the number of the directed edges in $\mathcal{G}(\mathcal{A})$. Noticeably, under the condition that the target (7.27) possesses a global attractive solution $\bar{s}(t)$, the consensus tracking in CNS (7.29) can be achieved asymptotically even when there is no coupling between any pair of neighboring nodes in $\mathcal{G}(\tilde{\mathcal{A}})$. In the sequel, it is assumed

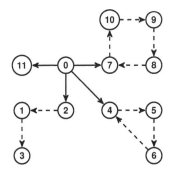

Figure 7.1 The augmented network $\mathcal{G}(\tilde{\mathcal{A}})$. Agent 0 in $\mathcal{G}(\tilde{\mathcal{A}})$ is used to represent the target (7.27). The dashed lines indicate the neighboring relationship of agents in $\mathcal{G}(\mathcal{A})$ while the solid lines represent the pinning links.

that the target (7.27) does not possess a global attractive solution. Then, it is not hard to verify that consensus tracking in the CNS (7.29) can not be ensured if the augmented graph $\mathcal{G}(\tilde{\mathcal{A}})$ does not contain any directed spanning tree. Next, we demonstrate how to use Algorithm 7.3 to find the agents to be pinned in a given directed graph such that the augmented graph contains a directed spanning tree. Suppose that the network topology $\mathcal{G}(\mathcal{A})$ contains 12 agents as shown in Figure 7.1. According to step (1) of Algorithm 7.3, one has that there are two agents in $\mathcal{G}(\mathcal{A})$ with zero in-degree, agents 2 and 11, and two strongly connected components. According to step (2) of Algorithm 7.3, one knows that agents 2 and 11 should be pinned. By steps (3)–(6) of Algorithm 7.3, one obtains that two distinct agents selected from agent sets $\{7, 8, 9, 10\}$ and $\{4, 5, 6\}$ should be pinned, respectively. Then, select agents 4 and 7 to be pinned. One may observe that the augmented network $\mathcal{G}(\tilde{\mathcal{A}})$ contains a directed spanning tree rooted at agent 0.

7.3.3 Main results

In this subsection, consensus tracking for the CNS (7.29) with $\omega_i(t) \equiv \mathbf{0}_n$, $i = 1, \ldots, N$, is firstly studied. The $\mathcal{H}_\infty$ consensus tracking for the CNS (7.29) in the presence of external disturbances is then addressed.

Based on the above discussions, one can establish the following theorem which summarizes the main results on consensus tracking for the CNS (7.29) with $\omega(t) \equiv \mathbf{0}_{Nn}$. For notational brevity, asterisk '*' in a symmetric matrix denotes the entry implied by symmetry.

Theorem 7.3 *Suppose that Assumptions 7.2 and 7.3 hold, and $\omega(t) \equiv \mathbf{0}_{Nn}$. Then, consensus tracking in the CNS (7.29) can be achieved if there exist a scalar $\varsigma > 0$, positive definite matrices $P, Q \in \mathbb{R}^{N \times N}$, positive semi-definite matrices $X \in \mathbb{R}^{3N \times 3N}$, and $N_i \in \mathbb{R}^{N \times N}$, $i = 1, 2, 3$, such that*

$$\Lambda = \begin{bmatrix} \Lambda_{11} + \varsigma\rho^2 & \Lambda_{12} & \Lambda_{13} \\ * & \Lambda_{22} & \Lambda_{23} \\ * & * & \Lambda_{33} - \tau I_N \end{bmatrix} < 0, \tag{7.32}$$

$$\Xi = \begin{bmatrix} X_{11} & X_{12} & X_{13} & N_1 \\ * & X_{22} & X_{23} & N_2 \\ * & * & X_{33} & N_3 \\ * & * & * & Q \end{bmatrix} \geq 0, \tag{7.33}$$

where $\Lambda_{11} = N_1 + N_1^T + hX_{11}$, $\Lambda_{12} = -\alpha P\overline{\mathcal{L}} + N_2^T - N_1 + hX_{12}$, $\Lambda_{13} = P + N_3^T + hX_{13}$, $\Lambda_{22} = -N_2 - N_2^T + hX_{22} + \alpha^2 h\overline{\mathcal{L}}^T Q\mathcal{L}$, $\Lambda_{23} = -N_3^T + hX_{23} - \alpha h\mathcal{L}^T Q$, and $\Lambda_{33} = hX_{33} + hQ$, with $X_{ij} \in \mathbb{R}^{N\times N}$, $i, j = 1, 2, 3$, and $X = \begin{bmatrix} X_{11} & X_{12} & X_{13} \\ * & X_{22} & X_{23} \\ * & * & X_{33} \end{bmatrix} \in \mathbb{R}^{3N\times 3N}$.

Proof 7.3 *Construct the following piecewise differentiable Lyapunov-Krasovskii functional for error system (7.31):*

$$V(t) = e^T(t)(P \otimes I_n)e(t) + (t_{k+1} - t)\int_{t_k}^t \dot{e}^T(\iota)(Q\otimes I_n)\dot{e}(\iota)d\iota,$$

where $P > 0$, $Q > 0$, $t \in [t_k, t_{k+1})$, $k \in \mathbb{N}$. Then, taking the time derivative of $V(t)$ along the trajectory of (7.31) gives that

$$\dot{V}(t) = 2e^T(t)(P \otimes I_n)\dot{e}(t) + (t_{k+1} - t)\dot{e}^T(t)(Q \otimes I_n)\dot{e}(t)$$
$$- \int_{t_k}^t \dot{e}^T(\iota)(Q \otimes I_n)\dot{e}(\iota)d\iota. \tag{7.34}$$

From the Newton-Leibnitz formula, the following equation holds for any given matrices $N_i \in \mathbb{R}^{N\times N}$,

$$2\left[e^T(t)(N_1\otimes I_n) + e^T(t - d_k(t))(N_2\otimes I_n) + f^T(x(t); s(t))(N_3\otimes I_n)\right]$$
$$\times \left[e(t) - e(t-d_k(t)) - \int_{t-d_k(t)}^t \dot{e}(\iota)d\iota\right] = 0. \tag{7.35}$$

On the other hand, for any given $X = \begin{bmatrix} X_{11} & X_{12} & X_{13} \\ * & X_{22} & X_{23} \\ * & * & X_{33} \end{bmatrix} \in \mathbb{R}^{3N\times 3N}$, the following inequality holds:

$$h\eta_1^T(t)(X\otimes I_n)\eta_1(t) - \int_{t-d_k(t)}^t \eta_1^T(\iota)(X\otimes I_n)\eta_1(\iota)d\iota \geq 0, \tag{7.36}$$

where $\eta_1(t) = \left[e^T(t), e^T(t - d_k(t)), f^T(x(t); s(t))\right]^T$. It thus follows from (7.34) to (7.36) that

$$\dot{V}(t) \leq 2e^T(t)(P \otimes I_n)\left[f(x(t); s(t)) - \alpha(\overline{\mathcal{L}} \otimes I_n)e(t - d_k(t))\right]$$
$$+ h\dot{e}^T(t)(Q \otimes I_n)\dot{e}(t) - \int_{t-d_k(t)}^t \dot{e}^T(\iota)(Q\otimes I_n)\dot{e}(\iota)d\iota$$
$$+ 2\left[e^T(t)(N_1\otimes I_n) + e^T(t - d_k(t))(N_2 \otimes I_n)\right.$$

$$+ f^T(x(t); s(t))(N_3 \otimes I_n)] \left[e(t) - e(t - d_k(t)) - \int_{t-d_k(t)}^t \dot{e}(\iota) d\iota \right]$$

$$+ h\eta_1^T(t)(X \otimes I_n)\eta_1(t) - \int_{t-d_k(t)}^t \eta_1^T(\iota)(X \otimes I_n)\eta_1(\iota) d\iota. \tag{7.37}$$

Some calculations give that

$$h\dot{e}^T(t)(Q \otimes I_n)\dot{e}(t)$$
$$= \alpha^2 h e^T(t - d_k(t))(\overline{\mathcal{L}}^T Q \overline{\mathcal{L}} \otimes I_n)e(t - d_k(t))$$
$$+ h f^T(x(t); s(t))(Q \otimes I_n)f(x(t); s(t))$$
$$- 2\alpha h e^T(t - d_k(t))(\overline{\mathcal{L}}^T Q \otimes I_n)f(x(t); s(t)). \tag{7.38}$$

Substituting (7.38) into (7.37) gives

$$\dot{V}(t) \leq \eta_1^T(t)(\widehat{\Lambda} \otimes I_n)\eta_1(t) - \int_{t-d_k(t)}^t \eta_2^T(t, \iota)(\Xi \otimes I_n)\eta_2(t, \iota) d\iota,$$

where $\eta_2(t, \iota) = [e^T(t), e^T(t - d_k(t)), f^T(x(t); s(t)), \dot{e}^T(\iota)]^T,$

$$\widehat{\Lambda} = \begin{bmatrix} \Lambda_{11} & \Lambda_{12} & \Lambda_{13} \\ * & \Lambda_{22} & \Lambda_{23} \\ * & * & \Lambda_{33} \end{bmatrix}, \quad \Xi = \begin{bmatrix} X_{11} & X_{12} & X_{13} & N_1 \\ * & X_{22} & X_{23} & N_2 \\ * & * & X_{33} & N_3 \\ * & * & * & Q \end{bmatrix},$$

$\Lambda_{11} = N_1 + N_1^T + hX_{11}, \quad \Lambda_{12} = -\alpha P\overline{\mathcal{L}} + N_2^T - N_1 + hX_{12}, \quad \Lambda_{13} = P + N_3^T + hX_{13},$
$\Lambda_{22} = -N_2 - N_2^T + hX_{22} + \alpha^2 h\overline{\mathcal{L}}^T Q\mathcal{L}, \quad \Lambda_{23} = -N_3^T + hX_{23} - \alpha h\mathcal{L}^T Q, \quad \Lambda_{33} = hX_{33} + hQ.$

According to (7.26), it is sufficient to show that $\dot{V}(t) < 0$ *if there exists a positive scalar* $\varsigma > 0$ *such that*

$$\eta_1^T(t)(\widehat{\Lambda} \otimes I_n)\eta_1(t) - \int_{t-d_k(t)}^t \eta_2^T(t, \iota)(\Xi \otimes I_n)\eta_2(t, \iota) d\iota$$
$$+ \varsigma\varrho^2 e^T(t)e(t) - \varsigma f^T(x(t); s(t))f(x(t); s(t)) < 0. \tag{7.39}$$

Let $\Lambda = \widehat{\Lambda} + \begin{bmatrix} \varsigma\rho^2 & 0 & 0 \\ * & 0 & 0 \\ * & * & -\varsigma I_N \end{bmatrix}$. *Noticeably, for each* $t \in [t_k, t_{k+1})$ *and* $k \in \mathbb{N}$, *one has*

$\dot{V}(t) < 0$ *if* $\Xi \geq 0$ *and* $\widehat{\Lambda} < 0$.

Furthermore, it can be verified that $\lim_{t \nearrow t_{k+1}} V(t) = V(t_{k+1}) = e^T(t_{k+1})(P \otimes I_n)e(t_{k+1})$, *for all* $k \in \mathbb{N}$. *By using some similar arguments as those in the standard proof of Lyapunov-Krasovskii stability theory [53], one gets that* $\|e(t)\|$ *will converge to zero asymptotically under conditions (7.32) and (7.33), which indicates that consensus tracking in the CNS (7.29) with* $\omega(t) \equiv \mathbf{0}_{Nn}$ *is achieved.*

Suppose that the conditions given in Theorem 7.3 can be ensured, i.e., consensus tracking can be ensured for some given sampling interval $h = h_0 > 0$. It is interesting to further study the maximum allowable sampling interval $h_{\max}$ guaranteeing consensus tracking in Theorem 7.3. For this purpose, the following algorithm is provided.

Algorithm 7.4 *The maximum allowable sampling interval h_{max} guaranteeing consensus tracking in Theorem 7.3 can be estimated by the following procedures:*

(1) Set $h_{max} = h_0$ and step size $\iota = \iota_0$, where $\iota_0 > 0$ is sufficiently small compared to h_0.

(2) Search matrices $P > 0$, $Q > 0$, $X \geq 0$, N_i, $i = 1, 2, 3$, and scalar $\varsigma > 0$ such that LMIs (7.32) and (7.33) hold. If the conditions are satisfied, set $h_{max} = h_{max} + \iota_0$ and re-perform step (2). Otherwise, stop and let h_{max} be the maximum allowable sampling interval.

Note that to obtain a less-conservative estimation on the maximum allowable sampling interval, the free-weighting matrices technique was employed in the proof of Theorem 7.3. It is also worth noting that the dimensions of the LMIs (7.32) and (7.33) are independent of those of the agents' states in the CNS (7.29). This 'decoupling' feature will be more desirable when each agent is a high-dimensional system. Alternatively, one may get the following corollary where the dimensions of the consensus criteria are dependent on those of the agents' states. Generally speaking, the consensus conditions given in the following corollary will be less conservative than those given in Theorem 7.3. However, it will be seen that solving the LMIs given in the following corollary is challenging.

Corollary 7.1 *Suppose that Assumptions 7.2 and 7.3 hold, and $\omega(t) \equiv \mathbf{0}_{Nn}$. Then, consensus tracking in the CNS (7.29) can be achieved if there exist a scalar $\overline{\varsigma} > 0$, positive definite matrices $\overline{P}$, $\overline{Q} \in \mathbb{R}^{Nn \times Nn}$, positive semi-definite matrix $\overline{X} \in \mathbb{R}^{3Nn \times 3Nn}$, and $\overline{N}_i \in \mathbb{R}^{Nn \times Nn}$, $i = 1, 2, 3$, such that*

$$\overline{\Lambda} = \begin{bmatrix} \overline{\Lambda}_{11} + \overline{\varsigma}\rho^2 & \overline{\Lambda}_{12} & \overline{\Lambda}_{13} \\ * & \overline{\Lambda}_{22} & \overline{\Lambda}_{23} \\ * & * & \overline{\Lambda}_{33} - \overline{\varsigma} I_{Nn} \end{bmatrix} < 0, \tag{7.40}$$

$$\overline{\Xi} = \begin{bmatrix} \overline{X}_{11} & \overline{X}_{12} & \overline{X}_{13} & \overline{N}_1 \\ * & \overline{X}_{22} & \overline{X}_{23} & \overline{N}_2 \\ * & * & \overline{X}_{33} & \overline{N}_3 \\ * & * & * & \overline{Q} \end{bmatrix} \geq 0, \tag{7.41}$$

where $\overline{\Lambda}_{11} = \overline{N}_1 + \overline{N}_1^T + h\overline{X}_{11}$, $\overline{\Lambda}_{12} = -\alpha\overline{P}(\overline{\mathcal{L}} \otimes I_n) + \overline{N}_2^T - \overline{N}_1 + h\overline{X}_{12}$, $\overline{\Lambda}_{13} = \overline{P} + \overline{N}_3^T + h\overline{X}_{13}$, $\overline{\Lambda}_{22} = -\overline{N}_2 - \overline{N}_2^T + h\overline{X}_{22} + \alpha^2 h(\overline{\mathcal{L}}^T \otimes I_n)\overline{Q}(\overline{\mathcal{L}} \otimes I_n)$, $\overline{\Lambda}_{23} = -\overline{N}_3^T + h\overline{X}_{23} - \alpha h(\overline{\mathcal{L}}^T \otimes I_n)\overline{Q}$, $\overline{\Lambda}_{33} = h\overline{X}_{33} + h\overline{Q}$, with $\overline{X}_{ij} \in \mathbb{R}^{Nn \times Nn}$, $i, j = 1, 2, 3$, and

$$\overline{X} = \begin{bmatrix} \overline{X}_{11} & \overline{X}_{12} & \overline{X}_{13} \\ * & \overline{X}_{22} & \overline{X}_{23} \\ * & * & \overline{X}_{33} \end{bmatrix} \in \mathbb{R}^{3Nn \times 3Nn}.$$

Proof 7.4 *Construct the following piecewise differentiable Lyapunov-Krasovskii functional for the error system (7.31):*

$$V(t) = e^T(t)\overline{P}e(t) + (t_{k+1} - t)\int_{t_k}^{t} \dot{e}^T(\iota)\overline{Q}\dot{e}(\iota)d\iota, \tag{7.42}$$

with $\overline{P} > 0$, $\overline{Q} > 0$, $t \in [t_k, t_{k+1})$, $k \in \mathbb{N}$. The corollary can be proved by following the steps in the proof of Theorem 7.3.

Remark 7.7 *Compared with the proof of Corollary 7.1, a special kind of Lyapunov-Krasovskii functional was employed in proving Theorem 7.3. Though the consensus conditions given in Theorem 7.3 have less complexity, they may be conservative in estimating the maximum allowable sampling interval for achieving consensus. However, the numerical studies indicate that the conservativeness introduced by employing a special kind of Lyapunov-Krasovskii functional in the proof of Theorem 7.3 is not severe.*

Based on Theorem 7.3, one may get the following theorem, which states that $\mathcal{H}_\infty$ consensus tracking for the CNS (7.29) with external disturbances can be ensured under some suitable conditions.

Theorem 7.4 *Suppose that Assumptions 7.2 and 7.3 hold. Then, $\mathcal{H}_\infty$ consensus tracking with performance index $\gamma > 0$ in the CNS (7.29) can be achieved if there exist a scalar $\varsigma > 0$, positive definite matrices S, $T \in \mathbb{R}^{N \times N}$, positive semi-definite matrix $Y \in \mathbb{R}^{4N \times 4N}$, and $W_i \in \mathbb{R}^{N \times N}$, $i = 1, \ldots, 4$, such that*

$$\Psi = \begin{bmatrix} \Psi_{11} + \varsigma\rho^2 + \gamma^{-1}I_N & \Psi_{12} & \Psi_{13} & \Psi_{14} \\ * & \Psi_{22} & \Psi_{23} & \Psi_{24} \\ * & * & \Psi_{33} - \varsigma I_N & \Psi_{34} \\ * & * & * & -\gamma I_N \end{bmatrix} < 0, \qquad (7.43)$$

$$\Omega = \begin{bmatrix} Y_{11} & Y_{12} & Y_{13} & Y_{14} & W_1 \\ * & Y_{22} & Y_{23} & Y_{24} & W_2 \\ * & * & Y_{33} & Y_{34} & W_3 \\ * & * & * & Y_{44} & W_4 \\ * & * & * & * & T \end{bmatrix} \geq 0, \qquad (7.44)$$

where $\Psi_{11} = W_1 + W_1^T + hY_{11}$, $\Psi_{12} = -\alpha S\overline{\mathcal{L}} + W_2^T - W_1 + hY_{12}$, $\Psi_{13} = S + W_3^T + hY_{13}$, $\Psi_{14} = S + W_4^T + hY_{14}$, $\Psi_{22} = -W_2 - W_2^T + hY_{22} + \alpha^2 h\overline{\mathcal{L}}^T T\overline{\mathcal{L}}$, $\Psi_{23} = -W_3^T + hY_{23} - \alpha h\overline{\mathcal{L}}^T T$, $\Psi_{24} = -W_4^T + hY_{24}$, $\Psi_{33} = hY_{33} + hT$, $\Psi_{34} = hY_{34}$, with $Y_{ij} \in \mathbb{R}^{N \times N}$,

$$i, j = 1, 2, 3, 4, \text{ and } Y = \begin{bmatrix} Y_{11} & Y_{12} & Y_{13} & Y_{14} \\ * & Y_{22} & Y_{23} & Y_{24} \\ * & * & Y_{33} & Y_{34} \\ * & * & * & Y_{44} \end{bmatrix} \in \mathbb{R}^{4N \times 4N}.$$

Proof 7.5 *According to Theorem 7.3, it follows from conditions (7.43) and (7.44) that consensus tracking in the CNS (7.29) with $\omega(t) \equiv \mathbf{0}_{Nn}$ can be achieved.*

Next, $\mathcal{H}_\infty$ consensus problem with performance index γ is studied. For $\vartheta \in [t_k, t_{k+1})$ and an arbitrarily given $k \in \mathbb{N}$, define

$$J_\vartheta = \int_0^\vartheta \left(\gamma^{-1}e^T(t)e(t) - \gamma\omega^T(t)\omega(t) \right) dt, \qquad (7.45)$$

where $\omega(t) = [\omega_1^T(t), \ldots, \omega_N^T(t)]^T$ with $\omega_i(t) \in \mathbb{L}_2[0, +\infty)$. By the zero initial condition $e(t) \equiv \mathbf{0}_{Nn}$ for $t \in [-h, 0]$, one gets

$$J_\vartheta = \int_0^\vartheta \left(\gamma^{-1} e^T(t) e(t) - \gamma \omega^T(t) \omega(t) + \dot{V}(t) \right) dt - V(\vartheta)$$

$$\leq \int_0^\vartheta \left(\gamma^{-1} e^T(t) e(t) - \gamma \omega^T(t) \omega(t) + \dot{V}(t) \right) dt, \tag{7.46}$$

where

$$V(t) = e^T(t)(S \otimes I_n) e(t) + (t_{k+1} - t) \int_{t_k}^t \dot{e}^T(\iota)(T \otimes I_n) \dot{e}(\iota) d\iota.$$

Furthermore, from the Newton-Leibnitz formula, the following equation holds for any given matrices $W_i \in \mathbb{R}^{N \times N}$, $i = 1, \ldots, 4$:

$$2 \big[e^T(t)(W_1 \otimes I_n) + e^T(t - d_k(t))(W_2 \otimes I_n) + f^T(x(t); s(t))(W_3 \otimes I_n)$$

$$+ \omega^T(t)(W_4 \otimes I_n) \big] \left[e(t) - e(t - d_k(t)) - \int_{t - d_k(t)}^t \dot{e}(\iota) d\iota \right] = 0. \tag{7.47}$$

Based on the above analysis and by arguments similar to the proof of Theorem 7.3, one has

$$\dot{V}(t) \leq \hat{\eta}_1^T(t)(\widehat{\Psi} \otimes I_n)\hat{\eta}_1(t) - \int_{t - d_k(t)}^t \hat{\eta}_2^T(t, \iota)(\Omega \otimes I_n)\hat{\eta}_2(t, \iota) d\iota,$$

where $\hat{\eta}_1(t) = [e^T(t), e^T(t - d_k(t)), f^T(x(t); s(t)), \omega^T(t)]^T$, $\hat{\eta}_2(t, \iota) = [e^T(t), e^T(t - d_k(t)), f^T(x(t); s(t)), \omega^T(t), \dot{e}^T(\iota)]^T$,

$$\widehat{\Psi} = \begin{bmatrix} \Psi_{11} & \Psi_{12} & \Psi_{13} & \Psi_{14} \\ * & \Psi_{22} & \Psi_{23} & \Psi_{24} \\ * & * & \Psi_{33} & \Psi_{34} \\ * & * & * & \Psi_{44} \end{bmatrix}, \quad Y = \begin{bmatrix} Y_{11} & Y_{12} & Y_{13} & Y_{14} \\ * & Y_{22} & Y_{23} & Y_{24} \\ * & * & Y_{33} & Y_{34} \\ * & * & * & Y_{44} \end{bmatrix} \geq 0,$$

Ψ_{ij}, $i, j = 1, \ldots, 4$, *are defined in (7.43), Ω is defined in (7.44).*

On the other hand, one obtains from (7.26) that $\varsigma \varrho^2 e^T(t) e(t) - \varsigma f^T(x(t); s(t))$ $\cdot f(x(t); s(t)) > 0$ for all $\varsigma > 0$. It thus follows from conditions (7.43) and (7.44) that $J_\vartheta < 0$, i.e.,

$$\int_0^\vartheta e^T(\iota) e(\iota) d\iota < \gamma^2 \int_0^\vartheta \omega^T(\iota) \omega(\iota) d\iota, \tag{7.48}$$

for $\vartheta \in [t_k, t_{k+1})$ and $k \in \mathbb{N}$. By noticing that $\omega(t) \in \mathbb{L}_2$, integrating (7.48) from $\vartheta = 0$ to ∞ yields $\int_0^\infty e^T(\iota) e(\iota) d\iota < \gamma^2 \int_0^\infty \omega^T(\iota) \omega(\iota) d\iota$. This indicates that the $\mathcal{H}_\infty$ consensus tracking problem of the CNS (7.29) with a prescribed performance index γ is indeed achieved.

Remark 7.8 *For a prescribed $\gamma > 0$, the $\mathcal{H}_\infty$ consensus with a disturbance rejection level less than γ can be verified by checking the conditions in Theorem 7.4. However, it is practically important to know the allowable smallest disturbance rejection level $\gamma_{\min}$ for consensus in the CNS (7.29). Note that it is very hard or even impossible to calculate $\gamma_{\min}$ theoretically. But it can be numerically estimated by solving the following optimization problem:*

$$\text{minimize } \gamma$$

$$\text{subject to: } \Psi < 0, \ \Omega \geq 0,$$

where Ψ and Ω are respectively given in (7.43) and (7.44), $\varsigma > 0$, matrices $S > 0$, $T > 0$, $Y \geq 0$, and W_i, $i = 1, \ldots, 4$, with appropriate dimensions.

In Theorem 7.4, the dimensions of LMIs (7.43) and (7.44) are independent of those of the agents' states in the CNS (7.29). Alternatively, one may get the following corollary where the dimensions of the $\mathcal{H}_\infty$ consensus criteria are dependent on those of the agents' states.

Corollary 7.2 *Suppose that Assumptions 7.2 and 7.3 hold. Then, the $\mathcal{H}_\infty$ consensus tracking with performance index $\gamma > 0$ in the CNS (7.29) can be achieved if there exist a scalar $\overline{\varsigma} > 0$, positive definite matrices $\overline{S}, \overline{T} \in \mathbb{R}^{Nn \times Nn}$, positive semi-definite matrix $\overline{Y} \in \mathbb{R}^{4Nn \times 4Nn}$, and $\overline{W}_i \in \mathbb{R}^{Nn}$, $i = 1, \ldots, 4$, such that*

$$\overline{\Psi} = \begin{bmatrix} \overline{\Psi}_{11} + \overline{\varsigma}\rho^2 + \gamma^{-1}I_{Nn} & \overline{\Psi}_{12} & \overline{\Psi}_{13} & \overline{\Psi}_{14} \\ * & \overline{\Psi}_{22} & \overline{\Psi}_{23} & \overline{\Psi}_{24} \\ * & * & \overline{\Psi}_{33} - \overline{\varsigma}I_{Nn} & \overline{\Psi}_{34} \\ * & * & * & -\gamma I_{Nn} \end{bmatrix} < 0, \qquad (7.49)$$

$$\overline{\Omega} = \begin{bmatrix} \overline{Y}_{11} & \overline{Y}_{12} & \overline{Y}_{13} & \overline{Y}_{14} & \overline{W}_1 \\ * & \overline{Y}_{22} & \overline{Y}_{23} & \overline{Y}_{24} & \overline{W}_2 \\ * & * & \overline{Y}_{33} & \overline{Y}_{34} & \overline{W}_3 \\ * & * & * & \overline{Y}_{44} & \overline{W}_4 \\ * & * & * & * & \overline{T} \end{bmatrix} \geq 0, \qquad (7.50)$$

where $\overline{\Psi}_{11} = \overline{W}_1 + \overline{W}_1^T + h\overline{Y}_{11}$, $\overline{\Psi}_{12} = -\alpha\overline{S}(\overline{\mathcal{L}} \otimes I_n) + \overline{W}_2^T - \overline{W}_1 + h\overline{Y}_{12}$, $\overline{\Psi}_{13} = \overline{S} + \overline{W}_3^T + h\overline{Y}_{13}$, $\overline{\Psi}_{14} = \overline{S} + \overline{W}_4^T + h\overline{Y}_{14}$, $\overline{\Psi}_{22} = -\overline{W}_2 - \overline{W}_2^T + h\overline{Y}_{22} + \alpha^2 h(\overline{\mathcal{L}}^T \otimes I_n)\overline{T}(\overline{\mathcal{L}} \otimes I_n)$, $\overline{\Psi}_{23} = -\overline{W}_3^T + h\overline{Y}_{23} - \alpha h(\overline{\mathcal{L}}^T \otimes I_n)\overline{T}$, $\overline{\Psi}_{24} = -\overline{W}_4^T + h\overline{Y}_{24}$, $\overline{\Psi}_{33} = h\overline{Y}_{33} + h\overline{T}$,

$\overline{\Psi}_{34} = h\overline{Y}_{34}$, $\overline{Y}_{ij} \in R^{Nn \times Nn}$, $i, j = 1, 2, 3, 4$, and $\overline{Y} = \begin{bmatrix} \overline{Y}_{11} & \overline{Y}_{12} & \overline{Y}_{13} & \overline{Y}_{14} \\ * & \overline{Y}_{22} & \overline{Y}_{23} & \overline{Y}_{24} \\ * & * & \overline{Y}_{33} & \overline{Y}_{34} \\ * & * & * & \overline{Y}_{44} \end{bmatrix}$

$\in \mathbb{R}^{4Nn \times 4Nn}$.

7.3.4 Extension to $\mathcal{H}_\infty$ consensus of CNSs with directed switching topologies

The $\mathcal{H}_\infty$ consensus tracking for CNSs with directed fixed topology under sampled-data communications has been investigated in previous subsections. However, in reality, the network topology may be time-varying due to technological limitations of sensors, external disturbances or communication channel failures. From this observation, $\mathcal{H}_\infty$ consensus tracking of CNSs with directed switching topologies under sampled-data communications is further studied in this subsection.

Let $\{\mathcal{G}(\mathcal{A}^1), \ldots, \mathcal{G}(\mathcal{A}^\kappa)\}$, $\kappa \geq 2$ be the set of all possible topologies. Suppose that there exists an infinite sequence of uniformly bounded non-overlapping time intervals $[t_z, t_{z+1})$, $z \in \mathbb{N}$, with $t_0 = 0$, $0 < \tau_m < t_{z+1} - t_z$, and τ_m such that the underlying topology is time-invariant for all $t \in [t_z, t_{z+1})$. Let $\mathcal{G}(\mathcal{A}^{\sigma(t)})$ be the topology of the CNS (7.29) at time $t \geq 0$. On the other hand, the coupling force between any pair of neighboring agents is generated by employing sampling technique and a zero-order hold circuit, i.e., the coupling force acting on each agent i, $i = 1, \ldots, N$, is time-invariant for all $t \in [t_k, t_{k+1})$, $k \in \mathbb{N}$. This indicates that for each time interval $[t_k, t_{k+1})$, $k \in \mathbb{N}$, the interaction among the agents is determined only by the coupling strength α and the communication topology $\mathcal{G}(\mathcal{A}^{\sigma(t_k)})$. Let $\{\mathcal{G}(\widetilde{\mathcal{A}}^1), \ldots, \mathcal{G}(\widetilde{\mathcal{A}}^\kappa)\}$ be the set of all possible augmented network topologies. Thus, the closed-loop CNSs under pinning control can be described as:

$$\dot{x}_i(t) = f(x_i(t), t) + \alpha \sum_{j=1}^{N} a_{ij}^{\sigma(t_k)} (x_j(t_k) - x_i(t_k)) + \omega_i(t)$$
$$- \alpha a_{i0}^{\sigma(t_k)} (x_i(t_k) - s(t_k)), \quad t \in [t_k, t_{k+1}), \tag{7.51}$$

where $a_{i0}^{\sigma(t_k)} > 0$ if agent i is pinned at time t_k, and $a_{i0}^{\sigma(t_k)} = 0$ otherwise, $i = 1, \ldots, N$. Here, $\mathcal{A}^{\sigma(t_k)} = [a_{ij}^{\sigma(t_k)}]$ is the adjacency matrix of graph $\mathcal{G}(\mathcal{A}^{\sigma(t_k)})$, $k \in \mathbb{N}$. Taking the target system (7.27) as a virtual node labeled as 0 in the considered network, one then gets

$$\dot{e}(t) = f(x(t); s(t)) - \alpha(\overline{\mathcal{L}}^{\sigma(t_k)} \otimes I_n)e(t - d_k(t)) + \omega(t), \quad t \in [t_k, t_{k+1}), \tag{7.52}$$

where $e(\theta) \equiv e(t_0)$ for all $\theta = [-h, 0]$, $\overline{\mathcal{L}}^{\sigma(t_k)} = \mathcal{L}^{\sigma(t_k)} + \text{diag}\{a_{10}^{\sigma(t_k)}, \ldots, a_{N0}^{\sigma(t_k)}\}$, $\mathcal{L}^{\sigma(t_k)}$ is the Laplacian matrix of $\mathcal{G}(\mathcal{A}^{\sigma(t_k)})$. Furthermore, one has that $\widetilde{\mathcal{L}}^{\sigma(t_k)} = \begin{bmatrix} 0 & \mathbf{0}_N^T \\ \mathbf{P}^{\sigma(t_k)} & \overline{\mathcal{L}}^{\sigma(t_k)} \end{bmatrix}$ is the Laplacian matrix of the augmented graph $\mathcal{G}(\widetilde{\mathcal{A}}^{\sigma(t_k)})$. Here, $\mathbf{P}^{\sigma(t_k)} = -[a_{10}^{\sigma(t_k)}, \ldots, a_{N0}^{\sigma(t_k)}]^T \in \mathbb{R}^N$.

To derive the main results of this subsection, the following assumption is introduced.

Assumption 7.4 *Each augmented graph $\mathcal{G}(\widetilde{\mathcal{A}}^i)$, $i = 1, \ldots, \kappa$, contains a directed spanning tree with agent 0 being the root.*

From the above analysis, one gets the following two theorems, which summarize the main results of this subsection.

Theorem 7.5 *Suppose that Assumptions 7.2 and 7.4 hold, and $\omega(t) \equiv \mathbf{0}_{Nn}$. Then, the consensus tracking in the CNS (7.51) can be achieved if there exist a scalar $\varsigma > 0$, positive definite matrices P, $Q \in \mathbb{R}^{N \times N}$, positive semi-definite matrix $X \in \mathbb{R}^{3N \times 3N}$, and $N_i \in \mathbb{R}^{N \times N}$, $i = 1, 2, 3$, such that*

$$\Lambda_i = \begin{bmatrix} \Lambda_{11} + \varsigma\rho^2 & \Lambda_{12} & \Lambda_{13} \\ * & \Lambda_{22} & \Lambda_{23} \\ * & * & \Lambda_{33} - \varsigma I_N \end{bmatrix} < 0, \tag{7.53}$$

$$\begin{bmatrix} X_{11} & X_{12} & X_{13} & N_1 \\ * & X_{22} & X_{23} & N_2 \\ * & * & X_{33} & N_3 \\ * & * & * & Q \end{bmatrix} \geq 0, \tag{7.54}$$

where $\Lambda_{11} = N_1 + N_1^T + hX_{11}$, $\Lambda_{12} = -\alpha P\overline{\mathcal{L}}^i + N_2^T - N_1 + hX_{12}$, $\Lambda_{13} = P + N_3^T + hX_{13}$, $\Lambda_{22} = -N_2 - N_2^T + hX_{22} + \alpha^2 h \left(\overline{\mathcal{L}}^i\right)^T Q\mathcal{L}^i$, $\Lambda_{23} = -N_3^T + hX_{23} - \alpha h \left(\mathcal{L}^i\right)^T Q$, $\Lambda_{33} = hX_{33} + hQ$, $\overline{\mathcal{L}}^i = M\widetilde{\mathcal{L}}^i M^T$, $i = 1, \ldots, \kappa$, $M = [I_N, \mathbf{0}_N]$, $X_{ij} \in \mathbb{R}^{N \times N}$, $i, j = 1, 2, 3$, and $X = \begin{bmatrix} X_{11} & X_{12} & X_{13} \\ * & X_{22} & X_{23} \\ * & * & X_{33} \end{bmatrix} \in \mathbb{R}^{3N \times 3N}$.

Proof 7.6 *Construct the following common piecewise differentiable Lyapunov-Krasovskii functional for the error system (7.52):*

$$V(t) = e^T(t)(P \otimes I_n)e(t) + (t_{k+1} - t) \int_{t_k}^t \dot{e}^T(\iota)(Q \otimes I_n)\dot{e}(\iota)d\iota,$$

where $P > 0$, $Q > 0$, $t \in [t_k, t_{k+1})$, $k \in \mathbb{N}$. Then, the theorem can be proved by following the steps in the proof of Theorem 7.3.

Furthermore, one can get the following theorem on the $\mathcal{H}_\infty$ consensus tracking of the CNS (7.51) with directed switching topologies. The detailed proof is omitted for brevity.

Theorem 7.6 *Suppose that Assumptions 7.2 and 7.4 hold. Then, $\mathcal{H}_\infty$ consensus tracking with performance index $\gamma > 0$ in the CNS (7.51) can be achieved if there exist a scalar $\varsigma > 0$, positive definite matrices S, $T \in \mathbb{R}^{N \times N}$, positive semi-definite matrix $Y \in \mathbb{R}^{4N \times 4N}$, and $W_i \in \mathbb{R}^{N \times N}$, $i = 1, \ldots, 4$, such that*

$$\Psi_i = \begin{bmatrix} \Psi_{11} + \varsigma\rho^2 + \gamma^{-1}I_N & \Psi_{12} & \Psi_{13} & \Psi_{14} \\ * & \Psi_{22} & \Psi_{23} & \Psi_{24} \\ * & * & \Psi_{33} - \varsigma I_N & \Psi_{34} \\ * & * & * & \Psi_{44} - \gamma I_N \end{bmatrix} < 0, \tag{7.55}$$

$$\Omega = \begin{bmatrix} Y_{11} & Y_{12} & Y_{13} & Y_{14} & W_1 \\ * & Y_{22} & Y_{23} & Y_{24} & W_2 \\ * & * & Y_{33} & Y_{34} & W_3 \\ * & * & * & Y_{44} & W_4 \\ * & * & * & * & Q \end{bmatrix} \geq 0, \tag{7.56}$$

$\Psi_{11} = W_1 + W_1^T + hY_{11}$, $\Psi_{12} = -\alpha S \overline{\mathcal{L}}^i + W_2^T - W_1 + hY_{12}$, $\Psi_{13} = S + W_3^T + hY_{13}$, $\Psi_{14} = S + W_4^T + hY_{14}$, $\Psi_{22} = -W_2 - W_2^T + hY_{22} + \alpha^2 h \left(\overline{\mathcal{L}}^i\right)^T T \overline{\mathcal{L}}^i$, $\Psi_{23} = -W_3^T + hY_{23} - \alpha h \left(\overline{\mathcal{L}}^i\right)^T T$, $\Psi_{24} = -W_4^T + hY_{24}$, $\Psi_{33} = hY_{33} + hT$, $\Psi_{34} = hY_{34}$, $\overline{\mathcal{L}}^i = M \widetilde{\mathcal{L}}^i M^T$, $\widetilde{\mathcal{L}}^i$ is the Laplacian matrix of $\mathcal{G}(\widetilde{\mathcal{A}}^i)$, $i = 1, \ldots, \kappa$, $M = [I_N, 0_N]$, $Y_{ij} \in \mathbb{R}^{N \times N}$, $i, j = 1, 2, 3, 4$, and $Y = \begin{bmatrix} Y_{11} & Y_{12} & Y_{13} & Y_{14} \\ * & Y_{22} & Y_{23} & Y_{24} \\ * & * & Y_{33} & Y_{34} \\ * & * & * & Y_{44} \end{bmatrix} \in \mathbb{R}^{4N \times 4N}$.

Remark 7.9 *In the present section, the dynamical agents in the considered CNS are assumed to have a homogeneous time-varying sampling rate, i.e., the sampler embedded in each agent works at the same time instants: t_k, $k \in \mathbb{N}$. It is more interesting but challenging to further study how to achieve $\mathcal{H}_\infty$ consensus tracking in CNSs with directed fixed or switching topologies under heterogeneous time-varying sampling rates. Furthermore, the consensus criteria provided in the present section are dependent on the solvability of some high dimensional LMIs, which are thus inapplicable for CNSs of huge size.*

7.4 NUMERICAL SIMULATIONS

In this example, the $\mathcal{H}_\infty$ consensus tracking problem for CNS (7.51) is numerically studied. The topology is assumed to switch back and forth between $\mathcal{G}(\widetilde{\mathcal{A}}^1)$ and $\mathcal{G}(\widetilde{\mathcal{A}}^2)$ per 0.12 s. The possible topologies $\mathcal{G}(\widetilde{\mathcal{A}}^1)$ and $\mathcal{G}(\widetilde{\mathcal{A}}^2)$ are shown in Figure 7.2, where the weights are indicated on the edges. It can be verified that Assumption 7.4 holds. Let $f(x_i(t), t) = [0.5\sin(x_{i1}(t)), 0.5\cos(x_{i2}(t))]^T$, for $i = 0, 1, \ldots, 5$. Set $c = 0.5$. It can be then obtained from Theorem 7.5 that the maximum allowable sampling interval for achieving consensus in the CNS (7.51) is $h_{\max} = 0.10132$. The state trajectories of the CNS (7.51) with coupling strength $c = 0.5$ and sampling interval $h = 0.10$ are shown in Figs. 7.3 and 7.4. Use $\text{Error}(t) = \sqrt{\sum_{j=1}^5 \|x_j - x_0(t)\|^2}$ to denote the consensus error of the CNS. Figure 7.5 indicates that a faster convergence rate will be yielded when enlarging the sampling interval. Furthermore, it can be obtained from Theorem 7.5 that $\mathcal{H}_\infty$ consensus with performance index $\gamma = 0.45$ in the CNS (7.51) can be guaranteed. Choose $w_i(t) = [2\sin(it), 2\cos(it)]^T$, for $0 \leq t \leq 4$, and $w_i(t) = 0_2$ for $t > 4$. The energy trajectories of $e(t)$ and $w(t)$ are shown in Figure 7.6, which indicate that the $\mathcal{H}_\infty$ consensus tracking problem is solved. It can be observed from Figure 7.6 that the estimation of the $\mathcal{H}_\infty$ performance index, i.e., $\gamma = 0.45$, is not very conservative.

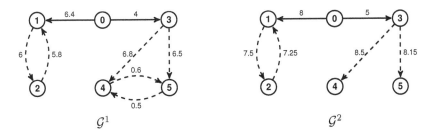

Figure 7.2 The communication graphs $\mathcal{G}^1$ and $\mathcal{G}^2$.

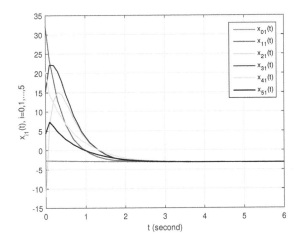

Figure 7.3 The agents' state trajectories $x_{i1}(t)$, $i = 0, 1, \ldots, 5$, of the CNS with $\omega(t) \equiv \mathbf{0}_{10}$.

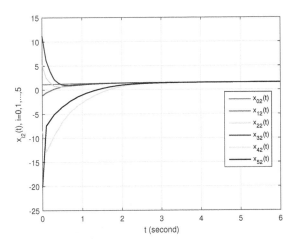

Figure 7.4 The agents' state trajectories $x_{i2}(t)$, $i = 0, 1, \ldots, 5$, of the CNS with $\omega(t) \equiv \mathbf{0}_{10}$.

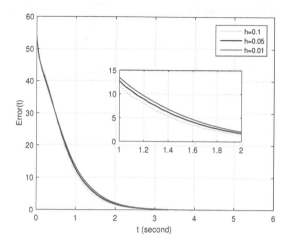

Figure 7.5 Trajectories of the consensus error $\mathrm{Error}(t)$ versus the sampling interval h.

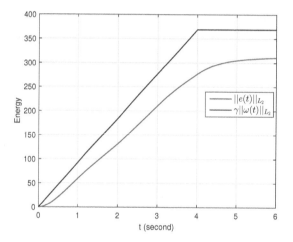

Figure 7.6 Energy trajectories of $e(t)$ and $\omega(t)$.

7.5 CONCLUSIONS

In this chapter, both $\mathcal{H}_\infty$ leaderless consensus and $\mathcal{H}_\infty$ consensus tracking problems for CNSs with disturbances under directed topology have been studied. To achieve $\mathcal{H}_\infty$ consensus in linear CNSs, a new consensus protocol has been constructed and analyzed in Section 7.2. By using tools from algebraic graph theory and switched systems theory, it has been theoretically proved that the $\mathcal{H}_\infty$ consensus problem in the closed-loop CNS can be solved if the control parameters of the protocol are appropriately designed. Furthermore, the interesting problem of consensus rate for nominal CNSs with directed switching topologies has been investigated. In Section 7.3, $\mathcal{H}_\infty$ consensus tracking in CNSs with directed fixed and switching topologies under aperiodic sampled-data communications has been investigated. By using a combined tool from input-delay approach, Lyapunov-Krasovskii stability analysis, and the LMI technique, some sufficient criteria for achieving consensus in CNSs directed fixed topology have been derived and discussed. The results are then extended to $\mathcal{H}_\infty$ consensus tracking with external disturbances and switching topologies. Here, it is assumed that the dynamic agents in the present network model have a common homogeneous sampling rate, that is, only global $\mathcal{H}_\infty$ consensus tracking for CNSs with directed switching topologies under synchronous sampling rate is addressed in this chapter. The $\mathcal{H}_\infty$ consensus problem for CNSs with asynchronous sampled-data communication and directed switching topologies is still in its infancy. It remains to be seen how to solve such a challenging problem in the future.

CHAPTER 8

Distributed tracking of nonlinear CNSs with directed switching topologies: An observer-based protocol

Unlike most existing works where the relative full state measurements of neighboring agents are used for controller design, this chapter proposes a relative outputs based observer type consensus controller for CNSs with Lipschitz nonlinear dynamics and directed switching topologies. This chapter begins by studying the case where the agents layer and the observer layer have the same communication topology. By using the MLFs based methods, we show that consensus can be achieved if each topology contains a directed spanning tree rooted at the leader and the dwell time for the switches among different topologies is less than a derived positive quantity. Then, this chapter extends the derived results to the case where the topologies of the agents layer and the observer layer are independent. Finally, this chapter presents some simulations to validate the obtained theoretical results.

8.1 INTRODUCTION

In [188], some preliminary results for consensus tracking of CNSs with Lipschitz nonlinear dynamics and directed switching topologies were provided under the assumption that the relative full state information of neighboring agents can be used for coordination. More recently, consensus tracking for CNSs with Lipschitz nonlinear dynamics and fixed topology was studied in [64] by proposing some dynamic protocols. Recently, the evolution behaviors of CNSs with nonlinear dynamics and switching topologies have attracted an increasing attention [16]. Despite the progresses, there have not been many results on consensus tracking of CNSs with Lipschitz nonlinear dynamics and directed switching topologies, especially under the scenario that only some partial state information of neighboring agents are available for protocol design. The above-mentioned concern motivates our study.

This chapter is concerned with the consensus tracking problem for CNSs with Lipschitz nonlinear dynamics and directed switching topologies where only the relative output information of neighboring agents can be used for protocol design. One challenge we face is how to reconstruct the full state information of each follower by using only the local relative output information. Motivated by the observer design approach for a single Lipschitz nonlinear system, a distributed observer is designed for each follower to estimate its full state information online. Next to reconstruction of the full state information of followers, there is a need for control algorithms that can make followers coordinate with each other under directed switching topologies. Motivated by this concern, a new class of communication protocols is proposed and employed. Since the separation principle does not hold for controller and observer design in the present model, a Lyapunov function-based design approach is adopted. By appropriately constructing MLFs, it is shown that consensus tracking in the CNSs can be ensured if each possible communication topology contains a directed spanning tree rooted at the leader and the control parameters are appropriately selected. The results are further extended to consensus tracking of CNSs for the case where the communication topologies among the observers are different from those of the agents.

8.2 PROBLEM FORMULATION

Consider a CNS consisting of a leader and N followers, where the leader is labelled as agent 0, and the followers are respectively labelled as agents $1, \ldots, N$. The dynamics of agent i, $i = 1, \ldots, N$, are described by

$$
\begin{aligned}
\dot{x}_i(t) &= Ax_i(t) + Bu_i(t) + Df(x_i(t), t), \\
y_i(t) &= Cx_i(t),
\end{aligned}
\tag{8.1}
$$

where $x_i(t) \in \mathbb{R}^n$ is the state, $u_i(t) \in \mathbb{R}^m$ is the control input, $y_i(t) \in \mathbb{R}^q$ is the output, $f(\cdot, \cdot) : \mathbb{R}^n \times [0, +\infty) \mapsto \mathbb{R}^n$ is a continuously differentiable vector-valued function satisfying the Lipschitz condition

$$
\|f(x, t) - f(y, t)\| \le \varrho \|x - y\|, \ \forall \, x, y \in \mathbb{R}^n, \ t \ge 0,
\tag{8.2}
$$

for some $\varrho \ge 0$, A, B, C, D are constant real matrices with compatible dimensions. It is assumed that the triple (A, B, C) is stabilizable and detectable. The dynamics of the leader are described by

$$
\begin{aligned}
\dot{x}_0(t) &= Ax_0(t) + Df(x_0(t), t), \\
y_0(t) &= Cx_0(t),
\end{aligned}
\tag{8.3}
$$

and $x_0(t) \in \mathbb{R}^n$ is the leader's state.

It is assumed that the communication topologies among the $N + 1$ agents switch among the graphs $\mathcal{G}^1, \ldots, \mathcal{G}^\kappa$, $\kappa \geq 1$. Let $t_0 = 0$, and let $t_1, t_2, \ldots$ be the switching sequence at which the communication topologies switch and it is assumed that $0 < \tau_m \leq t_{k+1} - t_k \leq \tau_M < +\infty$. The Laplacian matrix of $\mathcal{G}^{\sigma(t)}$ can be written as

$$\mathcal{L}^{\sigma(t)} = \begin{bmatrix} 0 & \mathbf{0}_N^T \\ \mathbf{P}^{\sigma(t)} & \overline{\mathcal{L}}^{\sigma(t)} \end{bmatrix},$$

$$\overline{\mathcal{L}}^{\sigma(t)} = \begin{bmatrix} \sum_{j \in \mathcal{N}_1} a_{1j}^{\sigma(t)} & -a_{12}^{\sigma(t)} & \cdots & -a_{1N}^{\sigma(t)} \\ -a_{21}^{\sigma(t)} & \sum_{j \in \mathcal{N}_2} a_{2j}^{\sigma(t)} & \cdots & -a_{2N}^{\sigma(t)} \\ \vdots & \vdots & \ddots & \vdots \\ -a_{N1}^{\sigma(t)} & -a_{N2}^{\sigma(t)} & \cdots & \sum_{j \in \mathcal{N}_N} a_{Nj}^{\sigma(t)} \end{bmatrix}, \tag{8.4}$$

where $\mathbf{P}^{\sigma(t)} = -[a_{10}^{\sigma(t)}, \cdots, a_{N0}^{\sigma(t)}]^T$.

In existing works on consensus tracking of CNSs with Lipschitz nonlinear dynamics, e.g., [79, 188], the relative full state of neighboring agents were used for protocol design. However, in practice, the full state may be unavailable due to technical limitations. Compared with the aforementioned works, the following consensus tracking protocol based only on the relative outputs between agent i and its neighbors is proposed:

$$\dot{\tilde{x}}_i(t) = A\tilde{x}_i(t) + Bu_i(t) + \alpha \sum_{j=0}^{N} a_{ij}^{\sigma(t)} F(\delta_j(t) - \delta_i(t)) + Df(\tilde{x}_i(t), t),$$

$$u_i(t) = \beta K \sum_{j=0}^{N} a_{ij}^{\sigma(t)} (\tilde{x}_j(t) - \tilde{x}_i(t)), \tag{8.5}$$

where $\tilde{x}_i(t) \in \mathbb{R}^n$ is the state of the observer embedded in follower i, $\delta_i(t) = \tilde{y}_i(t) - y_i(t)$ represents the output error between follower i and the associated observer, $\tilde{y}_i(t) = C\tilde{x}_i(t)$, $\alpha > 0$ and $\beta > 0$ are the coupling strengths to be selected later, $F \in \mathbb{R}^{n \times q}$ and $K \in \mathbb{R}^{m \times n}$ are the feedback matrices to be designed, and $\mathcal{A}^{\sigma(t)} = [a_{ij}^{\sigma(t)}]_{(N+1) \times (N+1)}$ is the adjacency matrix of $\mathcal{G}^{\sigma(t)}$. Since the leader in this chapter takes the role of a reference signal generator, it is assumed in (8.5) that $\tilde{x}_0(t) = x_0(t)$, i.e., the leader does not need to observe its state. Now, it follows from (8.1) and (8.5) that

$$\dot{x}_i(t) = Ax_i(t) + \beta BK \sum_{j=0}^{N} a_{ij}^{\sigma(t)} (\tilde{x}_j(t) - \tilde{x}_i(t)) + Df(x_i(t), t),$$

$$\dot{\tilde{x}}_i(t) = A\tilde{x}_i(t) + \beta BK \sum_{j=0}^{N} a_{ij}^{\sigma(t)} (\tilde{x}_j(t) - \tilde{x}_i(t)) + Df(\tilde{x}_i(t), t)$$

$$+ \alpha F \sum_{j=0}^{N} a_{ij}^{\sigma(t)} (\delta_j(t) - \delta_i(t)),$$

for $i = 1, \ldots, N$. By taking $x(t) = [x_1^T(t), \ldots, x_N^T(t)]^T$ and $\tilde{x}(t) = [\tilde{x}_1^T(t), \ldots, \tilde{x}_N^T(t)]^T$, one obtains

$$\dot{x}(t) = (I_N \otimes A)x(t) - \beta \left(\widehat{\widetilde{\mathcal{L}}}^{\sigma(t)} \otimes BK \right) \widehat{\tilde{x}}(t) + (I_N \otimes D)f(x(t); t)$$

$$\dot{\tilde{x}}(t) = (I_N \otimes A)\tilde{x}(t) - \beta \left(\widehat{\widetilde{\mathcal{L}}}^{\sigma(t)} \otimes BK \right) \widehat{\tilde{x}}(t) \qquad (8.6)$$

$$+ (I_N \otimes D)f(\tilde{x}(t); t) - \alpha \left(\overline{\mathcal{L}}^{\sigma(t)} \otimes F \right) \delta(t),$$

where $\widehat{\widetilde{\mathcal{L}}}^{\sigma(t)} = [\mathbf{P}^{\sigma(t)}, \overline{\mathcal{L}}^{\sigma(t)}] \in \mathbb{R}^{N \times (N+1)}$, $\overline{\mathcal{L}}^{\sigma(t)}$ and $\mathbf{P}^{\sigma(t)}$ are defined in (8.4), $\widehat{\tilde{x}}(t) = [\tilde{x}_0^T(t), \tilde{x}^T(t)]^T$, $\delta(t) = [\delta_1^T(t), \ldots, \delta_N^T(t)]^T$, $f(x(t); t) = [f^T(x_1(t), t), \ldots, f^T(x_N(t), t)]^T$, and $f(\tilde{x}(t); t) = [f^T(\tilde{x}_1(t), t), \ldots, f^T(\tilde{x}_N(t), t)]^T$.

To facilitate the subsequent analysis, the following assumption is made.

Assumption 8.1 *Each graph $\mathcal{G}^j$ contains a directed spanning tree with agent 0 (i.e., the leader) as the root, $j = 1, \ldots, \kappa$.*

If Assumption 8.1 holds, it can be obtained from Lemma 2.15 that there exists a positive definite diagonal matrix $\Phi^j = \text{diag}\{\phi_1^j, \ldots, \phi_N^j\}$ such that $(\overline{\mathcal{L}}^j)^T \Phi^j + \Phi^j \overline{\mathcal{L}}^j > 0$, where Φ^j satisfies $(\overline{\mathcal{L}}^j)^T [\phi_1^j, \ldots, \phi_N^j]^T = \mathbf{1}_N$, $j = 1, \ldots, \kappa$.

8.3 MAIN RESULTS

This section presents the main theoretical results. For notational convenience, one may let $\hat{e}(t) = [\tilde{e}^T(t), e^T(t)]^T$, $\tilde{e}^T(t) = [\tilde{e}_1^T(t), \ldots, \tilde{e}_N^T(t)]^T$, $e^T(t) = [e_1^T(t), \ldots, e_N^T(t)]^T$, $\tilde{e}_i(t) = x_i(t) - \tilde{x}_i(t)$, and $e_i(t) = x_i(t) - x_0(t)$, $i = 1, \ldots, N$. By using the properties of Kronecker product, it can be obtained from (8.6) that

$$\dot{\tilde{e}}(t) = \left[I_N \otimes A - \alpha \left(\overline{\mathcal{L}}^{\sigma(t)} \otimes FC \right) \right] \tilde{e}(t) + (I_N \otimes D)(f(x(t); t) - f(\tilde{x}(t); t)),$$

$$\dot{e}(t) = (I_N \otimes A)e(t) - \beta \left(\widehat{\widetilde{\mathcal{L}}}^{\sigma(t)} \otimes BK \right) \widehat{\tilde{x}}(t)$$

$$+ (I_N \otimes D)[f(x(t); t) - (\mathbf{1}_N \otimes f(x_0(t), t))], \qquad (8.7)$$

where $\widehat{\widetilde{\mathcal{L}}}^{\sigma(t)} = [\mathbf{P}^{\sigma(t)}, \overline{\mathcal{L}}^{\sigma(t)}] \in \mathbb{R}^{N \times (N+1)}$, $\overline{\mathcal{L}}^{\sigma(t)}$ and $\mathbf{P}^{\sigma(t)}$ are defined in (8.4), $\widehat{\tilde{x}}(t) = [\tilde{x}_0^T(t), \tilde{x}^T(t)]^T$. On the other hand, it can be verified that

$$\left(\widehat{\widetilde{\mathcal{L}}}^{\sigma(t)} \otimes BK \right) \widehat{\tilde{x}}(t) = \left(\widehat{\widetilde{\mathcal{L}}}^{\sigma(t)} \otimes BK \right) \left[\widehat{\tilde{x}}(t) - (\mathbf{1}_{N+1} \otimes x_0(t)) \right]$$

$$= \left(\overline{\mathcal{L}}^{\sigma(t)} \otimes BK \right) [\tilde{x}(t) - (\mathbf{1}_N \otimes x_0(t))]$$

$$= \left(\overline{\mathcal{L}}^{\sigma(t)} \otimes BK \right) (e(t) - \tilde{e}(t)). \qquad (8.8)$$

It thus follows from (8.7) and (8.8) that

$$\dot{\hat{e}}(t) = \widehat{A}\hat{e}(t) + (I_{2N} \otimes D)\widehat{f}(\hat{e}(t); t), \qquad (8.9)$$

where $\widehat{e}(t) = [\widetilde{e}^T(t), e^T(t)]^T$ and

$$\widehat{A} = \begin{bmatrix} I_N \otimes A - \alpha \left(\overline{\mathcal{L}}^{\sigma(t)} \otimes FC \right) & 0 \\ \beta \left(\overline{\mathcal{L}}^{\sigma(t)} \otimes BK \right) & I_N \otimes A - \beta \left(\overline{\mathcal{L}}^{\sigma(t)} \otimes BK \right) \end{bmatrix},$$

$$\widehat{f}(\widehat{e}(t); t) = \begin{bmatrix} f(x(t); t) - f(\widetilde{x}(t); t) \\ f(x(t); t) - (\mathbf{1}_N \otimes f(x_0(t), t)) \end{bmatrix}.$$

Under Assumption 8.1, it can be obtained that all eigenvalues of $\overline{\mathcal{L}}^j + (\Phi^j)^{-1} \left(\overline{\mathcal{L}}^j \right)^T \Phi^j$, $j = 1, \ldots, \kappa$, are positive. For notational brevity, let

$$\lambda_0 = \min_{j \in \{1, \ldots, \kappa\}} \left\{ \lambda_{\min}(\overline{\mathcal{L}}^j + (\Phi^j)^{-1}(\overline{\mathcal{L}}^j)^T \Phi^j) \right\}. \tag{8.10}$$

Before moving forward, the following algorithm is presented for selecting the parameters in protocol (8.5).

Algorithm 8.1 *Suppose that (A, B, C) is stabilizable and detectable, and Assumption 8.1 holds. The consensus tracking protocol (8.5) can be designed as follows:*

(1) Select scalars $\gamma_1 > 0$, $\rho > 0$, and $c_1 > 0$. Solve the LMI

$$\begin{bmatrix} A^T Q + QA - \gamma_1 C^T C + (1/\rho)I_n + c_1 Q & \varrho\sqrt{\rho}QD \\ \varrho\sqrt{\rho}D^T Q^T & -I_n \end{bmatrix} < 0 \tag{8.11}$$

to get a positive definite matrix Q, where ϱ is defined in (8.1). Then, set $F = Q^{-1}C^T$.

(2) Select scalars $\gamma_2 > 0$, $\tilde{\rho} > 0$, and $c_2 > 0$. Solve the LMI

$$\begin{bmatrix} AP + PA^T - \gamma_2 BB^T + \varrho^2\tilde{\rho}DD^T + c_2 P & (1/\sqrt{\tilde{\rho}})P \\ (1/\sqrt{\tilde{\rho}})P & -I_n \end{bmatrix} < 0 \tag{8.12}$$

to get a positive definite matrix P. Then, take $K = B^T P^{-1}$.

(3) Choose the coupling strengths $\alpha > \gamma_1/\lambda_0$ and $\beta > \gamma_2/\lambda_0$, where λ_0 is defined in (8.10).

Theorem 8.1 *Suppose that (A, B, C) is stabilizable and detectable, and Assumption 8.1 holds. Then, the consensus tracking problem for CNS with followers given in (8.1) and a leader given in (8.3) can be solved by protocol (8.5) constructed in Algorithm 8.1 if the dwell time $\tau_m > \tau_{th}$, where $\tau_{th} = (\ln\mu)/c_0$, $c_0 = \min_{i \in \{1,2\}}\{c_i\}$, $c_1 > 0$ and $c_2 > 0$ are given respectively in LMIs (8.11) and (8.12), $\mu = \overline{\phi}/\underline{\phi}$, $\overline{\phi} = \max_{i,j}\phi_i^j$, $\underline{\phi} = \min_{i,j}\phi_i^j$, $i \in \{1, \ldots, N\}$, $j \in \{1, \ldots, \kappa\}$.*

Proof 8.1 *Construct the following MLFs for the error systems (8.9):*

$$V(t) = \tilde{e}^T(t) \left(\Phi^{\sigma(t)} \otimes Q \right) \tilde{e}(t) + \iota e^T(t) \left(\Phi^{\sigma(t)} \otimes P^{-1} \right) e(t),$$

where Q and P are the solutions of LMIs (8.11) and (8.12), respectively, ι is a positive scalar will be given later.

For $t \in [t_k, t_{k+1})$ and an arbitrarily given $k \in \mathbb{N}$, taking the time derivative of $V(t)$ along the trajectories of systems (8.9) yields

$$\dot{V}(t) = \Lambda(t) + \iota \Psi(t),$$

with

$$\Lambda(t) = \tilde{e}^T(t) \left[\Phi^{\sigma(t)} \otimes (QA + A^T Q) - 2\alpha \left(\Phi^{\sigma(t)} \overline{\mathcal{L}}^{\sigma(t)} \otimes QFC \right) \right] \tilde{e}(t)$$
$$+ 2\tilde{e}^T(t) \left(\Phi^{\sigma(t)} \otimes QD \right) (f(x(t); t) - f(\tilde{x}(t); t)), \quad (8.13)$$

$$\Psi(t) = 2\beta e^T(t) \left(\Phi^{\sigma(t)} \overline{\mathcal{L}}^{\sigma(t)} \otimes P^{-1} BK \right) \tilde{e}(t)$$
$$+ e^T(t) \left[\Phi^{\sigma(t)} \otimes (A^T P^{-1} + P^{-1} A) - 2\beta \left(\Phi^{\sigma(t)} \overline{\mathcal{L}}^{\sigma(t)} \otimes P^{-1} BK \right) \right] e(t)$$
$$+ 2e^T(t) \left(\Phi^{\sigma(t)} \otimes P^{-1} D \right) (f(x(t); t) - (\mathbf{1}_N \otimes f(x_0(t), t))). \quad (8.14)$$

Substituting $F = Q^{-1} C^T$ into (8.13) yields

$$\Lambda(t) \leq \tilde{e}^T(t) \left[\Phi^{\sigma(t)} \otimes (QA + A^T Q - \alpha \lambda_0 C^T C) \right] \tilde{e}(t)$$
$$+ 2\tilde{e}^T(t) \left(\Phi^{\sigma(t)} \otimes QD \right) (f(x(t); t) - f(\tilde{x}(t); t)). \quad (8.15)$$

According to the fact $\alpha > \gamma_1/\lambda_0$, it follows from (8.15) that

$$\Lambda(t) \leq \tilde{e}^T(t) \left[\Phi^{\sigma(t)} \otimes (QA + A^T Q - \gamma_1 C^T C + \varrho^2 \rho QDD^T Q^T + \frac{1}{\rho} I_n) \right] \tilde{e}(t) \quad (8.16)$$

for any given $\rho > 0$. By step (1) of Algorithm 8.1, one has that there exists a positive scalar $\tilde{c}_0$ $(0 < \tilde{c}_0 \ll c_1)$ such that

$$\Lambda(t) < -(c_1 + \tilde{c}_0) \tilde{e}^T(t) \left(\Phi^{\sigma(t)} \otimes Q \right) \tilde{e}(t). \quad (8.17)$$

Furthermore, substituting $K = B^T P^{-1}$ into (8.14) gives that

$$\Psi(t) \leq 2\beta e^T(t) \left(\Phi^{\sigma(t)} \overline{\mathcal{L}}^{\sigma(t)} \otimes P^{-1} BB^T P^{-1} \right) \tilde{e}(t) + e^T(t) \left[\Phi^{\sigma(t)} \otimes \Pi \right] e(t), \quad (8.18)$$

where $\Pi = A^T P^{-1} + P^{-1} A - \beta \lambda_0 P^{-1} BB^T P^{-1} + \varrho^2 \tilde{\rho} P^{-1} DD^T P^{-1} + (1/\tilde{\rho}) I_n$ for some given $\tilde{\rho} > 0$. According to the fact $\beta > \gamma_2/\lambda_0$, one has

$$\Pi < \widetilde{\Pi}, \quad (8.19)$$

where $\widetilde{\Pi} = A^T P^{-1} + P^{-1} A - \gamma_2 P^{-1} B B^T P^{-1} + \varrho^2 \tilde{\rho} P^{-1} D D^T P^{-1} + (1/\tilde{\rho}) I_n$. *By step (2) of Algorithm 8.1, one may get that there exists a positive scalar $\widehat{c}_0$ $(0 < \widehat{c}_0 \ll c_2)$ such that $P\widetilde{\Pi}P + (c_2 + \widehat{c}_0)P < 0$. This indicates that*

$$\Pi < -c_2 P^{-1} - \widehat{c}_0 P^{-1}.$$

According to the above analysis, one obtains

$$\Psi(t) \leq 2\beta e^T(t) \left(\Phi^{\sigma(t)} \overline{\mathcal{L}}^{\sigma(t)} \otimes P^{-1} B B^T P^{-1} \right) \widetilde{e}(t)$$
$$- e^T(t) \left[\Phi^{\sigma(t)} \otimes \left(c_2 P^{-1} + \widehat{c}_0 P^{-1} \right) \right] e(t). \tag{8.20}$$

It thus can be derived from (8.17) and (8.20) that

$$\dot{V}(t) < - c_1 \widetilde{e}^T(t) \left(\Phi^{\sigma(t)} \otimes Q \right) \widetilde{e}(t) - \iota c_2 e^T(t) \left(\Phi^{\sigma(t)} \otimes P^{-1} \right) e(t)$$
$$+ \widehat{e}^T(t) \widehat{\Pi}(t) \widehat{e}(t), \tag{8.21}$$

where $\widehat{\Pi}(t) = \begin{bmatrix} -\widetilde{c}_0 \Phi^{\sigma(t)} \otimes Q & * \\ \iota \Omega^{\sigma(t)} & -\iota \widehat{c}_0 \Phi^{\sigma(t)} \otimes P^{-1} \end{bmatrix}$, $\Omega^{\sigma(t)} = \beta \Phi^{\sigma(t)} \overline{\mathcal{L}}^{\sigma(t)} \otimes P^{-1} B B^T P^{-1}$.
Since $-\iota \widehat{c}_0 \Phi^{\sigma(t)} \otimes P^{-1} < 0$, *it can be obtained from Schur complement lemma that* $\widehat{\Pi} < 0$ *if and only if*

$$\widetilde{c}_0 \Phi^{\sigma(t)} \otimes Q > \iota \left(\Omega^{\sigma(t)} \right)^T \left(\widehat{c}_0 \Phi^{\sigma(t)} \otimes P^{-1} \right)^{-1} \Omega^{\sigma(t)}. \tag{8.22}$$

According to the fact that $(\Omega^{\sigma(t)})^T (\widehat{c}_0 \Phi^{\sigma(t)} \otimes P^{-1})^{-1} \Omega^{\sigma(t)}$ *is positive semi-definite and* $\widetilde{c}_0 \Phi^{\sigma(t)} \otimes Q$ *is positive-definite, one gets that all the eigenvalues of* $(\widetilde{c}_0 \Phi^{\sigma(t)} \otimes Q)^{-1} (\Omega^{\sigma(t)})^T (\widehat{c}_0 \Phi^{\sigma(t)} \otimes P^{-1}) \Omega^{\sigma(t)}$ *are real and not less than 0. Based on the above analysis and according to the fact that* $\sigma(t) \in \{1, \ldots, \kappa\}$, *one may choose ι sufficiently small such that*

$$\iota < \min_{j \in \{1, \ldots, \kappa\}} \{\pi_j\}, \tag{8.23}$$

with $\pi_j = \dfrac{1}{\lambda_{\max} \left((\widetilde{c}_0 \Phi^j \otimes Q)^{-1} (\Omega^j)^T (\widehat{c}_0 \Phi^j \otimes P^{-1}) \Omega^j \right)}$ *and* $\Omega^j = \beta \Phi^j \overline{\mathcal{L}}^j \otimes P^{-1} B B^T P^{-1}$, $j = 1, \ldots, \kappa$. *According to (8.23), it can be obtained from Lemma 2.7 that (8.22) holds, i.e.,* $\widehat{\Pi}(t) < 0$ *for* $t \in [t_k, t_{k+1})$. *Based on the above analysis, it can be yielded from (8.21) that*

$$\dot{V}(t) < -c_0 V(t), \quad t \in [t_k, t_{k+1}), \tag{8.24}$$

for an arbitrarily given $k \in \mathbb{N}$, where $c_0 = \min_{i \in \{1,2\}} \{c_i\}$.

Note that the topology of CNS under consideration switches when $t = t_k$, $k \in \mathbb{N}$. It thus follows from (8.24) that

$$V(t) \leq \exp(-c_0(t - t_0)) V(t_0), \ \forall \ t \in [t_0, t_1), \tag{8.25}$$

and

$$V(t_1) \le \mu V(t_1^-) < \exp(-c_0\tau_m + \ln\mu)V(t_0), \tag{8.26}$$

where $\mu = \bar{\phi}/\underline{\phi}$, $\bar{\phi} = \max_{i,j}\phi_i^j$, $\underline{\phi} = \min_{i,j}\phi_i^j$, $i \in \{1,\ldots,N\}$, $j \in \{1,\ldots,\kappa\}$, *and* $V(t_1^-) = \lim_{t \nearrow t_1} V(t)$.

According to the fact that $\tau_m > (\ln\mu)/c_0$, *one gets that*

$$V(t_1) < \exp(-\nu\tau_m)V(t_0),$$

where $\nu = c_0 - (\ln\mu)/\tau_m > 0$. *For an arbitrarily given* $t \ge t_1$, *there exists a positive integer* $z \ge 1$ *such that* $t_z \le t < t_{z+1}$. *Furthermore, for an arbitrarily given* $z \in \mathbb{N}$, *one gets the following inequality by recursion:*

$$V(t_z) < \exp(-\nu\tau_m)V(t_{z-1}) < \exp(-z\nu\tau_m)V(t_0). \tag{8.27}$$

When $t \in (t_z, t_{z+1})$, *based on the above analysis, one gets*

$$
\begin{aligned}
V(t) &< \exp(-c_0(t - t_z))V(t_z) \\
&< \exp(-(c_0(t - t_z) + z\nu\tau_m))V(t_0) \\
&< \exp\left(-\frac{z\tau_m}{(z+1)\tau_M}\nu t\right) V(t_0) \\
&< \exp\left(-\frac{\tau_m\nu}{2\tau_M}t\right) V(t_0),
\end{aligned}
\tag{8.28}
$$

where the last inequality is derived by using the fact $\frac{z}{z+1} > 0.5$ *for all* $z \ge 1$.

When $t = t_z$, *one obtains*

$$V(t) < \exp\left(-\frac{\tau_m\nu}{\tau_M}t\right) V(t_0). \tag{8.29}$$

According to (8.28) and (8.29), one gets that $\|e(t)\| \to 0$ *as* $t \to +\infty$ *and thereby consensus is achieved.*

Remark 8.1 *It can be seen from Algorithm 8.1 and the proof of Theorem 8.1 that the intrinsic nonlinear dynamics of agents are regarded as perturbations in analyzing the consensus tracking behavior of closed-loop CNS (8.6). Furthermore, since the matrix* $\widehat{A}$ *in (8.9) possesses a favorable lower triangular form, in Algorithm 8.1, the observation gain matrix* F *and the feedback gain matrix* K *are designed separately, i.e., the Kalman separation principle for a single linear time-invariant system still holds for the considered CNS.*

Remark 8.2 *From Algorithm 8.1, one knows that the existence of protocol (8.5) relies on the feasibility of LMIs (8.11) and (8.12). The feasibility problem of LMI (8.11) is firstly analyzed as follows. Since the positive scalar* c_1 *is a free parameter in LMI (8.11), it can be obtained from the continuity theory that the feasibility of (8.11)* *is equivalent to the following feasible problem: There exist two positive scalars γ_1, ρ, and a positive definite matrix Q such that*

$$\begin{bmatrix} A^T Q + QA - \gamma_1 C^T C + \frac{1}{\rho} I_n & \varrho\sqrt{\rho}QD \\ \varrho\sqrt{\rho}D^T Q^T & -I_n \end{bmatrix} < 0. \tag{8.30}$$

Noticeably, LMI (8.30) holds if and only if there exist two positive scalars γ_1, ρ, and a positive definite matrix Q such that the following algebraic Riccati inequality (ARI) holds:

$$A^T Q + QA - \gamma_1 C^T C + \frac{1}{\rho} I_n + \varrho^2 \rho QDD^T Q < 0. \tag{8.31}$$

According to Finsler's lemma, one obtains that ARI (8.31) holds for some scalars $\gamma_1 > 0$, $\rho > 0$, and $Q > 0$ if and only if there exist a scalar $\rho > 0$, matrices $\widehat{F} \in \mathbb{R}^{n \times q}$, and $Q > 0$ such that

$$A^T Q + QA - \widehat{F}C - C^T \widehat{F}^T + \frac{1}{\rho} I_n + \varrho^2 \rho QDD^T Q < 0. \tag{8.32}$$

Without loss of generality, let $\widehat{F} = QL$. Then, one gets that LMI (8.32) is feasible if and only if there exist scalar $\rho > 0$, $L \in \mathbb{R}^{n \times q}$, $Q > 0$, such that

$$(A - LC)^T Q + Q(A - LC) + \frac{1}{\rho} I_n + \varrho^2 \rho QDD^T Q < 0. \tag{8.33}$$

According to the bounded real lemma, one obtains (8.33) holds if and only if there exists a matrix $L \in \mathbb{R}^{n \times q}$ such that

$$\left\| \gamma \left[sI - (A - LC) \right]^{-1} D \right\|_\infty < 1,$$

i.e.,

$$\left\| \left[sI - (A - LC) \right]^{-1} D \right\|_\infty < \frac{1}{\gamma}, \tag{8.34}$$

where $s \in \mathbb{C}$ and $\mathbb{C}$ represents the set of complex numbers. Thus, LMI (8.11) is feasible if and only if there exist a matrix $L \in \mathbb{R}^{n \times q}$ such that (8.34) holds. Also, it can be observed from the above analysis that (A, C) is detectable is a necessary condition for the feasibility of LMI (8.11). Similarly, one obtains that LMI (8.12) is feasible if and only if there exists a matrix $H \in \mathbb{R}^{m \times n}$ such that

$$\left\| D^T \left[sI - (A - BH)^T \right]^{-1} \right\|_\infty < \frac{1}{\gamma}. \tag{8.35}$$

Noticeably, one necessary condition for the feasibility of LMI (8.12) is that the matrix pair (A, B) is stabilizable.

Theorem 8.1 indicates that consensus tracking in the closed-loop CNS (8.6) under observer-type protocol constructed from Algorithm 8.1 can be ensured if the time interval between consecutive topology switchings is sufficiently large. However, it will be seen in the following theorem that this condition can be further relaxed by using the idea of ADT. Then, one may get the following theorem where some ADT based consensus tracking criteria are provided.

Theorem 8.2 *Suppose that* (A, B, C) *is stabilizable and detectable, and Assumption 8.1 holds. Then, the consensus tracking problem for CNS with followers given in (8.1) and a leader given in (8.3) can be solved by protocol (8.5) constructed by Algorithm 8.1 if the ADT* $\tau_a > \tau_{th}$, *where* τ_{th} *is defined in Theorem 8.1.*

Proof 8.2 *For any given* $t > 0$, *suppose that* t_i, $i = 1, \ldots, t_{N_\sigma(0,t)}$, *be the switching points of* $\sigma(t)$ *over the time interval* $(0, t)$, *where* $t_0 < t_1 < \ldots < t_{N_\sigma(0,t)}$. *Construct the same MLF* $V(t)$ *as this in the proof of Theorem 8.1. By noticing that* $V(t_i) \leq \mu V(t_i^-)$ *for each* $i \in \{1, \ldots, N_\sigma(0, t)\}$, *one obtains by induction that*

$$V(t) < \exp\left(-c_0(t - t_{N_\sigma(0,t)})\right) V(t_{N_\sigma(0,t)})$$
$$\leq \mu \exp\left(-c_0(t - t_{N_\sigma(0,t)})\right) V(t_{N_\sigma(0,t)}^-).$$

By induction, one obtains

$$V(t) < \mu^{N_\sigma(0,t)} \exp\left(-c_0(t - t_0)\right) V(t_0),$$

i.e.,

$$V(t) < \exp\left(-c_0(t - t_0) + N_\sigma(0, t)\ln\mu\right) V(t_0), \tag{8.36}$$

where μ *is defined in Theorem 8.1. According to the fact* $N_\sigma(0, t) \leq N_0 + \frac{t-t_0}{\tau_a}$, *it can be obtained from (8.36) that*

$$V(t) < \mu^{N_0} \exp\left(-\left(c_0 - \frac{\ln\mu}{\tau_a}\right)(t - t_0)\right) V(t_0). \tag{8.37}$$

According to the condition that $\tau_a > (\ln\mu)/c_0$, *it can be concluded that the consensus tracking for CNS with followers given in (8.1)and a leader given in (8.3) is indeed solved by protocol (8.5) constructed by Algorithm 8.1.*

Remark 8.3 *By constructing some MLFs, it has been shown in Theorem 8.1 that consensus tracking in the closed-loop CNS (8.6) can be achieved if the ADT is larger than a derived threshold. It is worth noting that consensus tracking in such a CNS can be ensured under arbitrarily given ADT* $\tau_a > 0$ *and chatter bound* N_0 *if there is a common Lyapunov function for error systems (8.9). However, it is still a challenging issue to find a common Lyapunov function for switched systems (8.9). Note also that, for switched systems, common Lyapunov function only exists under few situations.*

8.4 CONSENSUS TRACKING PROTOCOL DESIGN: INDEPENDENT TOPOLOGY CASE

It has been shown in the last section that, consensus tracking problem of CNS with followers given in (8.1) and a leader given in (8.3) can be solved by appropriately designing protocol (8.5). Noticeably, in protocol (8.5), the communication topology for the observation errors $\delta_i(t)$, $i = 0, 1, \ldots, N$, is previously set the same as that for the states of observers $\tilde{x}_i(t)$, $i = 0, 1, \ldots, N$. However, in some real situations, it is favorable to allow them to be independent with each other. Motivated by this observation, the following consensus protocol is proposed:

$$
\dot{\widetilde{x}}_i(t) = A\widetilde{x}_i(t) + Bu_i(t) + \alpha \sum_{j=0}^{N} \breve{a}_{ij}^{\breve{\sigma}(t)} F(\delta_j(t) - \delta_i(t)) + Df(\widetilde{x}_i(t), t),
$$

$$
u_i(t) = \beta K \sum_{j=0}^{N} \bar{a}_{ij}^{\bar{\sigma}(t)} (\widetilde{x}_j(t) - \widetilde{x}_i(t)), \ i = 1, \ldots, N, \tag{8.38}
$$

where $\breve{\mathcal{A}}^{\breve{\sigma}(t)} = \left[\breve{a}_{ij}^{\breve{\sigma}(t)}\right]_{(N+1)\times(N+1)}$ is the adjacency matrix of the graph $\breve{\mathcal{G}}^{\breve{\sigma}(t)}$ describing the communication topology for observations errors $\delta_i(t)$ with $i = 0, 1, \ldots, N$, the piecewise constant function $\breve{\sigma}(t) : [0, +\infty) \mapsto \{1, \ldots, \breve{\kappa}\}$ is the switching signal, $\bar{\mathcal{A}}^{\bar{\sigma}(t)} = \left[\bar{a}_{ij}^{\bar{\sigma}(t)}\right]_{(N+1)\times(N+1)}$ is the adjacency matrix of the graph $\bar{\mathcal{G}}^{\bar{\sigma}(t)}$ describing the communication topology for the observers' states $\widetilde{x}_i(t)$ with $i = 0, 1, \ldots, N$, the piecewise constant function $\bar{\sigma}(t) : [0, +\infty) \mapsto \{1, \ldots, \bar{\kappa}\}$ is the switching signal, and the other parameters in (8.38) are defined as the same as those in (8.5). It is worth noting that the switching signals $\breve{\sigma}(t)$ and $\bar{\sigma}(t)$ may be totally asynchronous. Similar to the analysis in the last subsection, it is assumed in (8.38) that $\widetilde{x}_0(t) = x_0(t)$, i.e., the leader does not need to observe its state. The switching sequences of $\breve{\mathcal{G}}^{\breve{\sigma}(t)}$ and $\bar{\mathcal{G}}^{\bar{\sigma}(t)}$ are denoted, respectively, by $\breve{t}_1, \breve{t}_2, \ldots,$ and $\bar{t}_1, \bar{t}_2, \ldots$. Without loss of generality, it is assumed that $\breve{t}_0 = \bar{t}_0 = 0$. For expressional convenience, let $\widehat{t}_0 = 0$ and $\widehat{t}_i = \min_{j,k\in\mathbb{N}}\left\{\breve{t}_j, \bar{t}_k : \breve{t}_j > \widehat{t}_{i-1}, \bar{t}_k > \widehat{t}_{i-1}\right\}$ for $i = 1, 2, \ldots$. To facilitate analysis, it is further assumed that $0 < \widehat{\tau}_m \leq \widehat{t}_{k+1} - \widehat{t}_k \leq \widehat{\tau}_M < +\infty, \ k \in \mathbb{N}$. Let $\breve{\mathcal{L}}^{\breve{\sigma}(t)}$ and $\bar{\mathcal{L}}^{\bar{\sigma}(t)}$ be respectively the Laplacian matrices of $\breve{\mathcal{G}}^{\breve{\sigma}(t)}$ and $\bar{\mathcal{G}}^{\bar{\sigma}(t)}$. Since the leader has no neighbors, one gets that $\breve{\mathcal{L}}^{\breve{\sigma}(t)}$ and $\bar{\mathcal{L}}^{\bar{\sigma}(t)}$ can be partitioned respectively into

$$
\breve{\mathcal{L}}^{\breve{\sigma}(t)} = \begin{bmatrix} 0 & \mathbf{0}_N^T \\ \breve{\mathbf{P}}^{\breve{\sigma}(t)} & \breve{\mathcal{L}}^{\breve{\sigma}(t)} \end{bmatrix}, \quad \bar{\mathcal{L}}^{\bar{\sigma}(t)} = \begin{bmatrix} 0 & \mathbf{0}_N^T \\ \bar{\mathbf{P}}^{\bar{\sigma}(t)} & \bar{\mathcal{L}}^{\bar{\sigma}(t)} \end{bmatrix}, \tag{8.39}
$$

where $\breve{\mathbf{P}}^{\breve{\sigma}(t)} = -[\breve{a}_{10}^{\breve{\sigma}(t)}, \ldots, \breve{a}_{N0}^{\breve{\sigma}(t)}]^T \in \mathbb{R}^N$, $\bar{\mathbf{P}}^{\bar{\sigma}(t)} = -[\bar{a}_{10}^{\bar{\sigma}(t)}, \ldots, \bar{a}_{N0}^{\bar{\sigma}(t)}]^T \in \mathbb{R}^N$, $\breve{\mathcal{A}}^{\breve{\sigma}(t)} = \left[\breve{a}_{ij}^{\breve{\sigma}(t)}\right]_{(N+1)\times(N+1)}$ and $\bar{\mathcal{A}}^{\bar{\sigma}(t)} = \left[\bar{a}_{ij}^{\bar{\sigma}(t)}\right]_{(N+1)\times(N+1)}$ are, respectively, the adjacency matrices of $\breve{\mathcal{G}}^{\breve{\sigma}(t)}$ and $\bar{\mathcal{G}}^{\bar{\sigma}(t)}$,

$$
\breve{\mathcal{L}}^{\breve{\sigma}(t)} = \begin{bmatrix} \sum_{j\in\breve{\mathcal{N}}_1} \breve{a}_{1j}^{\breve{\sigma}(t)} & -\breve{a}_{12}^{\breve{\sigma}(t)} & \cdots & -\breve{a}_{1N}^{\breve{\sigma}(t)} \\ -\breve{a}_{21}^{\breve{\sigma}(t)} & \sum_{j\in\breve{\mathcal{N}}_2} \breve{a}_{2j}^{\breve{\sigma}(t)} & \cdots & -\breve{a}_{2N}^{\breve{\sigma}(t)} \\ \vdots & \vdots & \ddots & \vdots \\ -\breve{a}_{N1}^{\breve{\sigma}(t)} & -\breve{a}_{N2}^{\breve{\sigma}(t)} & \cdots & \sum_{j\in\breve{\mathcal{N}}_N} \breve{a}_{Nj}^{\breve{\sigma}(t)} \end{bmatrix},
$$

$$
\bar{\mathcal{L}}^{\bar{\sigma}(t)} = \begin{bmatrix} \sum_{j\in\bar{\mathcal{N}}_1} \bar{a}_{1j}^{\bar{\sigma}(t)} & -\bar{a}_{12}^{\bar{\sigma}(t)} & \cdots & -\bar{a}_{1N}^{\bar{\sigma}(t)} \\ -\bar{a}_{21}^{\bar{\sigma}(t)} & \sum_{j\in\bar{\mathcal{N}}_2} \bar{a}_{2j}^{\bar{\sigma}(t)} & \cdots & -\bar{a}_{2N}^{\bar{\sigma}(t)} \\ \vdots & \vdots & \ddots & \vdots \\ -\bar{a}_{N1}^{\bar{\sigma}(t)} & -\bar{a}_{N2}^{\bar{\sigma}(t)} & \cdots & \sum_{j\in\bar{\mathcal{N}}_N} \bar{a}_{Nj}^{\bar{\sigma}(t)} \end{bmatrix}.
$$

Under this scenario, the error system (8.9) can be rewritten as

$$\dot{\hat{e}}(t) = \widetilde{\widetilde{A}}\hat{e}(t) + (I_{2N} \otimes D)\widehat{f}(\hat{e}(t); t), \tag{8.40}$$

where

$$\widetilde{\widetilde{A}} = \begin{bmatrix} I_N \otimes A - \alpha \left(\widecheck{\mathcal{L}}^{\breve{\sigma}(t)} \otimes FC \right) & 0 \\ \beta \left(\bar{\mathcal{L}}^{\bar{\sigma}(t)} \otimes BK \right) & I_N \otimes A - \beta \left(\bar{\mathcal{L}}^{\bar{\sigma}(t)} \otimes BK \right) \end{bmatrix}$$

and the other symbols are defined the same as those in (8.9).

To derive the main results of this section, the following assumption is needed.

Assumption 8.2 *For each $i \in \{1, \ldots, \widecheck{\kappa}\}$ and $j \in \{1, \ldots, \bar{\kappa}\}$, both $\widecheck{\mathcal{G}}^i$ and $\bar{\mathcal{G}}^j$ contain a directed spanning tree with agent 0 (i.e., the leader) as the root.*

Under Assumption 8.2, we can obtain that there exist positive definite diagonal matrices $\widecheck{\Phi}^i = \text{diag}\left\{ \widecheck{\phi}_1^i, \ldots, \widecheck{\phi}_N^i \right\}$ and $\bar{\Phi}^j = \text{diag}\left\{ \bar{\phi}_1^j, \ldots, \bar{\phi}_N^j \right\}$ such that

$$\widecheck{\Phi}^i \widecheck{\mathcal{L}}^i + \left(\widecheck{\mathcal{L}}^i \right)^T \widecheck{\Phi}^i > 0,$$

$$\bar{\Phi}^j \bar{\mathcal{L}}^j + \left(\bar{\mathcal{L}}^j \right)^T \bar{\Phi}^j > 0,$$

where $\widecheck{\phi}^i = \left[\widecheck{\phi}_1^i, \ldots, \widecheck{\phi}_N^i \right]^T \in \mathbb{R}^N$ and $\bar{\phi}^j = \left[\bar{\phi}_1^j, \ldots, \bar{\phi}_N^j \right]^T \in \mathbb{R}^N$ satisfy $\left(\widecheck{\mathcal{L}}^i \right)^T \widecheck{\phi}^i = \mathbf{1}_N$ and $\left(\bar{\mathcal{L}}^j \right)^T \bar{\phi}^j = \mathbf{1}_N$, respectively, $i = 1, \ldots, \widecheck{\kappa}$, $j = 1, \ldots, \bar{\kappa}$.

For notational brevity, let

$$\widecheck{\chi} = \min_{i \in \{1, \ldots, \widecheck{\kappa}\}} \lambda_{\min} \left(\widecheck{\mathcal{L}}^i + \left(\widecheck{\Phi}^i \right)^{-1} \left(\widecheck{\mathcal{L}}^i \right)^T \widecheck{\Phi}^i \right),$$

$$\bar{\chi} = \min_{j \in \{1, \ldots, \bar{\kappa}\}} \lambda_{\min} \left(\bar{\mathcal{L}}^j + \left(\bar{\Phi}^j \right)^{-1} \left(\bar{\mathcal{L}}^j \right)^T \bar{\Phi}^j \right). \tag{8.41}$$

Before moving on, the following constructive algorithm is presented for selecting the parameters in protocol (8.38) to achieve consensus tracking.

Algorithm 8.2 *Suppose that (A, B, C) is stabilizable and detectable, and Assumption 8.2 holds. The consensus tracking protocol (8.38) can be designed as follows:*

(1) Choose appropriate scalars $\widecheck{\gamma}_1 > 0$, $\widecheck{\varrho} > 0$, and $\widecheck{c}_1 > 0$, solve the LMI

$$\begin{bmatrix} A^T \widecheck{Q} + \widecheck{Q}A - \widecheck{\gamma}_1 C^T C + (1/\widecheck{\varrho})I_n + \widecheck{c}_1 \widecheck{Q} & \varrho\sqrt{\widecheck{\varrho}}\widecheck{Q}D \\ \varrho\sqrt{\widecheck{\varrho}}D^T \widecheck{Q}^T & -I_n \end{bmatrix} < 0 \tag{8.42}$$

to get a positive definite matrix $\widecheck{Q}$, where ϱ is defined in (8.1). Then, set $F = \widecheck{Q}^{-1}C^T$.

(2) Choose appropriate scalars $\check{\gamma}_2 > 0$, $\check{\rho} > 0$, and $\check{c}_2 > 0$, solve the LMI

$$\begin{bmatrix} A\check{P} + \check{P}A^T - \check{\gamma}_2 BB^T + \varrho^2\check{\rho}DD^T + \check{c}_2 P & (1/\sqrt{\check{\rho}})\check{P} \\ (1/\sqrt{\check{\rho}})\check{P} & -I_n \end{bmatrix} < 0 \qquad (8.43)$$

to get a positive definite matrix $\check{P}$. Then, take $K = B^T \check{P}^{-1}$.

(3) Choose the coupling strengths $\alpha > \check{\gamma}_1/\check{\chi}$ and $\beta > \check{\gamma}_2/\bar{\chi}$, where $\check{\chi}$ and $\bar{\chi}$ are defined in (8.41).

Let

$$\check{\mu}_0 = \check{\phi}_{\max}/\check{\phi}_{\min},$$
$$\bar{\mu}_0 = \bar{\phi}_{\max}/\bar{\phi}_{\min}, \qquad (8.44)$$

where $\check{\phi}_{\max} = \max_{k,i}\left\{\check{\phi}_k^i\right\}$, $\check{\phi}_{\min} = \min_{k,i}\left\{\check{\phi}_k^i\right\}$, $\bar{\phi}_{\max} = \max_{k,j}\left\{\bar{\phi}_k^j\right\}$, $\bar{\phi}_{\min} = \min_{k,j}\left\{\bar{\phi}_k^j\right\}$, $i \in \{1,\ldots,\check{\kappa}\}$, $j \in \{1,\ldots,\bar{\kappa}\}$, and $k \in \{1,\ldots,N\}$. Then, one may get the following theorem which summarizes the main results of this section.

Theorem 8.3 *Suppose that (A, B, C) is stabilizable and detectable, and Assumption 8.2 holds. Then, the consensus tracking problem for CNS with followers given in (8.1) and a leader given in (8.3) can be solved by protocol (8.38) constructed by Algorithm 8.2 if the dwell time $\hat{\tau}_m > \hat{\tau}_{th}$, where $\hat{\tau}_{th} = (\ln\hat{\mu}_0)/\check{c}_0$, $\check{c}_0 = \min_{i\in\{1,2\}}\{\check{c}_i\}$, $\check{c}_1 > 0$ and $\check{c}_2 > 0$ are given respectively in LMIs (8.42) and (8.43), $\hat{\mu}_0 = \max\{\check{\mu}_0, \bar{\mu}_0\}$, $\check{\mu}_0$ and $\bar{\mu}_0$ are defined in (8.44).*

Proof 8.3 *Constructing the following MLF for the error system (8.40):*

$$\check{V}(t) = \hat{e}^T(t)\left(\check{\Phi}^{\check{\sigma}(t)} \otimes \check{Q}\right)\hat{e}(t) + \check{\iota}e^T(t)\left(\bar{\Phi}^{\bar{\sigma}(t)} \otimes \check{P}^{-1}\right)e(t),$$

where $\hat{e}(t) = [\hat{e}^T(t), e^T(t)]^T \in \mathbb{R}^{2Nn}$, $\hat{e}(t), e(t) \in \mathbb{R}^{Nn}$, $\check{Q}$ and $\check{P}$ are the solutions of LMIs (8.42) and (8.43), respectively, $\check{\iota}$ is a positive scalar will be determined later.
For $t \in [\hat{t}_k, \hat{t}_{k+1})$ and an arbitrarily given $k \in \mathbb{N}$, taking the time derivative of $\check{V}$ along the trajectories of systems (8.38) yields

$$\dot{\check{V}}(t) = \check{\Phi}(t) + \check{\iota}\check{\Psi}(t), \qquad (8.45)$$

with

$$\check{\Phi}(t) = \hat{e}^T(t)\left[\check{\Phi}^{\check{\sigma}(t)} \otimes (\check{Q}A + A^T\check{Q}) - 2\alpha\left(\check{\Phi}^{\check{\sigma}(t)}\check{\mathcal{L}}^{\check{\sigma}(t)} \otimes \check{Q}FC\right)\right]\hat{e}(t)$$
$$+ 2\hat{e}^T(t)\left(\check{\Phi}^{\check{\sigma}(t)} \otimes \check{Q}D\right)(f(x(t);t) - f(\tilde{x}(t);t)), \qquad (8.46)$$

$$\check{\Psi}(t) = 2\beta e^T(t)\left(\bar{\Phi}^{\bar{\sigma}(t)}\bar{\mathcal{L}}^{\bar{\sigma}(t)} \otimes \check{P}^{-1}BK\right)\hat{e}(t)$$

$$+ e^T(t)\Big[\bar{\Phi}^{\bar{\sigma}(t)} \otimes (A^T \breve{P}^{-1} + \breve{P}^{-1} A) - 2\beta \bar{\Phi}^{\bar{\sigma}(t)} \bar{\bar{\mathcal{L}}}^{\bar{\sigma}(t)} \otimes \breve{P}^{-1} B K\Big] e(t)$$
$$+ 2e^T(t)\left(\bar{\Phi}^{\bar{\sigma}(t)} \otimes \breve{P}^{-1} D\right)(f(x(t);t) - (\mathbf{1}_N \otimes f(x_0(t),t))). \qquad (8.47)$$

Substituting $F = \breve{Q}^{-1} C^T$ *into (8.46) yields*

$$\breve{\Phi}(t) \le \tilde{e}^T(t)\Big[\breve{\Phi}^{\breve{\sigma}(t)} \otimes (\breve{Q}A + A^T \breve{Q} - \alpha \breve{\chi} C^T C)\Big]\tilde{e}(t)$$
$$+ 2\tilde{e}^T(t)\left(\breve{\Phi}^{\breve{\sigma}(t)} \otimes \breve{Q}D\right)(f(x(t);t) - f(\tilde{x}(t);t)), \qquad (8.48)$$

where $\breve{\chi}$ *is defined in (8.41). According to the fact* $\alpha > \breve{\gamma}_1/\breve{\chi}$, *it follows from (8.48) that*

$$\breve{\Phi}(t) \le \tilde{e}^T(t)\Big[\breve{\Phi}^{\breve{\sigma}(t)} \otimes (\breve{Q}A + A^T \breve{Q} - \breve{\gamma}_1 C^T C + \varrho^2 \breve{\varrho} \breve{Q} D D^T \breve{Q}^T + \frac{1}{\varrho} I_n)\Big]\tilde{e}(t) \qquad (8.49)$$

for any given $\breve{\varrho} > 0$. *By step (1) of Algorithm 8.2, one has that there exists a positive scalar* $\breve{\tilde{c}}_0$ $(0 < \breve{\tilde{c}}_0 \ll \breve{c}_1)$ *such that*

$$\breve{\Phi}(t) < -(\breve{c}_1 + \breve{\tilde{c}}_0)\tilde{e}^T(t)\left(\breve{\Phi}^{\breve{\sigma}(t)} \otimes \breve{Q}\right)\tilde{e}(t). \qquad (8.50)$$

Furthermore, substituting $K = B^T \breve{P}^{-1}$ *into (8.47) yields*

$$\breve{\Psi}(t) \le 2\beta e^T(t)\left(\bar{\Phi}^{\bar{\sigma}(t)} \bar{\bar{\mathcal{L}}}^{\bar{\sigma}(t)} \otimes \breve{P}^{-1} B B^T \breve{P}^{-1}\right)\tilde{e}(t) + e^T(t)\left(\bar{\Phi}^{\bar{\sigma}(t)} \otimes \breve{\Pi}\right)e(t), \qquad (8.51)$$

where $\breve{\Pi} = A^T \breve{P}^{-1} + \breve{P}^{-1} A - \beta \breve{\chi} \breve{P}^{-1} B B^T \breve{P}^{-1} + \varrho^2 \breve{\rho} \breve{P}^{-1} D D^T \breve{P}^{-1} + (1/\breve{\rho}) I_n$ *for some given* $\breve{\rho} > 0$. *According to the fact* $\beta > \breve{\gamma}_2/\breve{\chi}$, *one has*

$$\breve{\Pi} < \breve{\tilde{\Pi}}, \qquad (8.52)$$

where $\breve{\tilde{\Pi}} = A^T \breve{P}^{-1} + \breve{P}^{-1} A - \breve{\gamma}_2 \breve{P}^{-1} B B^T \breve{P}^{-1} + \varrho^2 \breve{\rho} \breve{P}^{-1} D D^T \breve{P}^{-1} + (1/\breve{\rho}) I_n$. *By step (2) of Algorithm 8.2, one may get that there exists a positive scalar* $\breve{\tilde{c}}_0$ $(0 < \breve{\tilde{c}}_0 \ll \breve{c}_2)$ *such that*

$$\breve{P} \breve{\tilde{\Pi}} \breve{P} + (\breve{c}_2 + \breve{\tilde{c}}_0)\breve{P} < 0,$$

which indicates that

$$\breve{\Pi} < -\breve{c}_2 \breve{P}^{-1} - \breve{\tilde{c}}_0 \breve{P}^{-1}. \qquad (8.53)$$

According to the above analysis, one obtains

$$\breve{\Psi}(t) \le 2\beta e^T(t)\left(\bar{\Phi}^{\bar{\sigma}(t)} \bar{\bar{\mathcal{L}}}^{\bar{\sigma}(t)} \otimes \breve{P}^{-1} B B^T \breve{P}^{-1}\right)\tilde{e}(t)$$
$$- e^T(t)\Big[\bar{\Phi}^{\bar{\sigma}(t)} \otimes \left(\breve{c}_2 \breve{P}^{-1} + \breve{\tilde{c}}_0 \breve{P}^{-1}\right)\Big]e(t). \qquad (8.54)$$

It thus can be derived from (8.50) and (8.54) that

$$\dot{V}(t) < -\check{c}_1 \tilde{e}^T(t)\left(\check{\Phi}^{\check{\sigma}(t)} \otimes \check{Q}\right)\tilde{e}(t) - \check{\iota}\check{c}_2 e^T(t)\left(\bar{\Phi}^{\bar{\sigma}(t)} \otimes \check{P}^{-1}\right)e(t)$$
$$+ \hat{e}^T(t)\check{\hat{\Pi}}(t)\hat{e}(t), \tag{8.55}$$

where

$$\check{\hat{\Pi}}(t) = \begin{bmatrix} -\check{\tilde{c}}_0 \check{\Phi}^{\check{\sigma}(t)} \otimes \check{Q} & * \\ \check{\iota}\bar{\Omega}^{\bar{\sigma}(t)} & -\check{\iota}\check{\tilde{c}}_0 \bar{\Phi}^{\bar{\sigma}(t)} \otimes \check{P}^{-1} \end{bmatrix},$$

of which $\bar{\Omega}^{\bar{\sigma}(t)} = \beta\bar{\Phi}^{\bar{\sigma}(t)}\bar{\mathcal{L}}^{\bar{\sigma}(t)} \otimes \check{P}^{-1}BB^T\check{P}^{-1}$. Since $-\check{\iota}\check{\tilde{c}}_0 \bar{\Phi}^{\bar{\sigma}(t)} \otimes \check{P}^{-1} < 0$, it can be obtained from Schur complement lemma that $\check{\hat{\Pi}} < 0$ if and only if

$$\check{\tilde{c}}_0 \check{\Phi}^{\check{\sigma}(t)} \otimes \check{Q} > \check{\iota}\left(\bar{\Omega}^{\bar{\sigma}(t)}\right)^T \left(\check{\tilde{c}}_0 \bar{\Phi}^{\bar{\sigma}(t)} \otimes \check{P}^{-1}\right)^{-1} \bar{\Omega}^{\bar{\sigma}(t)}. \tag{8.56}$$

Noticeably, $\left(\bar{\Omega}^{\bar{\sigma}(t)}\right)^T \left(\check{\tilde{c}}_0 \bar{\Phi}^{\bar{\sigma}(t)} \otimes \check{P}^{-1}\right)^{-1} \bar{\Omega}^{\bar{\sigma}(t)}$ is semi-positive definite and $\check{\tilde{c}}_0 \check{\Phi}^{\check{\sigma}(t)} \otimes \check{Q}$ is positive definite, one thus gets that all the eigenvalues of $\left(\check{\tilde{c}}_0 \check{\Phi}^{\check{\sigma}(t)} \otimes \check{Q}\right)^{-1} \left(\bar{\Omega}^{\bar{\sigma}(t)}\right)^T$ $\cdot \left(\check{\tilde{c}}_0 \bar{\Phi}^{\bar{\sigma}(t)} \otimes \check{P}^{-1}\right) \bar{\Omega}^{\bar{\sigma}(t)}$ are real and not less than 0. According to the fact $\check{\sigma}(t) \in \{1, \ldots, \check{\kappa}\}$ and $\bar{\sigma}(t) \in \{1, \ldots, \bar{\kappa}\}$, where $\check{\kappa}$ and $\bar{\kappa}$ are two given positive natural numbers, one may choose $\check{\iota}$ sufficiently small such that

$$\check{\iota} < \min_{i \in \{1, \ldots, \check{\kappa}\}, j \in \{1, \ldots, \bar{\kappa}\}}\{\check{\pi}_{i,j}\}, \tag{8.57}$$

with $\check{\pi}_{i,j} = \dfrac{1}{\lambda_{\max}\left(\left(\check{\tilde{c}}_0 \check{\Phi}^i \otimes \check{Q}\right)^{-1}(\bar{\Omega}^j)^T \left(\check{\tilde{c}}_0 \bar{\Phi}^j \otimes \check{P}^{-1}\right)\bar{\Omega}^j\right)}$ and $\bar{\Omega}^j = \beta\bar{\Phi}^j\bar{\mathcal{L}}^j \otimes \check{P}^{-1}BB^T\check{P}^{-1}$. According to (8.57), it can be obtained from the Schur complement lemma that (8.56) holds, which indicates that $\check{\hat{\Pi}}(t) < 0$ for $t \in \left[\hat{t}_k, \hat{t}_{k+1}\right)$.

Then, this theorem can be proved by following the steps in the proof of Theorem 8.1.

Remark 8.4 The solvability analysis of LMIs (8.42) and (8.43) is similar to that of LMIs (8.11) and (8.12), respectively. Then, by following the steps in Remark 8.2, the solvability conditions of LMIs (8.42) and (8.43) can be obtained without any difficulty. It is also worth noting that the results given in Theorem 8.3 can be extended to consensus tracking in CNS with only ADT constraints, which are omitted for brevity.

Remark 8.5 The present protocols (8.5) and (8.38) are partly motivated by the observer-type protocols for linear CNSs in the work of [76]. However, the observer design here is indeed different from that in [76]. It is also worth noting that the observers' states in the protocols given in [76] can not converge to the states of agents, while the observers' states in the present protocols will converge to those of the agents asymptotically (see numerical simulations in Section 8.5 for more details).

Remark 8.6 *Distributed consensus tracking problem for CNSs with Lipschitz nonlinear dynamics and directed switching topologies is solved in this chapter by designing some observer-type protocols. However, it is still an open issue of how to ensure distributed consensus tracking in CNSs with directed switching topologies and unknown continuous time or discrete time nonlinear dynamics which do not satisfy the Lipschitz condition. The design methods provided in [91, 174] might be helpful for studying this issue.*

8.5 NUMERICAL SIMULATIONS

Numerical simulations are provided in this section to verify the effectiveness of the theoretical results obtained in this chapter.

The inherent dynamics of each agent in simulations are assumed to be a linear satellite launch vehicle (SLV) model subject to nonlinear disturbances. According to [28], the dynamics of the i-th SLV for a typical experimental condition can be described by (8.1), with $x_i(t) = [x_{i1}(t), x_{i2}(t), x_{i3}(t), x_{i4}(t), x_{i5}(t), x_{i6}(t)]^T \in \mathbb{R}^6$,

$$A = \begin{bmatrix} 0 & 1 & 0 & 0 & 0 & 0 \\ 0.7066 & 0 & 1.87 \times 10^{-5} & 0 & 0 & 0 \\ 0 & 0 & 0 & 1 & 0 & 0 \\ 2.71 \times 10^{-5} & 0 & 0.4379 & 0 & 0 & 0 \\ 0 & 0 & 0 & 0 & 0 & 1 \\ 5.71 \times 10^{-4} & 0 & 5.468 \times 10^{-4} & 0 & 0 & 0 \end{bmatrix},$$

$$B = \begin{bmatrix} 0 & 0 & 0 & 0 & 0 \\ 1.64 & 1.64 & 1.64 & 1.64 & 0.5567 \\ 0 & 0 & 0 & 0 & 0 \\ 1.4755 & 1.4755 & 1.4755 & 1.4755 & 0.2136 \times 10^{-4} \\ 0 & 0 & 0 & 0 & 0 \\ 4.0425 & 4.0425 & 4.0425 & 4.0425 & 4.5012 \times 10^{-4} \end{bmatrix}$$

$$\begin{bmatrix} 0 & 0 & 0 \\ 1.6592 \times 10^{-4} & 0.5567 & 1.6592 \times 10^{-4} \\ 0 & 0 & 0 \\ 0.5009 & 0.2136 \times 10^{-4} & 0.5009 \\ 0 & 0 & 0 \\ 1.3721 & 4.5012 \times 10^{-4} & 1.3721 \end{bmatrix},$$

$$C = \begin{bmatrix} 1 & 0 & 0 & 0 & 0 & 0 \\ 0 & 0 & 1 & 0 & 0 & 0 \\ 0 & 0 & 0 & 0 & 1 & 0 \\ 0 & 0 & 0 & 0 & 0 & 1 \end{bmatrix},$$

$$D = \begin{bmatrix} 0 & 0 & 0 & 0 \\ 7.82\times10^{-4} & -6.84\times10^{-6} & -4.34\times10^{-4} & -8.43\times10^{-8} \\ 0 & 0 & 0 & 0 \\ 3\times10^{-8} & 7.61\times10^{-3} & -1.665\times10^{-8} & 9.38\times10^{-5} \\ 0 & 0 & 0 & 0 \\ -6.32\times10^{-7} & 9.51\times10^{-6} & 3.509\times10^{-7} & 1.171\times10^{-7} \end{bmatrix},$$

and $f(x_i(t), t) = [\sin(x_{i1}(t)) + \sin(2t), x_{i4}(t), \cos(x_{i3}(t)), \cos(t)]^T \in \mathbb{R}^4$, of which $x_{i1}(t)$ is the angle of pitch, $x_{i2}(t)$ is the pitch rate, $x_{i3}(t)$ is the angle of yaw, $x_{i4}(t)$ is the yaw rate, $x_{i5}(t)$ is the angle of roll, and $x_{i6}(t)$ is the roll rate. Furthermore, $f(x_i(t), t)$ is the external disturbance that satisfies the Lipschitz condition with Lipschitz constant $\varrho = 1$. It can be observed that the evolution of each individual SLV is assumed to be affected by four external disturbances, i.e., the percentage differential thrust between main engine thrusters $\sin(x_{i1}(t)) + \sin(2t)$, percentage differential thrust between strap-ons $x_{i4}(t)$, center of gravity (CG) offset along yaw axis $\cos(x_{i3}(t))$, and CG offset along pitch axis $\cos(t)$. According to the classical PBH controllability criterion, one has that the triple (A, B, C) is completely controllable and observable, and thus is stabilizable and detectable.

In this example, it is assumed that the communication topology switches between $\mathcal{G}^1$ and $\mathcal{G}^2$ which are shown in Figure 8.1 every 0.7 s, where the positive numbers on the directed edges of $\mathcal{G}^1$ and $\mathcal{G}^2$ indicate the corresponding communication weights among neighboring agents. Direct calculation gives that $\lambda_0 = 1.0561$ and $\mu = 2$. To achieve consensus tracking, consensus protocol (8.5) will be designed by following the steps given in Algorithm 8.1. Set $\gamma_1 = 10$, $\rho = 5$, and $c_1 = 1$, by step (1) of Algorithm 8.1, one gets

$$F = \begin{bmatrix} 1.0058 & 0.0000 & 0.0001 & 0.0001 \\ 1.0609 & 0.0000 & 0.0001 & 0.0002 \\ 0.0000 & 0.7433 & 0.0001 & 0.0001 \\ 0.0000 & 0.7116 & 0.0000 & 0.0002 \\ 0.0001 & 0.0001 & 0.3321 & 0.0325 \\ 0.0001 & 0.0001 & 0.0325 & 0.2083 \end{bmatrix}.$$

Set $\gamma_2 = 10$, $\tilde{\rho} = 5$, and $c_2 = 1$, by step (2) of Algorithm 8.1, one has

$$K = \begin{bmatrix} 0.0538 & 0.0642 & 0.3486 & 0.5268 & -0.0447 & -0.1683 \\ 0.0538 & 0.0642 & 0.3486 & 0.5268 & -0.0447 & -0.1683 \\ 0.0538 & 0.0642 & 0.3486 & 0.5268 & -0.0447 & -0.1683 \\ 0.0538 & 0.0642 & 0.3486 & 0.5268 & -0.0447 & -0.1683 \\ 0.1991 & 0.2368 & -0.1745 & -0.2636 & -0.0001 & 0.0090 \\ -0.1808 & -0.2150 & 0.2928 & 0.4425 & -0.0151 & -0.0661 \\ 0.1991 & 0.2368 & -0.1745 & -0.2636 & -0.0001 & 0.0090 \\ -0.1808 & -0.2150 & 0.2928 & 0.4425 & -0.0151 & -0.0661 \end{bmatrix}.$$

According to the step (3) of Algorithm 8.1, one may choose $\alpha = \beta = 15 > 9.4688$. It thus follows from Theorem 8.1 that distributed consensus tracking in the considered

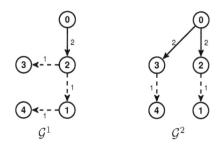

Figure 8.1 The communication graphs $\mathcal{G}^1$ and $\mathcal{G}^2$.

CNS equipped with protocol (8.5) designed above will be guaranteed if the dwell time is larger than 0.6931 s. The state trajectories of the closed-loop CNS are provided in Figs. 8.2–8.4, respectively. Profiles of the consensus tracking error $\|e(t)\|$ and observer error $\|\tilde{e}(t)\|$ are shown in Figure 8.5 which indicate that the distributed consensus tracking problem is solved.

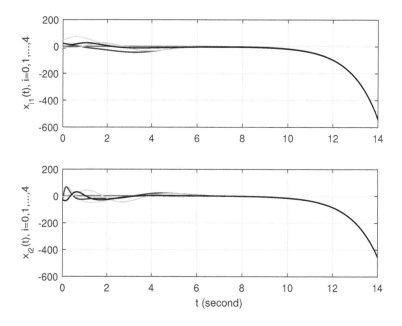

Figure 8.2 The agents' state trajectories $x_{i1}(t)$ and $x_{i2}(t)$, $i = 0, 1, \ldots, 4$.

8.6 CONCLUSIONS

We have addressed the consensus tracking for CNSs with Lipschitz nonlinear dynamics and directed switching topologies in this chapter. The challenge of only the relative output information of neighboring agents are available is successfully addressed by constructing a distributed observer. A Lyapunov function-based design approach has

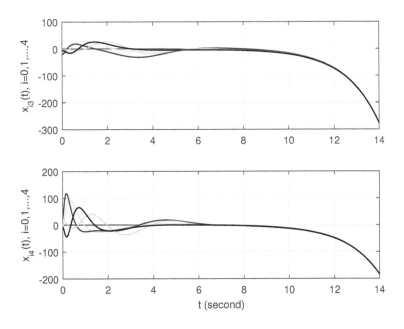

Figure 8.3 The agents' state trajectories $x_{i3}(t)$ and $x_{i4}(t)$, $i = 0, 1, \ldots, 4$.

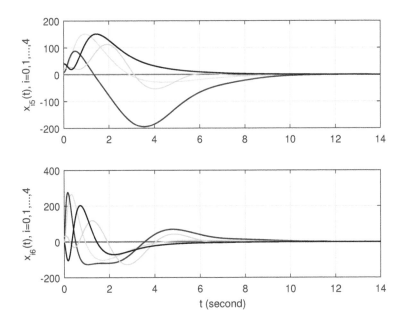

Figure 8.4 The agents' state trajectories $x_{i5}(t)$ and $x_{i6}(t)$, $i = 0, 1, \ldots, 4$.

been employed since we noticed that the separation principle is no longer satisfied in the considered controller and observer design scenario. It has been rigorously proved that consensus tracking could be achieved if each possible communication topology contains a spanning tree rooted at the leader, and the control parameters are suitably

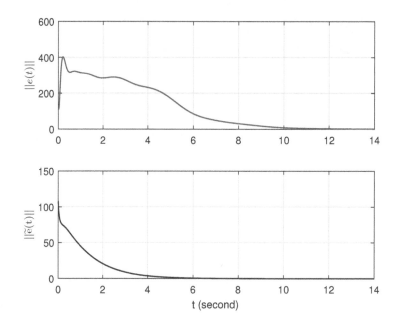

Figure 8.5 Trajectories of the consensus tracking error $\|e(t)\|$ and observer error $\|\tilde{e}(t)\|$.

selected. We have also considered a more general case where the distributed observers and controllers' communication configurations are independent and may be different.

CHAPTER 9

Cooperative tracking of CNSs with a high-dimensional leader and directed switching topologies

Unlike previous chapters where the leader and the followers have the same dimensional dynamics, this chapter studies the CNSs with a single high-dimensional leader, where the cooperative goal is that the states of the followers converge to the outputs of the leader. This chapter firstly proposes a reduced-order state observer for each follower to estimate the leader's dynamics, then the consensus controller is designed. Section 9.3 firstly studies the asymptotical consensus of CNSs without any external disturbance under directed switching topologies. Then we extend the results to the case with unknown external disturbances by showing that a finite $\mathbb{L}_2$-gain performance for tracking errors against external disturbances can be ensured. At last, the synthesis issue of designing an observer-based controller to achieve a prescribed $\mathbb{L}_2$-gain performance for consensus tracking is studied by using tools from $\mathcal{H}_\infty$ control theory, where the underlying topology is assumed to be undirected and fixed. Finally, this chapter presents some simulations to validate the theoretical results.

9.1 INTRODUCTION

In most practical applications, the evolution of CNSs is generally affected by unknown disturbances such as communication jamming, sensor noises, etc. Consensus tracking for CNSs with guaranteed $\mathbb{L}_2$-gain performance against disturbances has recently attracted much attention [172, 209]. Specifically, the $\mathbb{L}_2$-gain performance indexes for consensus tracking of disturbed CNSs with first-order and second-order dynamics are respectively addressed in [172] and [209]. In [184], consensus tracking and its $\mathbb{L}_2$-gain performance index are analyzed for a class of first-order CNSs with nonlinear node dynamics and external disturbances. The majority of consensus tracking problems currently studied in the literature are focused on homogeneous CNSs where the leader and followers are modeled by the same nominal systems. Comparatively speaking, consensus tracking problem of CNSs where the leader and followers have heterogeneously intrinsic dynamics has not received considerable attention. Particularly, we are at a loss in describing CNSs with a high-dimensional dynamic leader. Two typical such systems are the tractor-trailer systems consisting of a fire truck and several trailers [126], and the wireless sensor networks in the presence of a complex target to be monitored [232]. In [60], consensus tracking problem of first-order CNSs with a single high-dimensional dynamic leader has been investigated. Some necessary and sufficient criteria for achieving consensus in heterogeneous CNSs consisting of both first-order and second-order integrator-type agents have been derived in [230]. Containment control problem for a class of CNSs with multiple high-dimensional dynamic leaders has been studied in [25]. Then, practical consensus tracking of a class of uncertain CNSs with a high-dimensional dynamic leader has been addressed in [192], where the states of each follower asymptotically converge to some partial states of the leader with bounded residual errors.

This chapter is concerned with the distributed cooperative consensus tracking problem for a kind of general linear CNSs in the presence of a single high-dimensional leader. As the common assumption that the leader and followers have homogeneously nominal dynamics is removed from the present CNSs model, the considered CNSs model is general and thus broadly applicable to modeling some practical networking dynamical systems where the dynamics of the leader and the followers are heterogeneous. A class of distributed controllers associated with a reduced-order state observer are proposed for each follower to asymptotically track some partial states of the high-dimensional active leader under directed switching topologies. By using a combined tool from M-matrix theory and stability theory of switched systems, several efficient criteria are derived for consensus tracking of nominal CNSs. Theoretical analysis is then extended to consensus tracking of CNSs with unknown disturbances by showing that a finite $\mathbb{L}_2$-gain performance for tracking errors against external disturbances can be guaranteed if some suitable conditions are satisfied.

9.2 MODEL FORMULATION

Consider a CNS consisting of a leader with n-dimensional dynamics and N followers with m-dimensional dynamics, where $n > m$, $n, m \in \mathbb{N}$. We label the leader as agent 0 and the followers as agents $1, \dots, N$. The dynamics of the leader are described by:

$$\dot{\vartheta}(t) = \widetilde{A}\vartheta(t), \tag{9.1}$$

where $\vartheta(t) = [\vartheta_1(t), \dots, \vartheta_n(t)]^T \in \mathbb{R}^n$ is the state of the leader, $\widetilde{A} \in \mathbb{R}^{n \times n}$ is the system matrix. The dynamics of agent i, $i = 1, \dots, N$, are described by

$$\dot{x}_i(t) = A_i x_i(t) + \varpi_i(t) + u_i(t), \tag{9.2}$$

where $x_i(t) \in \mathbb{R}^m$ is the state, $A_i \in \mathbb{R}^{m \times m}$ is the system matrix, $\varpi_i(t) \in \mathbb{L}_2^m[0, +\infty)$ is the external disturbance, $\mathbb{L}_2^m[0, +\infty)$ denotes the m-dimensional square integrable function space, $u_i(t) \in \mathbb{R}^m$ is the control input to be designed later. In some practical leader-following systems, the followers need only to track some preselected but not all state variables of the leader. For example, the trailers need only to track the velocity of tractor in a tractor-trailer system consisting of a fire truck and several trailers. Furthermore, in the heterogeneous leader-following CNSs consisting of an unmanned aerial vehicle and several autonomous underwater vehicles, it is important to design distributed consensus tracking algorithms to make the velocity of the autonomous underwater vehicles track that of the unmanned aerial vehicle. Motivated by these observations, the control objective in this work is to drive the states of followers to asymptotically approach some preselected state variables of the leader, denoted by

$$\vartheta_{\text{out}}(t) = [\vartheta_{i_1}(t), \dots, \vartheta_{i_m}(t)]^T, \tag{9.3}$$

in the sense of $\lim_{t \to +\infty} \|x_i(t) - \vartheta_{\text{out}}(t)\| = 0$ under the condition that $\varpi_i(t) \equiv \mathbf{0}_m$, for $i = 1, \dots, N$, with $i_s \in \{1, \dots, n\}$, $\vartheta_{i_s}(t) \in \mathbb{R}$, for all $s = 1, \dots, m$, $i_j \neq i_k$ for $k \neq j$, $k, j \in \{1, \dots, m\}$; and guarantee a finite $\mathbb{L}_2$-gain performance index for consensus tracking of closed-loop CNSs against external disturbances for the case with external disturbances belonging to square integrable function space. For practicability, it is assumed that $\vartheta_{\text{out}}(t)$ can be only accessed by a subset of the followers. Before moving on, the following linear transformation is introduced:

$$\zeta(t) = T\vartheta(t), \tag{9.4}$$

where $T = \Pi_{i=1}^m T_i$, T_i, $i = 1, \dots, m$, are determined in the following Algorithm 9.1.

Algorithm 9.1 *Set $T_m = I(i_m, n)$. Here, $I(i_m, n)$ is the $n \times n$ elementary matrix determined by switching all matrix elements on row i_m with their counterparts on row n for I_n, i_m is given in (9.3). Set $\mu = m$.*

(1) Check the following condition: $\mu - 1 > 0$? If it does not hold, stop; else, go to step (2).

(2) Set $\vartheta^{[\mu]}(t) = T_\mu \vartheta(t)$. Without loss of any generality, one may let

$$\vartheta^{[\mu]}(t) = \left[\vartheta_{k_1^{[\mu]}}(t), \dots, \vartheta_{k_n^{[\mu]}}(t)\right]^T.$$

(3) Set $T_{\mu-1} = I(r^{[\mu-1]}, n-m+\mu-1)$, of which $r^{[\mu-1]} = \arg_{j \in \{1,2,\dots,n-m+\mu-1\}} k_j^{[\mu]} = i_{\mu-1}$, $I(r^{[\mu-1]}, n-m+\mu-1)$ is the $n \times n$ elementary matrix determined by switching all matrix elements on row $r^{[\mu-1]}$ with their counterparts on row $n-m+\mu-1$ for I_n, $i_{\mu-1} \in \{i_1, i_2, \dots, i_m\}$. Then, set $\mu = \mu - 1$ and go back to step (1).

According to Algorithm 9.1, one knows that, for each $i = 1, \dots, m$, T_i is an elementary matrix. Clearly, $T = \Pi_{k=1}^m T_i$ is invertible. According to (9.4), one obtains

$$\begin{cases} \dot{\zeta}(t) = \widehat{A}\zeta(t), \\ y(t) = \widehat{C}\zeta(t), \end{cases} \tag{9.5}$$

where $\zeta(t) = [\zeta_1(t), \ldots, \zeta_{n-m}(t), \vartheta_{\text{out}}^T(t)]^T$, $\zeta_i(t) \in \mathbb{R}$, for $i = 1, \ldots, n - m$, $\widehat{A} = T\tilde{A}T^{-1}$, $\widehat{C} = \begin{bmatrix} 0_{m \times (n-m)}, I_m \end{bmatrix} \in \mathbb{R}^{m \times n}$ is the output matrix, $y(t)$ is the output vector. Obviously, $y(t) = \vartheta_{\text{out}}(t)$ for all $t \geq 0$.

Let $\tilde{\zeta}(t) = [\zeta_1(t), \ldots, \zeta_{n-m}(t)]^T$, one gets that $\zeta(t) = [\tilde{\zeta}^T(t), \vartheta_{\text{out}}^T(t)]^T$. For the convenience of expression, one could partition $\widehat{A} = [\hat{a}_{ij}]_{n \times n}$ into the following compact

form: $\widehat{A} = \begin{bmatrix} \widehat{A}_{11} & \widehat{A}_{12} \\ \widehat{A}_{21} & \widehat{A}_{22} \end{bmatrix}$, of which $\widehat{A}_{11} \in \mathbb{R}^{(n-m) \times (n-m)}$, $\widehat{A}_{12} \in \mathbb{R}^{(n-m) \times m}$, $\widehat{A}_{21} \in \mathbb{R}^{m \times (n-m)}$, and $\widehat{A}_{22} \in \mathbb{R}^{m \times m}$. Note that $\vartheta_{\text{out}}(t)$ can be seen as the output vector of linear system (9.5). It is assumed that $\vartheta_{\text{out}}(t)$ is available to only a subset of followers. The followers, which can directly sense the output vector of linear system (9.5), are called as *informed agents*. Generally, the leader in practical CNSs takes the role of a command generator to yield desirable signals to be followed by the followers. It is thus supposed that the dynamics of the leader will not be affected by any followers.

Assumption 9.1 *The matrix pair $(\widehat{C}, \widehat{A})$ is detectable.*

The interaction topology of the considered CNS consisting a single leader given in (9.1) and N followers given in (9.2) is described by the directed switching graph with order $N+1$. This means that the communication relationship among the agents in the CNS under consideration is time-dependent but not fixed. Such a property provides the model with the ability to describe a great many CNSs communicated via wireless communication systems. Without loss of generality, we assume that the interaction topology of the considered CNS switches over a graph set $\widehat{\mathcal{G}}$ with $\widehat{\mathcal{G}} = \{\mathcal{G}^1, \ldots, \mathcal{G}^\kappa\}$, $\kappa \geq 2$. The switching time sequence is $t_1, t_2, \ldots$, with $t_1 > t_0 = 0$ and $\lim_{k \to +\infty} t_k = +\infty$, on which the topology switches. Note that the topology keeps fixed over each time interval $[t_k, t_{k+1})$, $k \in \mathbb{N}$. For expressional convenience, introduce the following switching function $\sigma(t) : [0, +\infty) \mapsto \{1, \ldots, \kappa\}$ to describe switching actions of the topology. One then gets that, at any given $t \geq 0$, $\mathcal{G}^{\sigma(t)}$ is the underlying interaction topology of the CNS. Clearly, for all $t \geq 0$, $\mathcal{G}^{\sigma(t)} \in \widehat{\mathcal{G}}$.

Since the evolution of the leader will not be affected any followers, the Laplacian matrix $\mathcal{L}^{\sigma(t)}$ of $\mathcal{G}^{\sigma(t)}$ can be rewritten as

$$\mathcal{L}^{\sigma(t)} = \begin{bmatrix} 0 & \mathbf{0}_N^T \\ \mathbf{P}^{\sigma(t)} & \overline{\mathcal{L}}^{\sigma(t)} \end{bmatrix},$$

$$\overline{\mathcal{L}}^{\sigma(t)} = \begin{bmatrix} \sum_{j \in \mathcal{N}_1} a_{1j}^{\sigma(t)} & -a_{12}^{\sigma(t)} & \cdots & -a_{1N}^{\sigma(t)} \\ -a_{21}^{\sigma(t)} & \sum_{j \in \mathcal{N}_2} a_{2j}^{\sigma(t)} & \cdots & -a_{2N}^{\sigma(t)} \\ \vdots & \vdots & \ddots & \vdots \\ -a_{N1}^{\sigma(t)} & -a_{N2}^{\sigma(t)} & \cdots & \sum_{j \in \mathcal{N}_N} a_{Nj}^{\sigma(t)} \end{bmatrix}, \quad (9.6)$$

where $\mathbf{P}^{\sigma(t)} = -[a_{10}^{\sigma(t)}, \ldots, a_{N0}^{\sigma(t)}]^T \in \mathbb{R}^N$. Note that, for all $t \geq 0$, $a_{i0}^{\sigma(t)} > 0$ if the output information $y(t)$ of the leader can be accessed by follower i and $a_{i0}^{\sigma(t)} = 0$ otherwise, $i = 1, \ldots, N$. The fact that the followers may only sense some partial state information about the leader prompts us to design an observer-based controller to complete the coordination goal. Specifically, for each follower i, the following local controller associated with a neighbor-based dynamic observer is proposed:

$$u_i(t) = \breve{A}x_i(t) + \widehat{A}_{21}\delta_i(t)$$

$$-cF_2\left[\sum_{j=1}^{N} a_{ij}^{\sigma(t)}(x_j(t) - x_i(t)) + a_{i0}^{\sigma(t)}(y(t) - x_i(t))\right], \qquad (9.7)$$

where $\breve{A} = \widehat{A}_{22} - A_i$, $F_2 \in \mathbb{R}^{m \times m}$ is the gain matrix to be determined later, $\delta_i(t) \in \mathbb{R}^{n-m}$ is the state of the dynamic observer embedded in follower i that is given by

$$\dot{\delta}_i(t) = \widehat{A}_{11}\delta_i(t) + \widehat{A}_{12}x_i(t)$$

$$- cF_1\left[\sum_{j=1}^{N} a_{ij}^{\sigma(t)}(x_j(t) - x_i(t)) + a_{i0}^{\sigma(t)}(y(t) - x_i(t))\right], \qquad (9.8)$$

where $F_1 \in \mathbb{R}^{(n-m) \times m}$ is the gain matrix to be selected later. For notational convenience, set $F = [F_1^T, F_2^T]^T$ and $\bar{x}_i(t) = [\delta_i^T(t), x_i^T(t)]^T$. Since the order of observer (9.8) is less than that of leader's state, it is thus a kind of reduced-order observer.

The following assumption is made.

Assumption 9.2 *Each graph $\mathcal{G}^j$, $j = 1, \ldots, \kappa$, contains a directed spanning tree rooted at agent 0 (i.e., the leader).*

Remark 9.1 *One interesting yet critical issue is how to select the agents to be informed such that, for each $j \in \{1, \ldots, \kappa\}$, the augmented graph $\mathcal{G}^j$ containing both the leader and N followers has a directed spanning tree rooted at node 0, i.e., Assumption 9.1 holds. Without loss of generality, let $\mathcal{G}(\widetilde{\mathcal{V}}, \widetilde{\mathcal{E}}^j, \widetilde{\mathcal{A}}^j)$ $(j \in \{1, \ldots, \kappa\})$ be the subgraph in $\mathcal{G}^j$ describing the interaction topology among N followers where $\widetilde{\mathcal{V}} = \{1, \ldots, N\}$, $\widetilde{\mathcal{E}}^j$ represents the set of communication edges among the N followers, and $\widetilde{\mathcal{A}}^j \in \mathbb{R}^{N \times N}$ indicates the weighted adjacency matrix of $\mathcal{G}(\widetilde{\mathcal{V}}, \widetilde{\mathcal{E}}^j, \widetilde{\mathcal{A}}^j)$. To make Assumption 9.2 holds, the candidate set $\widetilde{\mathcal{V}}_{inf}^j$ of nodes to be informed could be determined by employing the following algorithm.*

Algorithm 9.2 *Set $r = 0$, $h = 1$, and $\widetilde{\mathcal{V}}_{inf}^j = \emptyset$.*

(1) Use the depth-first search algorithm [47] to find all the strong connected components of $\mathcal{G}(\widetilde{\mathcal{V}}, \widetilde{\mathcal{E}}^j, \widetilde{\mathcal{A}}^j)$. Suppose that there are ι strong components, represented by $\mathcal{G}(\widetilde{\mathcal{V}}_1^j, \widetilde{\mathcal{E}}_1^j, \widetilde{\mathcal{A}}_1^j)$, ..., $\mathcal{G}(\widetilde{\mathcal{V}}_\iota^j, \widetilde{\mathcal{E}}_\iota^j, \widetilde{\mathcal{A}}_\iota^j)$, in $\mathcal{G}(\widetilde{\mathcal{V}}, \widetilde{\mathcal{E}}^j, \widetilde{\mathcal{A}}^j)$.

(2) Check whether there exists at least one node in $\widetilde{\mathcal{V}}_h^j$ which is reachable from a node belonging to the node set $\widetilde{\mathcal{V}} \backslash \widetilde{\mathcal{V}}_h^j$ in graph $\mathcal{G}(\widetilde{\mathcal{V}}, \widetilde{\mathcal{E}}^j, \widetilde{\mathcal{A}}^j)$. If yes, go to step (3); otherwise, go to step (4).

(3) Check the following condition: $h < \iota$? If it holds, let $h = h + 1$ and re-perform step (2); else stop.

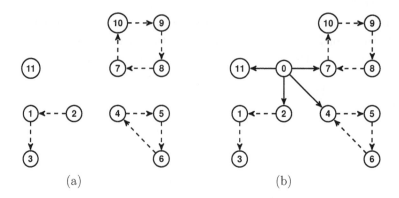

Figure 9.1 An illustration of Algorithm 9.2; (a): The interaction graph of 11 followers; (b): The augmented graph consisting of both the leader and 11 followers where node 0 represents the leader, the nodes located at the arrow heads of solid lines indicate the nodes being informed, respectively.

(4) Arbitrarily select one node v_h in $\widetilde{\mathcal{V}}_h^j$ to be informed and update the set $\widetilde{\mathcal{V}}_{inf}^j$, the value of r according to the rules: $\widetilde{\mathcal{V}}_{inf}^j = \widetilde{\mathcal{V}}_{inf}^j \cup \{v_h\}$ and $r = r + 1$, respectively. Then, check the following condition: $h < \iota$? If it holds, let $h = h + 1$ and go to step (2); else stop.

Note that the time complexity of depth-first search algorithm employed in step (1) of Algorithm 9.2 is $\mathcal{O}(N + |\widetilde{\mathcal{E}}^j|)$, with N and $|\widetilde{\mathcal{E}}^j|$ being respectively the numbers of nodes and edges of $\mathcal{G}(\widetilde{\mathcal{V}}, \widetilde{\mathcal{E}}^j, \widetilde{A}^j)$. Next, we illustrate how to use Algorithm 9.2 to determine the candidate set $\widetilde{\mathcal{V}}_{inf}^j$ of followers to be informed in a given directed graph such that the augmented graph contains a directed spanning tree. Assume that there are 11 followers in the considered CNS as shown in Figure 9.1(a) of which the dashed lines indicate the neighboring relationship among followers. According to steps (1)– (3) of Algorithm 9.2, we obtain that there are 4 strongly connected components in the directed graph describing the communication topology among followers. To make the augmented graph contains a directed spanning tree rooted at the leader, it can be obtained by performing steps (4) of Algorithm 9.2 that nodes 2, 11, two distinct nodes respectively selected from node sets $\{4, 5, 6\}$ and $\{7, 8, 9, 10\}$ should be informed. Set $\{2, 4, 7, 11\}$ as the candidate set of nodes to be informed, one gets that the augmented graph contains a directed spanning tree rooted at agent 0 (see Figure 9.1(b)).

Now, we are in the position to formulate the problems mathematically, which will be investigated in the present chapter.

Definition 9.1 The consensus tracking problem of CNS consisting of a high-dimensional leader (9.5) and followers (9.2) subject to $\varpi_i(t) = \mathbf{0}_m$ is said to be solved by controller (9.7) if $\lim_{t \to \infty} \|x_i(t) - y(t)\| = 0$ for each $i = 1, \ldots, N$.

Definition 9.2 The CNS consisting of a single high-dimensional active leader (9.5) and N followers (9.2) under controller (9.7), is said to have a finite $\mathbb{L}_2$-gain performance if the following two conditions simultaneously hold:

(1) The consensus tracking problem stated in Definition 9.1 is solved;

(2) There exists a positive scalar $\gamma > 0$ such that, under zero initial conditions $\zeta(0) = \mathbf{0}_n$, $\bar{x}_i(0) = \mathbf{0}_n$, $i = 1, \ldots, N$, the following condition is satisfied:

$$\int_0^{+\infty} \sum_{i=1}^N (\bar{x}_i(t) - \zeta(t))^T (\bar{x}_i(t) - \zeta(t)) \, dt \leq \gamma^2 \int_0^{+\infty} \varpi^T(t)\varpi(t)dt, \qquad (9.9)$$

where $\varpi(t) = [\varpi_1^T(t), \ldots, \varpi_N^T(t)]^T$.

9.3 CONSENSUS TRACKING AND ITS $\mathbb{L}_2$-GAIN PERFORMANCE OF CNSS WITH DIRECTED SWITCHING TOPOLOGIES

The main analytical results are presented and analyzed in this section.

Under the condition $\varpi_i(t) = \mathbf{0}_m$, for all $i = 1, \ldots, N$, consensus tracking of CNS with a single high-dimensional active leader given in (9.1) and followers given in (9.2) is firstly investigated in this section. Then, the $\mathbb{L}_2$-gain performance index of the considered CNS with external disturbances will be analyzed.

Let $e(t) = [e_1^T(t), \ldots, e_N^T(t)]^T$ with $e_i(t) = \bar{x}_i(t) - \zeta(t)$, $\bar{x}_i(t) = [\delta_i^T(t), x_i^T(t)]^T$, $i = 1, \ldots, N$. Then it can be derived from (9.5), (9.7), and (9.8) that

$$\dot{e}(t) = (I_N \otimes \hat{A} + c\overline{\mathcal{L}}^{\sigma(t)} \otimes F\hat{C})e(t), \qquad (9.10)$$

where $F = [F_1^T, F_2^T]^T$.

Suppose that Assumption 9.2 holds, one knows that, for each $j = 1, \ldots, \kappa$, $\overline{\mathcal{L}}^j$ is a nonsingular M-matrix and the following LMI

$$(\overline{\mathcal{L}}^j)^T \Phi^j + \Phi^j \overline{\mathcal{L}}^j > 0 \qquad (9.11)$$

is always feasible for some positive definite matrix $\Phi^j = \text{diag}\{\phi_1^j, \ldots, \phi_N^j\}$. Noticeably, for each $j \in \{1, \ldots, \kappa\}$, the selections of Φ^j for LMI (9.11) are not unique. Suppose that $\Phi^j = \text{diag}\{\phi_1^j, \ldots, \phi_N^j\}$ is a feasible solution for LMI (9.11), then for each $j \in \{1, \ldots, \kappa\}$, the following LMI holds:

$$\overline{\Phi}^j \overline{\mathcal{L}}^j + (\overline{\mathcal{L}}^j)^T \overline{\Phi}^j > 0, \qquad (9.12)$$

where $\overline{\Phi}^j = (1/\phi_{\max}^j)\Phi^j$, $\phi_{\max}^j = \max_{i=1,\ldots,N}\phi_i^j$. Obviously, the diagonal elements of $\overline{\Phi}^j$ belong to the interval $(0, 1]$, for all $j \in \{1, \ldots, \kappa\}$. The above statements indicate that, under Assumption 9.2, there always exist $\Psi^j = \text{diag}\{\psi_1^j, \ldots, \psi_N^j\}$ satisfying $(\overline{\mathcal{L}}^j)^T \Psi^j + \Psi^j \overline{\mathcal{L}}^j > 0$ and $\psi_i^j \in (0, 1]$ for all $j \in \{1, \ldots, \kappa\}$, $i \in \{1, \ldots, N\}$. Before moving forward, the following selection algorithm is provided.

Algorithm 9.3 *Under Assumption 9.2, one may select two positive scalars ϵ_1, ϵ_2 such that $\epsilon_1 << 1$ and $\epsilon_2 << 1$. Then, set $\epsilon_M = 1$ and $\epsilon_m = 0$.*

(1) Solve the LMIs:

$$(\overline{\mathcal{L}}^j)^T \Psi^j + \Psi^j \overline{\mathcal{L}}^j > 0, \ \ \Psi^j > \epsilon_m I_N, \ \ \Psi^j \leq \epsilon_M I_N \qquad (9.13)$$

to obtain some feasible solutions $\Psi^j = \text{diag}\{\psi_1^j, \ldots, \psi_N^j\}$, $j \in \{1, \ldots, \kappa\}$.

(2) Set $\epsilon_M = \epsilon_M - \epsilon_1$ and $\epsilon_m = \epsilon_m + \epsilon_2$. Check whether LMI (9.13) is feasible. If LMI (9.13) is feasible, solve LMI (9.13) to get some feasible solutions and return to the beginning of step (2); else, stop.

Remark 9.2 *As the common Lyapunov function exists for only a few CNSs with directed switching topologies, it is reasonable to alternatively construct MLFs for the error system of the considered CNSs. With the help of M-matrix theory and matrix inequality theory, a new class of MLFs will be constructed to analysis the evolution behavior of the error system. Note that the matrices Ψ^j determined in Algorithm 9.3 will be involved in constructing the MLFs for the tracking error system. Note that the update mechanisms for ϵ_m and ϵ_M given in step (2) of Algorithm 9.3 are designed to reduce the conservatism of the derived admissible ADT of switchings among different topologies for achieving consensus tracking.*

Furthermore, for notational brevity, let

$$\mu = \max_{i,j \in \{1,\dots,\kappa\}, i \neq j} \{\psi^i_{\max}/\psi^j_{\min}\}, \tag{9.14}$$

with $\psi^i_{\max} = \max_{k=1,\dots,N} \psi^i_k$, $\psi^j_{\min} = \min_{k=1,\dots,N} \psi^j_k$, for $i, j \in \{1,\dots,\kappa\}$. Obviously, $\mu \geq 1$. According to the fact $(\overline{\mathcal{L}}^j)^T \Psi^j + \Psi^j \overline{\mathcal{L}}^j > 0$, for each $j \in \{1,\dots,\kappa\}$, one knows that all the eigenvalues of matrix $(\Psi^j)^{-1}(\overline{\mathcal{L}}^j)^T \Psi^j + \overline{\mathcal{L}}^j$ are real and positive. For notational convenience, let λ^j_0 be the minimum eigenvalue of $(\Psi^j)^{-1}(\overline{\mathcal{L}}^j)^T \Psi^j + \overline{\mathcal{L}}^j$. Furthermore, set

$$\lambda_0 = \min_{j=1,\dots,\kappa} \lambda^j_0 > 0. \tag{9.15}$$

Theorem 9.1 *Under Assumptions 9.1 and 9.2, consensus tracking problem (stated in Definition 9.2) for CNS with a single high-dimensional active leader (9.1) and followers (9.2) subject to $\varpi_i(t) = \mathbf{0}_m$, $i = 1,\dots,N$, can be solved by controller (9.7) associated with state observer (9.8) with $F = -Q^{-1}\widehat{C}^T$ and $c > 2/\lambda_0$, if $\tau_a > (\ln\mu)/c_0$, where μ is defined in (9.14), $Q > 0$ and $c_0 > 0$ satisfy the LMI*

$$\widehat{A}^T Q + Q\widehat{A} - 2\widehat{C}^T \widehat{C} + c_0 Q < 0. \tag{9.16}$$

Proof 9.1 *Select the following MLFs for the error system (9.10):*

$$V(t) = e^T(t)(\Psi^{\sigma(t)} \otimes Q)e(t), \tag{9.17}$$

where $\Psi^{\sigma(t)} \in \{\Psi^1,\dots,\Psi^\kappa\}$, Ψ^j, $j \in \{1,\dots,\kappa\}$, are obtained by Algorithm 9.3, $Q > 0$ is given by (9.16).

For an arbitrarily given $T > 0$, let $t_0 = 0$ and t_i, $i = 1,\dots,N_\sigma[0,T)$, be the discontinuous time points of switching signal $\sigma(t)$ over $[0,T)$ with $t_1 < t_2 < \dots < t_{N_\sigma[0,T)}$. Then, for all $t \in [t_k, t_{k+1})$ and an arbitrarily given $k = 0,1,\dots,N_\sigma[0,T)-1$, calculating the time derivative of $V(t)$ on the trajectories of (9.10) gives that

$$\dot{V}(t) = e^T(t)[\Psi^{\sigma(t)} \otimes (Q\widehat{A} + \widehat{A}^T Q)]e(t) + 2ce^T(t)(\Psi^{\sigma(t)}\overline{\mathcal{L}}^{\sigma(t)} \otimes QF\widehat{C})e(t). \tag{9.18}$$

Substituting $F = -Q^{-1}\widehat{C}^T$ into (9.18) yields

$$\dot{V}(t) = e^T(t)\left[\Psi^{\sigma(t)} \otimes (Q\widehat{A} + \widehat{A}^T Q)\right] e(t)$$
$$- ce^T(t)\left\{\left[\Psi^{\sigma(t)}\overline{\mathcal{L}}^{\sigma(t)} + (\overline{\mathcal{L}}^{\sigma(t)})^T\Psi^{\sigma(t)}\right] \otimes \widehat{C}^T\widehat{C}\right\} e(t). \tag{9.19}$$

Since $\sigma(t) \in \{1,\ldots,\kappa\}$, then it follows that $\lambda_0\Psi^{\sigma(t)} \leq \Psi^{\sigma(t)}\overline{\mathcal{L}}^{\sigma(t)}+(\overline{\mathcal{L}}^{\sigma(t)})^T\Psi^{(\sigma(t))}$, where λ_0 is defined in (9.15). Based on the above analysis and by noticing the facts $\widehat{C}^T\widehat{C} \geq 0$, $c\lambda_0 > 2$, the following inequality can be obtained from (9.19):

$$\dot{V}(t) \leq e^T(t)\left[\Psi^{\sigma(t)} \otimes (Q\widehat{A} + \widehat{A}^T Q - 2\widehat{C}^T\widehat{C})\right] e(t).$$

It can be thus obtained from (9.16) and the above inequality that

$$\dot{V}(t) < -c_0 V(t) \tag{9.20}$$

for all $t \in [t_k, t_{k+1})$ and an arbitrarily given $k = 0, 1, \ldots, N_\sigma[0,T] - 1$. Furthermore, the following fact can be verified: $V(t_k) \leq \mu V(t_k^-)$, where μ is defined in (9.14), $k \in \{1, 2, \ldots, N_\sigma[0,T]\}$. Thus, one gets

$$V(T) \leq \exp\left(-c_0(T - t_{N_\sigma[0,T]})\right) V(t_{N_\sigma[0,T]})$$
$$\leq \mu \exp\left(-c_0(T - t_{N_\sigma[0,T]}^-)\right) V(t_{N_\sigma[0,T]}^-).$$

By induction, it can be derived from the above analysis that

$$V(T) \leq \mu^{N_\sigma[0,T]} \exp(-c_0 T)V(0). \tag{9.21}$$

Let τ_a and N_0 be respectively the ADT and the chatter bound of $\sigma(t)$, the following inequality can be obtained

$$V(T) \leq \mu^{N_\sigma[0,T]} \exp(-c_0 T)V(0)$$
$$\leq K_0 \exp\left(-(c_0 - \frac{\ln\mu}{\tau_a})T\right) V(0), \tag{9.22}$$

of which $K_0 = \mu^{N_0}$. Furthermore, the condition $\tau_a > (\ln\mu)/c_0$ implies $c_0 > (\ln\mu)/\tau_a$. Since T is an arbitrarily chosen positive number, one may conclude from (9.22) that cooperative consensus tracking problem for CNS with a single high-dimensional active leader (9.5) and followers (9.2) subject to $\varpi_i(t) = \mathbf{0}_m$, $i = 1,\ldots,N$, is indeed solved by controller (9.7).

Remark 9.3 *In Theorem 9.1, control parameter c in (9.7) and (9.8) is selected as $c > 2/\lambda_0$. Actually, this scalar can be selected as the following general form $c > \alpha_0/\lambda_0$, for any given $\alpha_0 > 0$. To possibly achieve cooperative consensus tracking in the considered CNS, the LMI (9.16) should be correspondingly revised*

$$\widehat{A}^T Q + Q\widehat{A} - \alpha_0\widehat{C}^T\widehat{C} + c_0 Q < 0. \tag{9.23}$$

However, it should be noted that, for some given $c_0 > 0$, the feasibility problems of LMIs (9.16) and (9.23) are the same. Furthermore, it can be observed that the conservativeness of consensus tracking criteria given in Theorem 9.1 is closely related to parameter c_0. Suppose that LMI (9.16) is always feasible, the larger the parameter c_0 is selected, the less the conservatism of the criteria is yielded. According to the Finsler's lemma and pole assignment theory [203], one knows that LMI (9.16) is always feasible for any given $c_0 > 0$ when the matrix pair $(\widehat{C}, \widehat{A})$ is completely observable. Suppose that Assumption 9.2 holds and the matrix pair $(\widehat{C}, \widehat{A})$ is completely observable, it can be obtained from Theorem 9.1 that, for an arbitrarily given switching signal $\sigma(t)$, the consensus tracking problem (stated in Definition 9.1) can be solved if the control parameters F, c are suitably designed.

Remark 9.4 *It can be obtained from Remark 9.3 that, for any given $c_0 > 0$, the LMI (9.16) is always feasible for some $Q > 0$ if the matrix pair $(\widehat{C}, \widehat{A})$ is completely observable. Generally, the above statement does not hold for the case where $(\widehat{C}, \widehat{A})$ is detectable but not completely observable. The feasibility of LMI (9.16) for the case that $(\widehat{C}, \widehat{A})$ is detectable but not completely observable is analyzed as follows. Suppose that $(\widehat{C}, \widehat{A})$ is detectable but not completely observable, let β_0 be the maximum real part of the unobservable eigenvalue of matrix pair $(\widehat{C}, \widehat{A})$. One hence gets $\beta_0 < 0$. One may then get that there exists a matrix L such that the real parts of all eigenvalues of $\widehat{A} + L\widehat{C}$ are not larger than β_0, i.e., the LMI*

$$(\widehat{A} + L\widehat{C})^T Q + Q(\widehat{A} + L\widehat{C}) - 2\beta_0 Q \leq 0 \tag{9.24}$$

is feasible for some $Q > 0$. Let $QL = W$, LMI (9.24) becomes

$$\widehat{A}^T Q + Q\widehat{A} + W\widehat{C} + \widehat{C}^T W^T - 2\beta_0 Q \leq 0. \tag{9.25}$$

According to (9.25) and by using Finsler's Lemma, one may observe that LMI

$$\widehat{A}^T Q + Q\widehat{A} - 2\widehat{C}^T \widehat{C} + c_0 Q < 0$$

is feasible for some $Q > 0$ under the condition that $c_0 < -2\beta_0$. This indicates that, under the condition that $(\widehat{C}, \widehat{A})$ is detectable but not completely observable, LMI (9.16) is feasible for each $c_0 \in (0, -2\beta_0)$.

Next, the finite $\mathbb{L}_2$-gain performance index of the considered CNS in the presence of external disturbances is analyzed. The control goal is to construct a suitable controller (9.7) and estimator (9.8) to reach consensus tracking under zero initial conditions, meanwhile maintaining a desirable disturbance rejection level. Note that $\mathbb{L}_2$-gain is an extension of $\mathcal{H}_\infty$ norm of linear time-invariant systems. In practice, $\mathbb{L}_2$-gain is an important index in assessing the quality of the closed-loop systems subject to external disturbances [149, 151, 184].

In this case, the following evolution equation can be derived from (9.10) as

$$\dot{e}(t) = (I_N \otimes \widehat{A} + c\overline{\mathcal{L}}^{\sigma(t)} \otimes F\widehat{C})e(t) + (I_N \otimes \widehat{C}^T)\varpi(t), \tag{9.26}$$

where $F = [F_1^T, F_2^T]^T$ and $\varpi(t) = [\varpi_1^T(t), \dots, \varpi_N^T(t)]^T$.

For notational brevity, let

$$\psi_m = \min_{j \in \{1,\dots,\kappa\}} \{\psi_{\min}^j\}, \tag{9.27}$$

of which $\psi_{\min}^j = \min_{k=1,\dots,N} \psi_k^j$, and $\Psi^j = \text{diag}\{\psi_1^j, \dots, \psi_N^j\}$, $j \in \{1, \dots, \kappa\}$, are determined in Algorithm 9.3.

Theorem 9.2 *Under Assumptions 9.1 and 9.2, the CNS with a single high-dimensional active leader given in (9.1) and followers given in (9.2) has a finite $\mathbb{L}_2$-gain performance under controller (9.7) associated with state observer (9.8) if the control parameters are appropriately selected such that $F = -\widehat{Q}^{-1}\widehat{C}^T$, $c > \chi_0/\lambda_0$, and $\tau_a > (\ln\mu)/c_1$, where λ_0 and μ are respectively defined in (9.14) and (9.15), $\widehat{Q} > 0$ is a feasible solution of the LMI*

$$\widehat{Q}\widehat{A} + \widehat{A}^T\widehat{Q} - \chi_0\widehat{C}^T\widehat{C} + (\rho_0 + c_1)\widehat{Q} + (1/\psi_m)I_n < 0, \tag{9.28}$$

where ψ_m is defined in (9.27), χ_0, c_1, and ρ_0 are positive scalars. Specifically, the following inequality holds:

$$\int_0^{+\infty} e^T(t)e(t)dt \leq \gamma^2 \int_0^{+\infty} \varpi^T(t)\varpi(t)dt, \tag{9.29}$$

where $e(t) = [e_1^T(t), \dots, e_N^T(t)]^T$, $e_i(t) = \bar{x}_i(t) - \zeta(t)$, $\varpi(t) = [\varpi_1^T(t), \dots, \varpi_N^T(t)]^T$,

$$\gamma = \sqrt{\frac{\widehat{\lambda}_M c_1 \exp(N_0 \ln\mu)}{\rho_0 c_2}}, \tag{9.30}$$

where $\widehat{\lambda}_M$ is the maximum eigenvalue of $\widehat{Q}$, $c_2 = c_1 - (\ln\mu)/\tau_a$, and $N_0 > 0$ is the chatter bound of the switching signal.

Proof 9.2 *Select the following MLFs for the system (9.26):*

$$\widehat{V}(t) = e^T(t)(\Psi^{\sigma(t)} \otimes \widehat{Q})e(t), \tag{9.31}$$

where $\widehat{Q} > 0$ is defined in (9.28). Calculating the right time derivative of $\widehat{V}(t)$ along the trajectories of (9.26) and invoking $F = -\widehat{Q}^{-1}\widehat{C}^T$ yield

$$\dot{\widehat{V}}(t) \leq e^T(t)[\Psi^{\sigma(t)} \otimes (\widehat{A}^T\widehat{Q} + \widehat{Q}\widehat{A} - \chi_0\widehat{C}^T\widehat{C})]e(t)$$
$$+ 2e^T(t)[\Psi^{\sigma(t)} \otimes (\widehat{Q}\widehat{C}^T)]\varpi(t), \tag{9.32}$$

where χ_0 is given in (9.28). Furthermore, the following inequality can be verified by using the properties of Kronecker product:

$$2e^T(t)(\Psi^{\sigma(t)} \otimes \widehat{Q}\widehat{C}^T)\varpi(t)$$
$$\leq \rho_0 e^T(t)(\Psi^{\sigma(t)} \otimes \widehat{Q})(I_N \otimes \widehat{Q}^{-1})(\Psi^{\sigma(t)} \otimes \widehat{Q})e(t)$$

$$+ \frac{1}{\rho_0} \varpi^T(t)(I_N \otimes \widehat{C})(I_N \otimes \widehat{Q})(I_N \otimes \widehat{C}^T)\varpi(t) \tag{9.33}$$

for any given $\rho_0 > 0$. According to Algorithm 9.3, one knows that the maximum eigenvalues of $\Psi^{\sigma(t)}$ are less than or equal to 1 for all t. Based on the previous analysis and by noticing the special structure of $\widehat{C}$, it can be yielded from (9.33) that

$$2e^T(t)(\Psi^{\sigma(t)} \otimes \widehat{Q}\widehat{C}^T)\varpi(t)$$
$$\leq \rho_0 e^T(t)(\Psi^{\sigma(t)} \otimes \widehat{Q})e(t) + (\widehat{\lambda}_M/\rho_0)\varpi^T(t)\varpi(t), \tag{9.34}$$

where $\widehat{\lambda}_M$ is the maximum eigenvalue of $\widehat{Q}$. Combining (9.32) and (9.34) gives

$$\dot{\widehat{V}}(t) \leq e^T(t)[\Psi^{\sigma(t)} \otimes (\widehat{A}^T\widehat{Q} + \widehat{Q}\widehat{A} - \chi_0\widehat{C}^T\widehat{C} + \rho_0\widehat{Q})]e(t) + \frac{\widehat{\lambda}_M}{\rho_0}\varpi^T(t)\varpi(t)$$
$$\leq e^T(t)\left[\Psi^{\sigma(t)} \otimes (\widehat{A}^T\widehat{Q} + \widehat{Q}\widehat{A} - \chi_0\widehat{C}^T\widehat{C} + \rho_0\widehat{Q} + (1/\psi_m)I_n)\right]e(t)$$
$$- e^T(t)e(t) + (\widehat{\lambda}_M/\rho_0)\varpi^T(t)\varpi(t), \tag{9.35}$$

where ψ_m is defined in (9.27). The following inequality can be thus derived from the above analysis and by using (9.28):

$$\dot{\widehat{V}}(t) \leq -c_1\widehat{V}(t) - [e^T(t)e(t) - (\widehat{\lambda}_M/\rho_0)\varpi^T(t)\varpi(t)]. \tag{9.36}$$

For any given $t > 0$, let $t_1, t_2, \ldots, t_{N_\sigma[0,t)}$ be the discontinuous time points of switching signal $\sigma(t)$ over $[0,t)$. By (9.36), one gets that

$$\widehat{V}(t) \leq \mu\widehat{V}(t_{N_\sigma[0,t)}^-) \exp(-c_1(t - t_{N_\sigma[0,t)}))$$
$$+ \int_{t_{N_\sigma[0,t)}}^t \exp(-c_1(t-s))(-e^T(s)e(s) + \frac{\widehat{\lambda}_M}{\rho_0}\varpi^T(s)\varpi(s))ds$$
$$\leq \mu\Big[\widehat{V}(t_{N_\sigma[0,t)-1}) \exp(-c_1(t_{N_\sigma[0,t)} - t_{N_\sigma[0,t)-1}))$$
$$+ \int_{t_{N_\sigma[0,t)-1}}^{t_{N_\sigma[0,t)}} \exp(-c_1(t_{N_\sigma[0,t)} - s))(-e^T(s)e(s)$$
$$+ \frac{\widehat{\lambda}_M}{\rho_0}\varpi^T(s)\varpi(s))ds\Big] \exp(-c_1(t - t_{N_\sigma[0,t)}))$$
$$+ \int_{t_{N_\sigma[0,t)}}^t \exp(-c_1(t-s))(-e^T(s)e(s) + \frac{\widehat{\lambda}_M}{\rho_0}\varpi^T(s)\varpi(s))ds, \tag{9.37}$$

where μ is defined in (9.14). By recursion, the following facts can be verified from (9.37) that

$$\widehat{V}(t) \leq \mu^{N_\sigma[0,t)} \exp(-c_1t)\widehat{V}(0)$$
$$+ \int_0^t \mu^{N_\sigma[s,t)} \exp(-c_1(t-s))\left(-e^T(s)e(s) + \frac{\widehat{\lambda}_M}{\rho_0}\varpi^T(s)\varpi(s)\right)ds$$
$$= \exp(-c_1t + N_\sigma[0,t)\ln\mu)\widehat{V}(0)$$

$$+ \int_0^t \exp(-c_1(t-s) + N_\sigma[s,t]\ln\mu) \left[-e^T(s)e(s) + \frac{\widehat{\lambda}_M}{\rho_0}\varpi^T(s)\varpi(s) \right] ds,$$

$$(9.38)$$

where $N_\sigma[s,t]$ represents the total number of discontinuous time points of switching signal $\sigma(t)$ over $[s,t)$ and satisfies

$$N_\sigma[s,t] \leq N_0 + (t-s)/\tau_a, \quad \forall s \leq t. \tag{9.39}$$

Then, based on the facts $\widehat{V}(0) = 0$ and $\widehat{V}(t) \geq 0$, it can be yielded from (9.38) and (9.39) that

$$\int_0^t \exp(-c_1(t-s))e^T(s)e(s)ds$$

$$\leq \frac{\widehat{\lambda}_M}{\rho_0} \int_0^t \exp\left(-c_1(t-s) + (N_0 + \frac{t-s}{\tau_a})\ln\mu\right) \varpi^T(s)\varpi(s)ds$$

$$= \frac{\widehat{\lambda}_M \exp(N_0\ln\mu)}{\rho_0} \int_0^t \exp(-c_2(t-s))\varpi^T(s)\varpi(s)ds, \tag{9.40}$$

where $c_2 = c_1 - (\ln\mu)/\tau_a > 0$. Thus, it can be derived from (9.40) that

$$\int_0^{+\infty} e^T(s)e(s)ds \leq \gamma^2 \int_0^{+\infty} \varpi^T(s)\varpi(s)ds, \tag{9.41}$$

where $\gamma = \sqrt{(\widehat{\lambda}_M c_1 \exp(N_0\ln\mu))/(\rho_0 c_2)}$.

Remark 9.5 *The finite $\mathbb{L}_2$-gain performance index of the considered CNSs with zero initial conditions is analyzed in Theorem 9.2. The relationships between the performance index γ and the system parameters are clearly provided in (9.30). Interestingly, it is found that a better disturbance rejection level will be yielded by enlarging ρ_0. Thus, though the free parameter ρ_0 is introduced to reduce the conservatism of (9.33), it can be actually seen as a tunable parameter for the performance index of closed-loop CNS. However, the larger the value of ρ_0, the lower the feasibility of ARI (9.28).*

9.4 CONSENSUS TRACKING AND ITS $\mathbb{L}_2$-GAIN PERFORMANCE OF CNSS WITH UNDIRECTED FIXED TOPOLOGY

Using tools from the stability theory of switched systems, consensus tracking and its $\mathbb{L}_2$-gain performance of CNS with a high-dimensional leader and directed switching topologies are analyzed in the last subsection. Note that the methodology employed in Theorem 9.2 is analyzing the performance index by using the MLFs-based approach. In practice, it is very important to know whether the CNS could attain a pre-specified disturbance rejection level by directly designing $\mathcal{H}_\infty$ controllers. Unlike the methodology utilized in Theorem 9.2 where the control parameters in the local controller and dynamic observer are constructed just for achieving consensus tracking qualitatively, the parameters in the local controller and dynamic observer developed in this section are designed by fully considering the pre-specified disturbance rejection performance. As a tradeoff, the interaction graph representing the communication relationship among the N followers in this section is restricted as an undirected graph as the decoupling technique will be employed in proving the main results. Actually, $\mathbb{L}_2$-gain performance synthesis problem is also called as the nonlinear $\mathcal{H}_\infty$ control problem.

Next, we give the following assumption, which will be employed in proving the analytical results of this subsection.

Assumption 9.3 *The interaction graph for the $N+1$ agents is fixed and the subgraph representing the communication relationship among the N followers is undirected.*

Without loss of any generality, denote by $\mathcal{G}$ the fixed interaction graph for the $N+1$ agents. The corresponding Laplacian matrix $\mathcal{L}$ can be written as:

$$\mathcal{L} = \begin{bmatrix} 0 & \mathbf{0}_N^T \\ \mathbf{P} & \overline{\mathcal{L}} \end{bmatrix}, \quad \overline{\mathcal{L}} = \begin{bmatrix} \sum_{j\in\mathcal{N}_1} a_{1j} & -a_{12} & \cdots & -a_{1N} \\ -a_{21} & \sum_{j\in\mathcal{N}_2} a_{2j} & \cdots & -a_{2N} \\ \vdots & \vdots & \ddots & \vdots \\ -a_{N1} & -a_{N2} & \cdots & \sum_{j\in\mathcal{N}_N} a_{Nj} \end{bmatrix}, \quad (9.42)$$

where $\mathbf{P} = -[a_{10}, \ldots, a_{N0}]^T$. Throughout this section, we suppose that Assumptions 9.2 and 9.3 hold. Hence, one may get that $\overline{\mathcal{L}}$ given in (9.42) is positive definite. Thus, one may let U be an orthogonal matrix such that $U^T \overline{\mathcal{L}} U = \text{diag}\{\overline{\lambda}_1, \ldots, \overline{\lambda}_N\}$, where $\overline{\lambda}_1, \ldots, \overline{\lambda}_N$ are the eigenvalues of $\overline{\mathcal{L}}$. Accordingly, the evolution of consensus tracking error system can be given as

$$\dot{e}(t) = \tilde{A}_c e(t) + (I_N \otimes \hat{C}^T)\varpi(t), \quad (9.43)$$

where $\tilde{A}_c = I_N \otimes \hat{A} + c\overline{\mathcal{L}} \otimes F\hat{C}$, $e(t) = [e_1^T(t), \ldots, e_N^T(t)]^T$, $e_i(t) = \bar{x}_i(t) - \zeta(t)$, $\varpi(t) = [\varpi_1^T(t), \ldots, \varpi_N^T(t)]^T$. The considered $\mathbb{L}_2$-gain performance synthesis problem of CNS with a single high-dimensional active leader given by (9.5) and followers given by (9.2) can be stated as follows. For a given $\gamma > 0$, find an appropriate controller (9.7) associated with state estimator (9.8) such that the following two conditions are satisfied: (1) The cooperative consensus tracking problem for the considered CNS is solved with $\varpi(t) = \mathbf{0}_{Nm}$, i.e., the zero fixed point of (9.43) with $\varpi(t) = \mathbf{0}_{Nm}$ is globally attractive, and (2)

$$\int_0^{+\infty} e^T(t)e(t)dt \leq \gamma^2 \int_0^{+\infty} \varpi^T(t)\varpi(t)dt. \quad (9.44)$$

Note that condition (1) holds if and only if $\tilde{A}_c$ is a Hurwitz matrix. Suppose that $\tilde{A}_c$ is a Hurwitz matrix, inequality (9.44) holds if and only if

$$\left\| (sI_{Nn} - \tilde{A}_c)^{-1}(I_N \otimes \hat{C}^T) \right\|_\infty < \gamma, \quad (9.45)$$

i.e.,

$$\sup_{\varpi\in\mathbb{R}} \overline{\sigma}[(i\varpi I_{Nn} - \tilde{A}_c)^{-1}(I_N \otimes \hat{C}^T)] < \gamma, \quad (9.46)$$

where s is a complex variable, i is the imaginary unit, $\|\cdot\|_\infty$ indicates the $\mathcal{H}_\infty$ norm. Furthermore, according to definition of $\mathcal{H}_\infty$ norm, one gets

$$\left\|(sI_{Nn} - \tilde{A}_c)^{-1}(I_N \otimes \hat{C}^T)\right\|_\infty$$

$$= \left\|(U^T \otimes I_n)(sI_{Nn} - \tilde{A}_c)^{-1}(I_N \otimes \hat{C}^T)(U \otimes I_n)\right\|_\infty$$

$$= \max_{i=1,\dots,N} \left\|\left[sI_n - (\hat{A} + c\tilde{\lambda}_i F \hat{C})\right]^{-1} \hat{C}^T\right\|_\infty. \tag{9.47}$$

Based upon the previous analysis, one may establish the following theorem.

Theorem 9.3 *Suppose that the communication topology of CNS with a high-dimensional active leader described by (9.5) and followers described by (9.2) is fixed, and Assumptions 9.2, 9.3 hold. Then, for any given $\gamma > 0$, cooperative consensus tracking problem of the considered CNS with $\mathbb{L}_2$-gain performance less than γ is solved by the controller (9.7) associated with state observer (9.8) if the control parameters are appropriately selected such that $F = -\hat{P}^{-1}\hat{C}^T$, $c > p_0/\tilde{\lambda}_{\min}$ for some $p_0 > 0$, where $\tilde{\lambda}_{\min} = \min_{i=1,\dots,N}\tilde{\lambda}_i$, $\hat{P}$ is a positive definite solution of ARI:*

$$\hat{P}\hat{A} + \hat{A}^T\hat{P} - 2p_0\hat{C}^T\hat{C} + I_n + \frac{1}{\gamma^2}\hat{P}\hat{C}^T\hat{C}\hat{P} < 0. \tag{9.48}$$

Proof 9.3 *According to the analysis given in (9.44)–(9.47), one knows that the cooperative consensus tracking problem of the considered CNS with a finite $\mathbb{L}_2$-gain performance index less than γ will be solved if and only if, for each $i = 1,\dots,N$, the following two conditions are simultaneously satisfied:*

(1) $\hat{A} + c\tilde{\lambda}_i F\hat{C}$ is a Hurwitz matrix;

(2) $\left\|\left[sI_n - (\hat{A} + c\tilde{\lambda}_i F\hat{C})\right]^{-1} \hat{C}^T\right\|_\infty < \gamma.$

Since $F = -\hat{P}^{-1}\hat{C}^T$ and $c > p_0/\tilde{\lambda}_{\min}$, one may get that, for each $i = 1,\dots,N$,

$$(\hat{A} + c\tilde{\lambda}_i F\hat{C})^T\hat{P} + \hat{P}(\hat{A} + c\tilde{\lambda}_i F\hat{C}) \leq \hat{A}^T\hat{P} + \hat{P}\hat{A}^T - 2p_0\hat{C}^T\hat{C} < 0, \tag{9.49}$$

where the last inequality is derived by using (9.48). Clearly, (9.49) implies that $\hat{A} + c\tilde{\lambda}_i F\hat{C}$ is a Hurwitz matrix, for each $i = 1,\dots,N$. Furthermore, according to the conditions $F = -\hat{P}^{-1}\hat{C}^T$ and $c > p_0/\tilde{\lambda}_{\min}$, the following inequalities can be obtained from (9.48):

$$(\hat{A} + c\tilde{\lambda}_i F\hat{C})^T\hat{P} + \hat{P}(\hat{A} + c\tilde{\lambda}_i F\hat{C}) + I_n + \frac{1}{\gamma^2}\hat{P}\hat{C}^T\hat{C}\hat{P}$$

$$\leq \hat{P}\hat{A} + \hat{A}^T\hat{P} - 2p_0\hat{C}^T\hat{C} + I_n + \frac{1}{\gamma^2}\hat{P}\hat{C}^T\hat{C}\hat{P} < 0. \tag{9.50}$$

According to the bounded real lemma, it can be derived from (9.50) that

$$\left\|\left[sI_n - (\hat{A} + c\tilde{\lambda}_i F\hat{C})\right]^{-1} \hat{C}^T\right\|_\infty < \gamma,$$

for each $i = 1,\dots,N$. This completes the proof.

Remark 9.6 *According to the Schur complement lemma, one knows that there is a positive definite matrix $\widehat{P}$ to make ARI (9.48) hold for some given $p_0 > 0$, $\gamma > 0$, if and only if there exists a positive definite matrix $\widehat{P}$ such that the following LMI holds:*

$$\begin{bmatrix} \widehat{P}\widehat{A} + \widehat{A}^T\widehat{P} - 2p_0\widehat{C}^T\widehat{C} + I_n & \frac{1}{\gamma}\widehat{P}\widehat{C}^T \\ \frac{1}{\gamma}\widehat{C}\widehat{P} & -I \end{bmatrix} < 0. \tag{9.51}$$

Thus, the best disturbance rejection index γ_{best} can be calculated by solving the following optimization problem:

Minimize γ
subject to LMI *(9.51)*, with $\widehat{P} > 0$, $p_0 > 0$.

Remark 9.7 *Under Assumptions 9.2 and 9.3, matrix $\overline{\mathcal{L}}$ defined in (9.42) is positive definite. In this case, $\overline{\mathcal{L}}$ is orthogonally similar to a positive definite diagonal matrix. Based on this fact, the cooperative consensus tracking problem with prescribed $\mathbb{L}_2$-gain performance specification has been equivalently converted to the $\mathcal{H}_\infty$ control problems of a set of decoupled systems with the same low dimensions.*

9.5 NUMERICAL SIMULATIONS

In this section, we validate the analytical results provided in the last sections by numerical example.

In this simulation example, the validity of theoretical results in Theorems 9.1 and 9.2 will be verified. The CNS has a Caltech Wireless Testbed vehicle (CWTV, acts as the leader) and 4 wheeled mobile robots (WMRs, act as the followers). The dynamics of the CWTV are described by (9.5) with $\zeta(t) = [\zeta_1(t), \ldots, \zeta_6(t)]^T \in \mathbb{R}^6$, and

$$\widehat{A} = \begin{bmatrix} 0 & 0 & 0 & 1 & 0 & 0 \\ 0 & 0 & 0 & 0 & 1 & 0 \\ 0 & 0 & 0 & 0 & 0 & 1 \\ 0 & 0 & -0.2003 & -0.2003 & 0 & 0 \\ 0 & 0 & 0.2003 & 0 & 0.2003 & 0 \\ 0 & 0 & 0 & 0 & 0 & -1.6129 \end{bmatrix},$$

$$\widehat{C} = \begin{bmatrix} 1 & 0 & 0 & 0 & 0 & 0 \\ 0 & 1 & 0 & 0 & 0 & 0 \end{bmatrix},$$

where $\zeta_1(t)$ and $\zeta_2(t)$ represent respectively the CWTV's positions along the x-axis and the y-axis, $\zeta_3(t)$ represents the CWTV's orientation, $\zeta_4(t) = \dot{\zeta}_1(t)$, $\zeta_5(t) = \dot{\zeta}_2(t)$, $\zeta_6(t) = \dot{\zeta}_3(t)$. It is obviously that $\vartheta_{\text{out}}(t) = [\zeta_1(t), \zeta_2(t)]^T \in \mathbb{R}^2$. Moreover, direct calculation gives rank$[\widehat{C} \ \widehat{C}\widehat{A} \ \ldots \ \widehat{C}\widehat{A}^5]^T = 6$ and thereby $(\widehat{C}, \widehat{A})$ is observable. Therefore Assumption 9.1 holds. The dynamics of the WMRs are described by (9.2) with $x_i(t) = [x_{i1}(t), x_{i2}(t)]^T$, $A_i = 0$, and $\varpi_i(t) = [0.1\varpi_{i1}(t), 0.1\varpi_{i2}(t)]^T \in \mathbb{R}^2$ in which

$$\varpi_{i1}(t) = \begin{cases} \sin(it), & 0 \leq t \leq 3, \\ 0, & \text{otherwise,} \end{cases} \quad \varpi_{i2}(t) = \begin{cases} \cos(it), & 0 \leq t \leq 3, \\ 0, & \text{otherwise,} \end{cases}$$

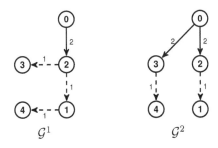

Figure 9.2 The communication graphs $\mathcal{G}^1$ and $\mathcal{G}^2$.

where $x_{i1}(t)$ and $x_{i2}(t)$ represent respectively the WMRs' positions along the x-axis and the y-axis, $i = 1, \ldots, 4$. Clearly, $\varpi_i(t) \in \mathbb{L}_2^2[0, +\infty)$.

The communication topologies are assumed to switch between $\mathcal{G}^1$ and $\mathcal{G}^2$ that are shown in Figure 9.2. Direct calculation gives that $\mu = 2$ and $\lambda_0 = 1.0561$. Since the matrix pair $(\widehat{C}, \widehat{A})$ is completely observable, it can be got from Remark 9.3 that LMI (9.16) is always feasible for any given positive scalar $c_0 > 0$. Set $c_0 = 1$, solving the LMI (9.16) gives that

$$
F = \begin{bmatrix} -2.5659 & 0.6863 & 4.8881 & -2.4646 & 1.7938 & -0.0264 \\ 0.6863 & -2.7996 & -3.4605 & 1.2614 & -3.3362 & 0.0154 \end{bmatrix}^T.
$$

Choose $c = 2 > 2/\lambda_0 = 1.8938$, it can be seen from Theorem 9.1 that cooperative consensus tracking of the closed-loop CNS without external disturbance can be ensured if the ADT $\tau_a > 0.6931\,\mathrm{s}$. In simulations, the underlying interaction graph is assumed to switch between graphs $\mathcal{G}^1$ and $\mathcal{G}^2$ every $0.7\,\mathrm{s}$. The trajectories of $\zeta_1(t)$, $x_{i1}(t)$ and $\zeta_2(t)$, $x_{i2}(t)$ are respectively plotted in Figs. 9.3 and 9.4, $i = 1, \ldots, 4$. The trajectories of the consensus tracking errors $\|e_i(t)\|$ are plotted in Figure 9.5 which indicate that the states of followers are able to asymptotically track the outputs of leader. This example validates the theoretical results in Theorem 9.1 very well.

9.6 CONCLUSIONS

In this chapter, consensus tracking for CNSs with a high-dimensional active leader and directed switching topologies has been studied, where the dynamics of followers may be subject to unknown disturbances. A class of cooperative controllers, together with local state estimators, has been designed to make the followers able to track the high-dimensional active leader. By using a combined tool from stability analysis of switched systems and nonsingular M matrix theory, sufficient criteria have been derived for achieving consensus tracking of CNSs without external disturbance under directed switching topologies. We further demonstrated that a finite $\mathbb{L}_2$-gain performance index for tracking errors against external disturbances can be ensured if some suitable conditions are satisfied. At last, the synthesis issue of designing a distributed controller to achieve a desired level of $\mathbb{L}_2$-gain performance for

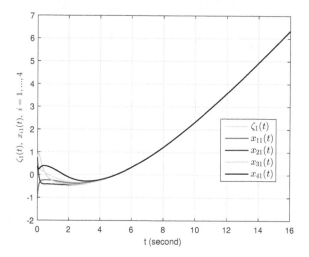

Figure 9.3 The agents' state trajectories $\zeta_1(t)$ and $x_{i1}(t)$, $i = 1, \ldots, 4$.

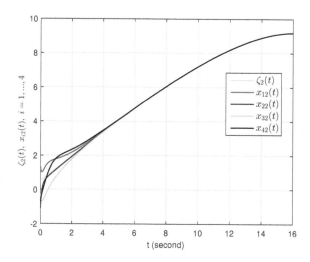

Figure 9.4 The agents' state trajectories $\zeta_2(t)$ and $x_{i2}(t)$, $i = 1, \ldots, 4$.

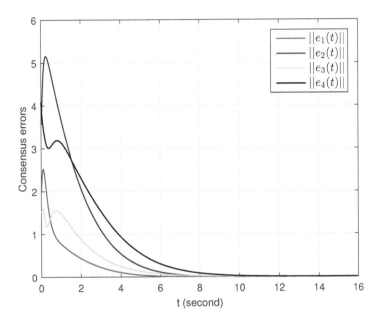

Figure 9.5 Trajectories of the consensus tracking errors $\|e_i(t)\|$, $i = 1, \ldots, 4$.

consensus tracking over an undirected fixed topology is addressed by using tools from $\mathcal{H}_\infty$ control theory.

CHAPTER 10

Neuro-adaptive consensus of CNSs with uncertain dynamics

This chapter studies the consensus problems for CNSs subject to uncertainties by using the neural network adaptive control approach. The achieved consensus is named as neuro-adaptive consensus. This chapter firstly presents the universal neural network approximation theory under which the uncertainties are linearized. Section 10.2 studies the practical consensus tracking of CNSs with a high-dimensional leader and directed switching topologies. And it is shown that the neuro-adaptive consensus error converges into a bounded set under the proposed controller. Section 10.3 solves the asymptotical consensus tracking problem for CNSs with a high-dimensional leader and directed fixed topology. Note that a favorability is that the neuro-adaptive consensus does not depend on any global information thereby the protocol is fully distributed. In Section 10.4, both practical and asymptotical neuro-adaptive containment problems for CNSs with multiple leaders under detail-balanced directed topology are studied. Note that the asymptotical neuro-adaptive containment is solved by proposing a second-order neuro-adaptive law.

10.1 INTRODUCTION

In most practical applications, uncertainties are inevitable due to the imprecise measurements, external disturbances, and interactions with the unknown environment. Hence, the question arises naturally: How to design the consensus algorithm for CNSs with uncertain dynamics? A major difficulty is to approximate the unknown uncertainties. In control engineering, neural networks (NNs) are usually employed as the function approximator to emulate the unknown function. And it has been proven that the radial basis function neural network (RBFNN) can approximate any continuous function over a compact set to arbitrary accuracy.

Theorem 10.1 (*NN universal approximation theorem*) *Let $f(x) : \mathbb{R}^n \mapsto \mathbb{R}^m$ be a continuous function. For a given compact set $\Omega \subset \mathbb{R}^n$ and any given positive*

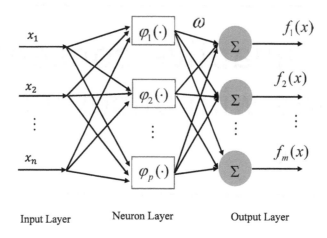

Figure 10.1 Structure of the RBFNN.

constant ϵ_M, there exists an ideal weight matrix $\omega \in \mathbb{R}^{p \times m}$ such that

$$f(x) = \omega^T \varphi(x) + \epsilon, \ \forall \ x \in \Omega, \tag{10.1}$$

where $\epsilon \in \mathbb{R}^m$ is the bounded function approximation error satisfying $\|\epsilon\| < \epsilon_M$, $\varphi(x) = [\varphi_1(x), \ldots, \varphi_p(x)]^T$ is the activation function.

A structure of the RBFNN is shown in Figure 10.1, where p denotes the number of neurons, $\varphi(x) = [\varphi_1(x), \ldots, \varphi_p(x)]^T$ is the activation function which can be selected as

$$\varphi_i(x) = \exp\left[\frac{-(x - \mu_i)^T (x - \mu_i)}{\nu_i}\right], \quad i = 1, \ldots, p, \tag{10.2}$$

where $\mu_i = [\mu_{i1}, \ldots, \mu_{in}]^T$ is the center of the receptive field and $\nu_i > 0$ is the width of the Gaussian function.

By using the NN universal approximation theorem, some distributed controllers together with adaptive NN weight updating laws were designed for uncertain CNSs [18, 24, 63]. And by using the Lyapunov stability theory, they showed that the consensus error was ultimately uniformly bounded (UUB) under undirected connected communication topology. Later, Sun and Geng [147] showed that asymptotical consensus tracking could be achieved in uncertain CNSs under undirected topology. But their approaches depend heavily on the 'linearity-in-parameters' (i.e., the NN approximation errors equal to 0) assumption. Inspired by the aforementioned works and some references therein, this chapter studies the neuro-adaptive consensus problem for uncertain CNSs with directed communication topology. This chapter is organized as follows: Section 10.2 studies the practical consensus tracking problem for uncertain CNSs with a high-dimensional leader under directed switching topologies. Section 10.3 studies the fully distributed consensus tracking problem for uncertain CNSs with a high-dimensional leader under directed fixed topology. Section 10.4 studies the containment problem for uncertain CNSs with multiple leaders under detail balanced directed topology.

10.2 PRACTICAL CONSENSUS TRACKING OF CNSS WITH A HIGH-DIMENSIONAL LEADER AND DIRECTED SWITCHING TOPOLOGIES

In this section, we discuss the neuro-adaptive consensus tracking problem for uncertain CNSs with a high-dimensional leader over directed fixed as well as directed switching topologies.

10.2.1 Model formulation

Consider a CNS consisting of a leader and N followers, where the leader is labelled as agent 0 and the N followers are, respectively, labelled as agents $1, \ldots, N$. The dynamics of the leader are described as

$$
\begin{aligned}
\dot{\theta}(t) &= \widehat{A}\theta(t) + g(\theta(t), t), \\
y(t) &= \widehat{C}\theta(t),
\end{aligned}
\tag{10.3}
$$

where $\theta(t) \in \mathbb{R}^{nm}$ is the state of the leader, n and m are two positive integers, $\widehat{A} \in \mathbb{R}^{nm \times nm}$ is the system matrix, nonlinear function $g(\cdot, \cdot) : \mathbb{R}^{nm} \times [0, +\infty) \mapsto \mathbb{R}^{nm}$ is a continuously differentiable vector-valued function satisfying the global Lipschitz condition

$$
\|g(\rho, t) - g(\varrho, t)\| \leq \gamma \|\rho - \varrho\|, \ \forall \ \rho, \varrho \in \mathbb{R}^{mn}, \ t \geq 0,
\tag{10.4}
$$

for some $\gamma \geq 0$. Suppose that the state of the leader can be partitioned as $\theta(t) = [\theta_1^T(t), \ldots, \theta_n^T(t)]^T$ with $\theta_i(t) \in \mathbb{R}^m$, $i = 1, \ldots, n$. Then, one may partition $\widehat{A} = [\hat{a}_{ij}]_{nm \times nm}$ into the following form:

$$
\widehat{A} = \begin{bmatrix}
\hat{A}_{11} & \hat{A}_{12} & \cdots & \hat{A}_{1n} \\
\hat{A}_{21} & \hat{A}_{22} & \cdots & \hat{A}_{2n} \\
\vdots & \vdots & \ddots & \vdots \\
\hat{A}_{n1} & \hat{A}_{n2} & \cdots & \hat{A}_{nn}
\end{bmatrix},
\tag{10.5}
$$

where $\hat{A}_{ij} \in \mathbb{R}^{m \times m}$, $i, j = 1, \ldots, n$. Furthermore, the nonlinear function $g(\theta(t), t)$ can be partitioned as $g(\theta(t), t) = [g_1^T(\theta_1(t), t), \ldots, g_n^T(\theta_n(t), t)]^T$, where $g_i(\theta_i(t), t) : \mathbb{R}^m \times [0, +\infty) \mapsto \mathbb{R}^m$ is a continuously differentiable vector-valued function. The vector $y(t) \in \mathbb{R}^m$ is the output of the leader which is assumed to be available to a subset of followers, the output matrix $\widehat{C} \in \mathbb{R}^{m \times nm}$. For the convenience of analysis, it is assumed that $\widehat{C} = [0, \ldots, 0, I_m] \in \mathbb{R}^{m \times nm}$ and the matrix pair $(\widehat{C}, \widehat{A})$ is detectable.

The dynamics of follower i, $i = 1, \ldots, N$, are given as

$$
\dot{x}_i(t) = Ax_i(t) + f_i(x_i(t)) + h_i(t) + u_i(t),
\tag{10.6}
$$

where $x_i(t) = [x_{i1}(t), \ldots, x_{im}(t)]^T \in \mathbb{R}^m$ is the state, $f_i(x_i(t)) : \mathbb{R}^m \mapsto \mathbb{R}^m$ is a continuous function that describe the unknown unmodeled dynamics, $h_i(t) \in \mathbb{R}^m$ are the unknown disturbances, $A \in \mathbb{R}^{m \times m}$ is a known matrix. By noticing that the nonlinear functions $f_i(x_i(t))$, $i = 1, \ldots, N$, in (10.6) are continuous, it can be thus obtained from the NN universal approximation theorem that $f_i(x_i(t))$ can be approximated on a compact set $\Omega \subset \mathbb{R}^m$ by

$$f_i(x_i(t)) = \omega_i^T \varphi_i(x_i(t)) + \epsilon_i, \ \forall \ x_i(t) \in \Omega, \tag{10.7}$$

where $\varphi_i(\cdot) : \mathbb{R}^m \mapsto \mathbb{R}^p$ is the activation function, $\omega_i \in \mathbb{R}^{p \times m}$ is the ideal NN weight matrix, $\epsilon_i \in \mathbb{R}^m$ is the NN approximation error. Suppose that $\max_{x_i(t) \in \Omega} \|\varphi_i(x_i(t))\|$ is finite, i.e., there exists a positive scalar $\varphi_{iM} > 0$ such that

$$\max_{x_i(t) \in \Omega} \|\varphi_i(x_i(t))\| \leq \varphi_{iM}, \tag{10.8}$$

for each $i = 1, \ldots, N$. Furthermore, one may let $\omega_{iM} = \|\omega_i\|_F$, where $\|\omega_i\|_F$ is the Frobenius norm of ω_i. Take $\varphi(x(t)) = [\varphi_1^T(x_1(t)), \ldots, \varphi_N^T(x_N(t))]^T$, $W = \text{diag}\{\omega_1, \ldots, \omega_N\}$, and $\epsilon = [\epsilon_1^T, \ldots, \epsilon_N^T]^T$, one then gets that there exist positive scalars φ_M, W_M, and ϵ_M such that $\|\varphi(x(t))\| \leq \varphi_M$, $\|W\|_F \leq W_M$, and $\|\epsilon\| \leq \epsilon_M$. The ideal approximating weight matrix ω_i in (10.7) is assumed to be unknown, i.e., ω_i can not be directly used for controller design. To compensate for unknown nonlinearities effectively, a new kind of NN-based estimators is designed for each follower to approximate the nonlinearities online. Define the local approximation of $f_i(x_i(t))$ as

$$\hat{f}_i(x_i(t)) = \hat{\omega}_i^T(t)\varphi_i(x_i(t)), \tag{10.9}$$

where $\hat{\omega}_i(t) \in \mathbb{R}^{p \times m}$ is the current estimate of the ideal weights for follower i, $i = 1, \ldots, N$. For notational brevity, let $\widehat{W}(t) = \text{diag}\{\hat{\omega}_1(t), \ldots, \hat{\omega}_N(t)\}$.

For practicability, it is assumed that the initial condition $\theta(0)$ is unknown to each follower and only the output information $y(t)$ of the leader can be measured by some designed followers. The control goal in the present section is to design some consensus tracking protocols $u_i(t)$, $i = 1, \ldots, N$, under which the states of the followers ultimately track the output of the single leader (given by (10.3)) with bounded residual errors.

In this section, the controllers $u_i(t)$, $i = 1, \ldots, N$, are designed as follows

$$u_i(t) = u_i^{ct}(t) + u_i^{ft}(t), \tag{10.10}$$

where $u_i^{ct}(t)$ and $u_i^{ft}(t)$ are, respectively, the observer-based compensation term and the local feedback term given as

$$u_i^{ct}(t) = \tilde{A}x_i(t) + g_n(z_i^n(t), t) + \sum_{j=1}^{n-1} \hat{A}_{nj} z_i^j(t) - \alpha F_n \eta_i(t),$$

$$u_i^{ft}(t) = -\beta(x_i(t) - z_i^n(t)) - \hat{\omega}_i^T(t)\varphi_i(x_i(t)), \tag{10.11}$$

where $\tilde{A} = \hat{A}_{nn} - A$, $z_i^j(t) \in \mathbb{R}^m$, $i = 1, \ldots, N$, $j = 1, \ldots, n$, $z_i(t) = [(z_i^1(t))^T, \ldots, (z_i^n(t))^T]^T \in \mathbb{R}^{nm}$ is the state of the estimator embedded at follower i that is given by

$$\dot{z}_i^k(t) = g_k(z_i^k(t), t) + \sum_{j=1}^{n} \hat{A}_{kj} z_i^j(t) - \alpha F_k \eta_i(t), \qquad (10.12)$$

where $\hat{A}_{kj}$, $k = 1, \ldots, n$, $j = 1, \ldots, n$, are given in (10.5), α and β are positive scalars to be selected, and

$$\eta_i(t) = \sum_{j=1}^{N} a_{ij}(z_j^n(t) - z_i^n(t)) + a_{i0}(y(t) - z_i^n(t)), \qquad (10.13)$$

$F = [F_1^T, \ldots, F_n^T]^T$ is the feedback gain matrix, and $\hat{\omega}_i(t)$ is the estimation of $\omega_i(t)$, $i = 1, \ldots, N$.

Since the dimensions of the leader's and the followers' dynamics are different, the CNS under consideration is indeed heterogeneous. Additionally, only some partial information of the leader is available for some followers. To make the states of each follower ultimately synchronize to the output of the leader with bounded residual errors, the controller $u_i(t)$ associated with an estimator (10.12) is constructed. As it will be seen that the estimator $z_i(t)$, for each $i = 1, \ldots, N$, can be regarded as a full-order observer of the high-dimensional leader. Typically, only the relative local information between agents and their neighbors can be used for controller design within the context of CNSs. In the present design framework, the vector $\eta_i(t)$ defined in (10.13) represents the relative local information available to follower i, $i = 1, \ldots, N$. However, to compensate the effect of unmeasurable states of high-dimensional leader on consensus tracking, the compensation term $u_i^{ct}(t)$ defined in (10.11) is utilized. Furthermore, the local feedback term $u_i^{ft}(t)$ is designed to achieve consensus tracking and reduce the effect of followers' unknown unmodeled dynamics on consensus tracking. It is also worth pointing out that another possible way to solve such a distributed consensus tracking problem with partial information of the leader is the distributed output regulation-based approach [61].

Before moving on, the following assumptions are made.

Assumption 10.1 *The graph $\mathcal{G}$ contains a directed spanning tree with agent 0 (i.e., the leader) being the root.*

Assumption 10.2 *There exists a positive scalar h_M such that $\|h(t)\| \leq h_M$ holds for all $t \geq 0$, where $h(t) = [h_1^T(t), \ldots, h_N^T(t)]^T$ and $h_i(t)$ represents the external disturbances given in (10.6).*

Assumption 10.3 *The leader's trajectory is in a bounded region, i.e., $\|\theta(t)\| < \theta_M$ holds for all $t \geq 0$ and a scalar $\theta_M > 0$.*

Since the leader has no neighbors, one has that $a_{0j} = 0$, for all $j = 1, \ldots, N$.

Thus, the Laplacian matrix $\mathcal{L}$ of topology $\mathcal{G}$ can be partitioned as

$$\mathcal{L} = \begin{bmatrix} 0 & \mathbf{0}_N^T \\ \mathbf{P} & \overline{\mathcal{L}} \end{bmatrix}, \quad \overline{\mathcal{L}} = \begin{bmatrix} \sum_{j\in\mathcal{N}_1} a_{1j} & -a_{12} & \cdots & -a_{1N} \\ -a_{21} & \sum_{j\in\mathcal{N}_2} a_{2j} & \cdots & -a_{2N} \\ \vdots & \vdots & \ddots & \vdots \\ -a_{N1} & -a_{N2} & \cdots & \sum_{j\in\mathcal{N}_N} a_{Nj} \end{bmatrix},$$

where $\mathbf{P} = -[a_{10}, \ldots, a_{N0}]^T \in \mathbb{R}^N$. Here, $a_{i0} > 0$ if and only if follower i can directly sense the output information of the leader, $i = 1, \ldots, N$. Under Assumption 10.1, it can be obtained from Lemma 2.15 that there exists a positive definite diagonal matrix $\Phi = \text{diag}\{\phi_1, \ldots, \phi_N\}$ such that $\overline{\mathcal{L}}^T \Phi + \Phi \overline{\mathcal{L}} > 0$, where Φ satisfies $\overline{\mathcal{L}}^T [\phi_1, \ldots, \phi_N]^T = \mathbf{1}_N$. For notational brevity, let

$$\lambda_0 = \lambda_{\min}(\overline{\mathcal{L}} + \Phi^{-1}\overline{\mathcal{L}}^T \Phi). \tag{10.14}$$

Note that $\lambda_0 > 0$ if Assumption 10.1 holds.

We are now in a position to formulate the consensus tracking problem considered in this section.

Definition 10.1 *The consensus tracking problem with residual error $\varpi > 0$ for the CNS consisting of the high-dimensional leader (10.3) and the followers (10.6) equipped with the controller (10.10) is said to be achieved if $\lim_{t\to\infty}\|x_i(t) - y(t)\| \leq \varpi$ for all $i = 1, \ldots, N$.*

It is worth noting that, despite that many significant achievements have been made on achieving consensus tracking/pinning synchronization for homogeneous networking systems, we continue to lack tools to efficiently guarantee consensus tracking for CNSs with unknown dynamics and a high-dimensional leader. The main challenges are twofold: Firstly, to obtain the consensus tracking error system, one needs to reconstruct the unmeasurable dynamics of a high-dimensional leader by designing distributed observers; and secondly, the effects of unmodeled dynamics and unknown external disturbances on consensus tracking should be coped with online while only local relative information is available.

10.2.2 CNSs with fixed topology

In this subsection, neuro-adaptive consensus of CNSs with a fixed communication topology is addressed.

Let $e(t) = [e_1^T(t), \ldots, e_N^T(t)]^T$, $e_i(t) = z_i(t) - \theta(t)$, $i = 1, \ldots, N$, it can be obtained from (10.3) and (10.12) that

$$\dot{e}(t) = \left(I_N \otimes \widehat{A} + \alpha \overline{\mathcal{L}} \otimes F\widehat{C}\right) e(t) + \tilde{g}(e(t), t), \tag{10.15}$$

where $\tilde{g}(e(t), t) = [(g(z_1(t), t))^T, \ldots, (g(z_N(t), t))^T]^T - \mathbf{1}_N \otimes g(\theta(t), t)$. Furthermore, some calculations give that

$$\dot{\delta}_i(t) = \widehat{A}_{nn}\delta_i(t) + f_i(x_i(t), t) - \hat{\omega}_i^T(t)\varphi_i(x_i(t)) - \beta\delta_i(t) + h_i(t)$$

$$= \hat{A}_{nn}\delta_i(t) - \tilde{\omega}_i^T(t)\varphi_i(x_i(t)) - \beta\delta_i(t) + \epsilon_i + h_i(t), \tag{10.16}$$

where $\delta_i(t) = x_i(t) - z_i^n(t)$, $\tilde{\omega}_i(t) = \hat{\omega}_i(t) - \omega_i$, $i = 1, \ldots, N$. Set $\delta(t) = [\delta_1^T(t), \ldots, \delta_N^T(t)]^T$, one has

$$\dot{\delta}(t) = (I_N \otimes \hat{A}_{nn})\delta(t) - \widetilde{W}^T(t)\varphi(x(t)) - \beta\delta(t) + \epsilon + h(t), \tag{10.17}$$

where $\widetilde{W}(t) = \text{diag}\{\tilde{\omega}_1(t), \ldots, \tilde{\omega}_N(t)\}$, $\varphi(x(t)) = [(\varphi_1(x_1(t)))^T, \ldots, (\varphi_N(x_N(t)))^T]^T$, $\epsilon = [\epsilon_1^T, \ldots, \epsilon_N^T]^T$, $h(t) = [h_1^T(t), \ldots, h_N^T(t)]^T$.

Before moving forward, a multi-step design procedure is given to select the control parameters of controller (10.10) and observer (10.12) under the fixed topology $\mathcal{G}$.

Algorithm 10.1 *Under Assumptions 10.1–10.3, and the condition that $(\hat{C}, \hat{A})$ is detectable, the control parameters of controller (10.10) and observer (10.12) can be designed as follows.*

(1) Select $\alpha > 0$ and $c_0 > 0$. Then, solve the LMI

$$\begin{bmatrix} \hat{A}^T P + P\hat{A} - \alpha\lambda_0\hat{C}^T\hat{C} + I_{nm} + c_0 P & \gamma P \\ \gamma P & -I_{nm} \end{bmatrix} < 0 \tag{10.18}$$

to get a matrix $P > 0$, where λ_0 is defined in (10.14). Then, choose $F = -P^{-1}\hat{C}^T$.

(2) Select $c_1 > 0$, $c_2 > 0$, and $\beta > (\chi_{\max} + c_2/2)$, where $\chi_{\max}$ represents the maximal real part of the eigenvalues of $\hat{A}_{nn}$. Solve the LMI

$$\hat{A}_{nn}^T Q + Q\hat{A}_{nn} - (2\beta - c_2)Q < 0 \tag{10.19}$$

to get a matrix $Q > 0$. Then, set the neuro-adaptive tuning law as

$$\dot{\hat{\omega}}_i(t) = \Gamma_{\omega_i}\left[\varphi_i(x_i(t))\delta_i^T(t)Q - c_1\hat{\omega}_i(t)\right], \tag{10.20}$$

where $\Gamma_{\omega_i} > 0$ and $\delta_i(t) = x_i(t) - z_i^n(t)$, $i = 1, \ldots, N$.

Then, one can establish the following theorem which summarizes the main results of this subsection.

Theorem 10.2 *Suppose that Assumptions 10.1–10.3 hold, LMIs (10.18) and (10.19) admit some feasible solutions. Then, the consensus tracking problem defined in Definition 10.1 for the CNS with leader given by (10.3) and followers given by (10.6) can be solved by the neuro-adaptive protocol (10.10) associated with the observer (10.12) with control parameters designed by Algorithm 10.1. Specially, for each $i = 1, \ldots, N$,*

$$\lim_{t\to\infty}\|x_i(t) - y(t)\| \leq \varpi, \tag{10.21}$$

where $\varpi = \sqrt{\frac{V_2(t_0)}{\Upsilon_{\min}}} + \frac{\|\nu\|\Upsilon_{\max}}{\tilde{c}\Upsilon_{\min}}$, $V_2(t_0) = \delta^T(t_0)(I_N \otimes Q)\delta(t_0) + \text{tr}(\widetilde{W}^T(t_0)\Gamma_W^{-1}\widetilde{W}(t_0))$, $\Upsilon_{\min} = \min_{i=1,\ldots,N}\{\lambda_{\min}(Q), \Gamma_{\omega_i}^{-1}\}$, $\Upsilon_{\max} = \max_{i=1,\ldots,N}\{\lambda_{\max}(Q), \Gamma_{\omega_i}^{-1}\}$, $\nu = [2\kappa_0\|Q\|_2, 2c_1 W_M]$, $\kappa_0 = \epsilon_M + h_M$, $\tilde{c} = \min\{c_2\lambda_{\min}(Q), 2c_1\}$, and matrix $Q > 0$ is a feasible solution of (10.19).

Proof 10.1 *Construct the following Lyapunov function for the observer error system (10.15):*

$$V_1(t) = e^T(t)(\Phi \otimes P)e(t), \tag{10.22}$$

where P is a solution of (10.18). Taking the time derivative of $V_1(t)$ along the trajectories of (10.15) yields

$$\dot{V}_1(t) = e^T(t)\left[\Phi \otimes (\widehat{A}^T P + P\widehat{A}) + \alpha(\Phi\overline{\mathcal{L}} + \overline{\mathcal{L}}^T\Phi) \otimes PF\widehat{C}\right]e(t)$$
$$+ 2e(t)^T(\Phi \otimes P)\widetilde{g}(e(t), t), \tag{10.23}$$

where $\widetilde{g}(e(t), t) = [(g(z_1(t), t))^T, \ldots, (g(z_N(t), t))^T]^T - \mathbf{1}_N \otimes g(\theta(t), t)$. By Lemma 2.10, it can be obtained from (10.4), (10.23), and the fact $F = -P^{-1}\widehat{C}^T$ that

$$\dot{V}_1(t) \le e^T(t)\left[\Phi \otimes (\widehat{A}^T P + P\widehat{A}) - \alpha(\Phi\overline{\mathcal{L}} + \overline{\mathcal{L}}^T\Phi) \otimes \widehat{C}^T\widehat{C}\right]e(t)$$
$$+ \sum_{i=1}^N \phi_i e_i^T(t)(\gamma^2 P^2 + I_{nm})e_i(t). \tag{10.24}$$

By using Lemma 2.7 and noticing the fact that $\widehat{C}^T\widehat{C}$ is a semi-positive definite matrix, it can be derived from (10.24) that

$$\dot{V}_1(t) \le e^T(t)[\Phi \otimes (\widehat{A}^T P + P\widehat{A} - \alpha\lambda_0\widehat{C}^T\widehat{C} + \gamma^2 P^2 + I_{nm})]e(t), \tag{10.25}$$

where $\lambda_0 = \lambda_{\min}(\overline{\mathcal{L}} + \Phi^{-1}\overline{\mathcal{L}}^T\Phi)$. According to step (1) of Algorithm 10.1, it can be thus obtained from (10.25) that

$$\dot{V}_1(t) \le -c_0 V_1(t), \tag{10.26}$$

where $c_0 > 0$ is given in Algorithm 10.1. By the fact that $\|\theta(t)\| \le \theta_M$, it can be concluded from (10.26) that the trajectory $z(t)$ for any given initial condition $z(t_0)$ is uniformly bounded. Furthermore, it can be further obtained that

$$\|e(t)\| \le \sqrt{\frac{V_1(t_0)}{\phi_{\min}\lambda_{\min}(P)}} \exp\left(-\frac{c_0(t - t_0)}{2}\right), \ \forall\, t \ge t_0, \tag{10.27}$$

where $\phi_{\min} = \min_{i=1,2,\ldots,N}\phi_i$, $\lambda_{\min}(P)$ represents the minimal eigenvalue of P. The above analysis indicates that $z_i^n(t)$ will converge to $y(t)$ with an exponential decay rate $c_0/2$.

Construct the following Lyapunov function for the consensus tracking error system (10.17):

$$V_2(t) = \delta^T(t)(I_N \otimes Q)\delta(t) + \text{tr}(\widetilde{W}^T(t)\Gamma_W^{-1}\widetilde{W}(t)), \tag{10.28}$$

where Q is a solution of (10.18). Taking the time derivative of $V_2(t)$ along the trajectories of (10.17) gives

$$\dot{V}_2(t) = \delta^T(t)\left[I_N \otimes (\widehat{A}_{nn}^T Q + Q\widehat{A}_{nn} - 2\beta Q)\right]\delta(t)$$
$$+ 2\delta^T(t)(I_N \otimes Q)\left(-\widetilde{W}^T(t)\varphi(x(t)) + \epsilon + \omega(t)\right)$$

$$+ 2\mathrm{tr}(\widetilde{W}^T(t)\Gamma_W^{-1}\dot{\widehat{W}}(t)), \qquad (10.29)$$

where $\Gamma_W^{-1} = \mathrm{diag}\{\Gamma_{\omega_1}^{-1}, \ldots, \Gamma_{\omega_N}^{-1}\}$. Substituting (10.20) into (10.29) and considering (10.19), one gets

$$\dot{V}_2(t) \leq - c_2\delta^T(t)(I_N \otimes Q)\delta(t) + 2\delta^T(t)(I_N \otimes Q)(\epsilon + \omega(t))$$
$$- 2c_1\mathrm{tr}(\widetilde{W}^T(t)\widehat{W}(t)),$$

where c_1 and c_2 are defined in step (2) of Algorithm 10.1. Since $\|\epsilon\| \leq \epsilon_M$, $\|h(t)\| \leq h_M$, $\|W\|_F \leq W_M$, and $\|\widetilde{W}(t)\|_F = \sqrt{\mathrm{tr}(\widetilde{W}^T(t)\widetilde{W}(t))}$, some calculations give that

$$\dot{V}_2(t) \leq - c_2\lambda_{\min}(Q)\|\delta(t)\|^2 + 2\kappa_0\|\delta(t)\|\|Q\|_2$$
$$- 2c_1\|\widetilde{W}(t)\|_F^2 + 2c_1\|\widetilde{W}(t)\|_F W_M, \qquad (10.30)$$

where $\kappa_0 = \epsilon_M + h_M$. Let $\varsigma(t) = [\|\delta(t)\|, \|\widetilde{W}(t)\|_F]^T$, it can be obtained from (10.30) that

$$\dot{V}_2(t) \leq -\varsigma^T(t)\Psi\varsigma(t) + \nu\varsigma(t), \qquad (10.31)$$

where $\Psi = \begin{bmatrix} c_2\lambda_{\min}(Q) & 0 \\ 0 & 2c_1 \end{bmatrix}$ and $\nu = [2\kappa_0\|Q\|_2, 2c_1W_M]$. Take $\tilde{c} = \min\{c_2\lambda_{\min}(Q), 2c_1\}$, it thus can be derived from (10.31) that

$$\dot{V}_2(t) \leq -\tilde{c}\|\varsigma(t)\|^2 + \|\nu\|\|\varsigma(t)\|. \qquad (10.32)$$

Clearly, $\tilde{c}$ in (10.32) is larger than zero since both $c_2\lambda_{\min}(Q)$ and $2c_1$ are positive scalars. By the fact $\Upsilon_{\min}\|\varsigma(t)\|^2 \leq V_2(t) \leq \Upsilon_{\max}\|\varsigma(t)\|^2$, with $\Upsilon_{\min} = \min_{i=1,\ldots,N}\{\lambda_{\min}(Q), \Gamma_{W_i}^{-1}\}$ and $\Upsilon_{\max} = \max_{i=1,2,\ldots,N}\{\lambda_{\max}(Q), \Gamma_{W_i}^{-1}\}$, one has

$$\dot{V}_2(t) \leq -\hat{c}V_2(t) + \bar{c}\sqrt{V_2(t)}, \qquad (10.33)$$

i.e.,

$$\frac{d}{dt}(\sqrt{V_2(t)}) \leq -\hat{c}\sqrt{V_2(t)} + \bar{c}, \qquad (10.34)$$

where $\hat{c} = \tilde{c}/\Upsilon_{\max}$ and $\bar{c} = \|\nu\|/\sqrt{\Upsilon_{\min}}$. Integrating both sides of (10.34) from t_0 to t yields

$$\sqrt{V_2(t)} \leq \sqrt{V_2(t_0)} \exp\left(-\frac{\hat{c}}{2}(t - t_0)\right) + \frac{\bar{c}}{\hat{c}}\left(1 - \exp\left(-\frac{\hat{c}}{2}(t - t_0)\right)\right)$$
$$\leq \sqrt{V_2(t_0)} + \frac{\bar{c}}{\hat{c}}.$$

This means that $\|\delta(t)\|$ is uniformly bounded for any given $V_2(t_0)$. Noticing that $\|\theta(t)\|$ is also uniformly bounded, it can be obtained that $y(t)$ and $x_i(t)$, for all $i = 1, \ldots, N$, are uniformly bounded. On the other hand, it can be obtained from (10.28) that

$\delta^T(t)(I_N \otimes Q)\delta(t) \leq V_2(t)$, *for all* $t \geq 0$. *Then, based on the above analysis, one may get*

$$\|\delta(t)\| \leq \varpi, \tag{10.35}$$

where $\varpi = \sqrt{\frac{V_2(t_0)}{\Upsilon_{\min}}} + \frac{\|\nu\|\Upsilon_{\max}}{\tilde{c}\Upsilon_{\min}}$, $\Upsilon_{\min} = \min_{i=1,\dots,N}\{\lambda_{\min}(Q),\Gamma_{\omega_i}^{-1}\}$, $\Upsilon_{\max} = \max_{i=1,\dots,N}\{\lambda_{\max}(Q),\Gamma_{\omega_i}^{-1}\}$, $\nu = [2\kappa_0\|Q\|_2, 2c_1 W_M]$, $\kappa_0 = \epsilon_M + h_M$, *and* $\tilde{c} = \min\{c_2\lambda_{\min}(Q), 2c_1\}$.

Combining (10.27) and (10.35) give that $\lim_{t\to\infty}\|x_i(t) - y(t)\| \leq \varpi$.

Remark 10.1 *From the proof of Theorem 10.2, it can be seen that the states of all followers are uniformly bounded for an arbitrarily given initial condition. It is thus reasonable to assume that* $x_i(t)$, $i = 1,\dots,N$, *are contained in a compact set* $\Omega \subset \mathbb{R}^m$. *Furthermore, it is also worth noting that the size of the compact set* Ω *is not needed for the protocol design which can be set as large as needed in practice. Also, the positives scalars* W_M, ϵ_M, h_M, *and* θ_M, *could be actually unknown since none of them is involved in the protocol design.*

Remark 10.2 *To solve the consensus tracking problem in CNS consisting of the leader (10.3) and the followers (10.6), each follower needs to maintain a distributed observer (10.12) to estimate the states of the high-dimensional leader. The local feature of the observer (10.12), as reflected in fact that only some local information has been used for observer design, is much favorable in some practical applications. This indicates that it is unnecessary to employ a centralized estimator which having the ability to communicate with all followers for solving such a consensus tracking problem (stated in Definition 10.1).*

Remark 10.3 *The feasibility problems of LMIs (10.18) and (10.19) are analyzed as follows. The feasibility condition of LMI (10.18) is first provided. Since the positive scalars* α *and* c_0 *are free parameters in LMI (10.18), it can be obtained from the continuity theory that the feasibility of (10.18) is equivalent to the following feasible problem: There exist a positive scalar* α *and a positive definite matrix* P *such that the following algebraic Riccati inequality (ARI) holds:*

$$\widehat{A}^T P + P\widehat{A} - \alpha\widehat{C}^T\widehat{C} + I_{nm} + \gamma^2 P^2 < 0, \tag{10.36}$$

where γ *is the Lipschitz constant given in (10.4). On the other hand, it can be obtained from Finsler's lemma that there exist a positive scalar* α *and a matrix* $P > 0$ *such that ARI (10.36) holds if and only if there exist matrices* $\widehat{F} \in \mathbb{R}^{nm\times m}$ *and* $P > 0$ *such that*

$$\widehat{A}^T P + P\widehat{A} - \widehat{F}\widehat{C} - \widehat{C}^T\widehat{F}^T + I_{nm} + \gamma^2 P^2 < 0. \tag{10.37}$$

Without loss of any generality, let $\widehat{F} = PS$. *Then, one gets that LMI (10.18) is feasible if and only if there exist a matrix* $S \in \mathbb{R}^{nm\times m}$ *and* $P > 0$, *such that the following ARI holds:*

$$(\widehat{A} - S\widehat{C})^T P + P(\widehat{A} - S\widehat{C}) + I_{nm} + \gamma^2 P^2 < 0. \tag{10.38}$$

By the bounded real lemma in $\mathcal{H}_\infty$ control theory, one has ARI (10.38) holds if and only if there exists a matrix $S \in \mathbb{R}^{nm \times m}$ such that

$$\left\| \gamma \left[sI_{nm} - (\widehat{A} - S\widehat{C}) \right]^{-1} \right\|_\infty < 1, \tag{10.39}$$

i.e.,

$$\left\| \left[sI_{nm} - (\widehat{A} - S\widehat{C}) \right]^{-1} \right\|_\infty < \frac{1}{\gamma}. \tag{10.40}$$

This means that LMI (10.18) is feasible if and only if there exists a matrix $S \in \mathbb{R}^{nm \times m}$ such that (10.40) holds. It can be also derived from the above analysis that the detectability of $(\widehat{C}, \widehat{A})$ is a necessary but not sufficient condition for the feasibility of LMI (10.18). Next, the feasibility of LMI (10.19) is analyzed. According to fact that $\beta > (\chi_{\max} + c_2/2)$ for some $c_2 > 0$ and $\chi_{\max}$ is the maximal real part of the eigenvalues of $\widehat{A}_{nn}$, one knows that LMI (10.19) is always feasible. Indeed, one feasible solution for LMI (10.19) is $Q = I_m$.

10.2.3 CNSs with switching topologies

Based on the analysis provided in the last subsection, neuro-adaptive consensus tracking problem for CNS (10.3) and (10.6) with periodic switching and directed communication topology is studied in this subsection.

Suppose that there exists an infinite sequence of uniformly bounded non-overlapping time intervals $[t_k, t_{k+1})$, $k \in \mathbb{N}$, with $t_0 = 0$, $t_{k+1} - t_k = \tau > 0$, over which the communication topology is time-invariant. The time sequence $t_1, t_2, \ldots,$ is called the switching sequence, at which the communication topology switches. For expressional convenience, introduce a switching signal $\sigma(t) : [0, +\infty) \mapsto \{1, \ldots, \kappa\}$. Then, let $\mathcal{G}^{\sigma(t)}$ be the interaction graph of the considered CNS at time t. For the convenience of analysis, it is assumed that the switching graph $\mathcal{G}^{\sigma(t)} \in \widehat{\mathcal{G}}$ for all $t \geq 0$, where $\widehat{\mathcal{G}} = \{\mathcal{G}^1, \ldots, \mathcal{G}^\kappa\}$, $\kappa > 1$.

Assumption 10.4 *Each graph $\mathcal{G}^i$ contains a directed spanning tree with node 0 (i.e., the leader) being the root, $i = 1, \ldots, \kappa$.*

Similar to the above subsection, the Laplacian matrix $\mathcal{L}^{\sigma(t)}$ of graph $\mathcal{G}^{\sigma(t)}$ can be written as

$$\mathcal{L}^{\sigma(t)} = \begin{bmatrix} 0 & \mathbf{0}_N^T \\ \mathbf{P}^{\sigma(t)} & \overline{\mathcal{L}}^{\sigma(t)} \end{bmatrix},$$

$$\overline{\mathcal{L}}^{\sigma(t)} = \begin{bmatrix} \sum_{j \in \mathcal{N}_1} a_{1j}^{\sigma(t)} & -a_{12}^{\sigma(t)} & \cdots & -a_{1N}^{\sigma(t)} \\ -a_{21}^{\sigma(t)} & \sum_{j \in \mathcal{N}_2} a_{2j}^{\sigma(t)} & \cdots & -a_{2N}^{\sigma(t)} \\ \vdots & \vdots & \ddots & \vdots \\ -a_{N1}^{\sigma(t)} & -a_{N2}^{\sigma(t)} & \cdots & \sum_{j \in \mathcal{N}_N} a_{Nj}^{\sigma(t)} \end{bmatrix},$$

where $\mathbf{P}^{\sigma(t)} = -[a_{10}^{\sigma(t)}, \ldots, a_{N0}^{\sigma(t)}]^T$. Under Assumption 10.4, one has that there exist positive vectors $\phi^{\sigma(t)} = \left((\overline{\mathcal{L}}^{\sigma(t)})^T \right)^{-1} 1_N$ such that

$$\Phi^{\sigma(t)} \overline{\mathcal{L}}^{\sigma(t)} + \left(\overline{\mathcal{L}}^{\sigma(t)} \right)^T \Phi^{\sigma(t)} > 0, \tag{10.41}$$

where $\Phi^{\sigma(t)} = \text{diag}\{\phi_1^{\sigma(t)}, \dots, \phi_N^{\sigma(t)}\}$ and $\phi^{\sigma(t)} = [\phi_1^{\sigma(t)}, \dots, \phi_N^{\sigma(t)}]^T$. Based on the above analysis, one has that all the eigenvalues of $\overline{\mathcal{L}}^{\sigma(t)} + (\Phi^{\sigma(t)})^{-1} (\overline{\mathcal{L}}^{\sigma(t)})^T \Phi^{\sigma(t)}$ are real and positive. For notational convenience, let

$$\check{\lambda}_0 = \min_{i=1,\dots,\kappa} \check{\lambda}_0^i, \tag{10.42}$$

where $\check{\lambda}_0^i = \lambda_{\min} \left(\overline{\mathcal{L}}^i + (\Phi^i)^{-1} \left(\overline{\mathcal{L}}^i \right)^T \Phi^i \right)$ for $i = 1, \dots, \kappa$. The observer error system and the consensus tracking error system can be respectively obtained as

$$\dot{e}(t) = \left(I_N \otimes \widehat{A} + \alpha \overline{\mathcal{L}}^{\sigma(t)} \otimes F\widehat{C} \right) e(t) + \tilde{g}(e(t), t), \tag{10.43}$$

$$\dot{\delta}(t) = (I_N \otimes \widehat{A}_{nn})\delta(t) - \widetilde{W}^T(t)\varphi(x(t)) - \beta\delta(t) + \epsilon + h(t), \tag{10.44}$$

where the notations are defined the same as those in the last subsection.

To proceed on, a multi-step design procedure is given for selecting the control parameters of controller (10.10) and observer (10.12) under switching topologies.

Algorithm 10.2 *Under Assumptions 10.1, 10.2, and 10.4, and the condition that $(\widehat{C}, \widehat{A})$ is detectable, the control parameters of controller (10.10) and observer (10.12) are designed as follows.*

(1) Select $\alpha > 0$ and $\check{c}_0 > 0$. Solve the LMI

$$\begin{bmatrix} \widehat{A}^T P + P\widehat{A} - \alpha\check{\lambda}_0 \widehat{C}^T \widehat{C} + I_{nm} + \check{c}_0 P & \gamma P \\ \gamma P & -I_{nm} \end{bmatrix} < 0 \tag{10.45}$$

to get $P > 0$, where $\check{\lambda}_0$ is defined in (10.42). Then, set $F = -P^{-1}\widehat{C}^T$.

(2) Select $\check{c}_1 > 0$, $\check{c}_2 > 0$, and $\beta > (\chi_{\max} + \check{c}_2/2)$, where $\chi_{\max}$ represents the maximal real part of the eigenvalues of $\widehat{A}_{nn}$. Solve the following LMI:

$$\widehat{A}_{nn}^T Q + Q\widehat{A}_{nn} - (2\beta - \check{c}_2)Q < 0 \tag{10.46}$$

to get $Q > 0$. Then, design the neuro-adaptive tuning law as

$$\dot{\hat{\omega}}_i(t) = \Gamma_{\omega_i} \left[\varphi_i(x_i(t))\delta_i(t)^T Q - \check{c}_1 \hat{\omega}_i(t) \right], \tag{10.47}$$

where Γ_{ω_i} is a positive scalar and $\delta_i(t) = x_i(t) - z_i^n(t)$, $i = 1, \dots, N$.

Then, one may get the following theorem which states the main results of this subsection.

Theorem 10.3 *Suppose that Assumptions 10.1, 10.2, and 10.4 hold, LMIs (10.45) and (10.46) admit some feasible solutions. Then, the consensus tracking problem defined in Definition 10.1 for the CNS (10.3) and (10.6) with directed switching topologies can be solved by the neuro-adaptive protocol (10.10) associated with the observer (10.12) with control parameters designed by Algorithm 10.2 if the switching periodic τ satisfies the following condition: $\tau > (\ln\mu)/\check{c}_0$, where $\mu = \max_{i,j}\phi_i^j/\min_{i,j}\phi_i^j$, ϕ_i^j, $i = 1,\ldots,N$, $j = 1,\ldots,\kappa$, are defined in (10.41). Specially, for each $i = 1,\ldots,N$,*

$$\lim_{t\to\infty}\|x_i(t) - y(t)\| \le \tilde{\varpi}, \tag{10.48}$$

where $\tilde{\varpi} = \sqrt{\dfrac{\check{V}_2(t_0)}{\tilde{\Upsilon}_{\min}}} + \dfrac{\|\tilde{\nu}\|\tilde{\Upsilon}_{\max}}{\tilde{d}\tilde{\Upsilon}_{\min}}$, $\tilde{\Upsilon}_{\min} = \min_{i=1,\ldots,N}\{\lambda_{\min}(Q),\Gamma_{\omega_i}^{-1}\}$, $\tilde{\Upsilon}_{\max} = \max_{i=1,\ldots,N}\{\lambda_{\max}(Q),\ \Gamma_{\omega_i}^{-1}\}$, $\tilde{\nu} = [2\kappa_0\|Q\|_2, 2\check{c}_1\omega_M]$, $\kappa_0 = \epsilon_M + h_M$, $\tilde{d} = \min\{\check{c}_2\lambda_{\min}(Q), 2\check{c}_1\}$, *and*

$$\check{V}_2(t_0) = \delta^T(t_0)(I_N \otimes Q)\delta(t_0) + \mathrm{tr}(\widetilde{W}^T(t_0)\Gamma_W^{-1}\widetilde{W}(t_0)).$$

Proof 10.2 *Choose the following MLFs for the switched observer error systems (10.43):*

$$\check{V}_1(t) = e^T(t)(\Phi^{\sigma(t)} \otimes P)e(t), \tag{10.49}$$

where $\Phi^{\sigma(t)}$ is defined in (10.41), $P > 0$ is a solution of (10.45).

Noticing the fact that the communication topology $\mathcal{G}^{\sigma(t)}$ is fixed for $t \in [t_k, t_{k+1})$, for any given $k \in \mathbb{N}$, and considering (10.43), one gets

$$\dot{\check{V}}_1(t) \le e^T(t)[\Phi^{\sigma(t)} \otimes (\widehat{A}^T P + P\widehat{A} - \alpha\check{\lambda}_0\widehat{C}^T\widehat{C} + \gamma^2 P^2 + I_{nm})]e(t), \tag{10.50}$$

where $\check{\lambda}_0$ is defined in (10.42). According to (10.45), it can be thus obtained from (10.50) that

$$\dot{\check{V}}_1(t) \le -\check{c}_0\check{V}_1(t) \tag{10.51}$$

for all $t \in [t_k, t_{k+1})$. Note that the closed-loop observer error system (10.43) switches at $t = t_{k+1}$. It then follows from the above analysis and (10.49) that

$$\check{V}_1(t_{k+1}^-) < \check{V}_1(t_k)\exp(-\check{c}_0\tau). \tag{10.52}$$

On the other hand, it can be verified that

$$\check{V}_1(t_{k+1}) \le e^T(t_{k+1})(\phi_M I_N \otimes P)\,e(t_{k+1}),$$
$$\check{V}_1(t_{k+1}^-) \ge e^T(t_{k+1})(\phi_m I_N \otimes P)\,e(t_{k+1}),$$

where $\phi_M = \max_{i,j}\phi_i^j$, $\phi_m = \min_{i,j}\phi_i^j$, ϕ_i^j, $i = 1,\ldots,N$, $j = 1,\ldots,\kappa$, are defined in (10.41). Let $\mu = \phi_M/\phi_m$, one obtains $\check{V}_1(t_{k+1}) < \mu\check{V}_1(t_{k+1}^-)$ for each $k \in \mathbb{N}$. By the fact $\tau > (\ln\mu)/\check{c}_0$, one obtains

$$\check{V}_1(t_{k+1}) < \check{V}_1(t_k)\exp(-\widehat{c}_0\tau), \ \forall\ k \in \mathbb{N}, \tag{10.53}$$

where $\widehat{c}_0 = \check{c}_0 - (\ln\mu)/\tau > 0$. According to the fact $t_0 = 0$, it can be yielded by recursion that

$$\check{V}_1(t_k) < \check{V}_1(t_0)\exp(-\widehat{c}_0 k\tau). \tag{10.54}$$

For an arbitrarily given $t > t_1$, there exists a positive integer $k \geq 2$ such that $t_k \leq t < t_{k+1}$. When $t \in (t_k, t_{k+1})$, based on the above analysis, one gets

$$\check{V}_1(t) < \exp(-\check{c}_0(t - t_k))\check{V}_1(t_k)$$

$$< \exp\left(-\frac{k}{k+1}\widehat{c}_0 t\right)\check{V}_1(t_0)$$

$$\leq \exp\left(-\frac{\widehat{c}_0}{2}t\right)\check{V}_1(t_0). \tag{10.55}$$

For the case of $t = t_p$, one has

$$\check{V}_1(t) < \exp(-\widehat{c}_0 t)\check{V}_1(t_0). \tag{10.56}$$

According to (10.55) and (10.56), one may conclude that $\|e(t)\|$ converges to zero with an exponential decay rate.

Construct the following Lyapunov function for the consensus tracking error system (10.44):

$$\check{V}_2(t) = \delta^T(t)(I_N \otimes Q)\delta(t) + \text{tr}(\widetilde{W}^T(t)\Gamma_W^{-1}\widetilde{W}(t)), \tag{10.57}$$

where Q is a solution of (10.47). Using some similar analysis as those employed in the proof of Theorem 10.2, one may get that

$$\|\delta(t)\| \leq \widetilde{\varpi}, \tag{10.58}$$

where $\widetilde{\varpi} = \sqrt{\frac{\check{V}_2(t_0)}{\widetilde{\Upsilon}_{\min}}} + \frac{\|\widetilde{\nu}\|\widetilde{\Upsilon}_{\max}}{\widetilde{d}\widetilde{\Upsilon}_{\min}}$, $\widetilde{\Upsilon}_{\min} = \min_{i=1,\ldots,N}\{\lambda_{\min}(Q), \Gamma_{\omega_i}^{-1}\}$, $\widetilde{\Upsilon}_{\max} = \max_{i=1,\ldots,N}\{\lambda_{\max}(Q), \Gamma_{\omega_i}^{-1}\}$, $\widetilde{\nu} = [2\kappa_0\|Q\|_2, 2\check{c}_1 W_M]$, $\kappa_0 = \epsilon_M + h_M$, and $\widetilde{d} = \min\{\check{c}_2\lambda_{\min}(Q), 2\check{c}_1\}$.

Combining (10.55), (10.56), and (10.58) give that $\lim_{t\to\infty}\|x_i(t) - y(t)\| \leq \widetilde{\varpi}$.

Remark 10.4 *To achieve consensus tracking in CNS (10.6) and (10.3) with directed switching topologies, MLFs have been proposed to prove the exponential stability of the zero equilibrium point of switched systems (10.43). It can be seen from Theorem 10.3 that, under some suitable conditions, practical consensus tracking may be ensured if the switching period τ is larger than a threshold value. It should be noted that consensus tracking in CNS (10.6) and (10.3) with directed switching topologies can be achieved for an arbitrarily given positive switching period τ if there exists a common Lyapunov function for the switched systems (10.43). However, it is still a challenging yet unsolved problem to find a common Lyapunov function for switched systems. Furthermore, it is assumed in the present section that the communication channels among neighboring agents work perfectly and the actuators embedded in agents and network topology switch synchronously. It is interesting yet important to further consider how to ensure consensus tracking for CNS in the presence of packet dropouts [171], actuator failure [151], and asynchronous switching [180].*

Remark 10.5 *By using some similar manipulations as used in Remark 10.3, the solvability conditions for LMIs (10.45) and (10.46) can be obtained. Since the observer error system (10.43) becomes a switched system, a class of MLFs has been constructed to analyze the convergence property of its zero equilibrium point. It is also worth noting that the results given in Theorem 10.3 can be directly extended to the case of CNSs with aperiodic switching directed topologies by using dwell-time-based analysis approach.*

Remark 10.6 *It is well known in literature on NN-based control that the size of the compact set in which the states of the control plant are uniformly bounded is not needed in NN-based controller design. Specifically, it can be seen from Theorems 10.2 and 10.3 that the bound of the compact set Ω in (10.7) is not involved in consensus protocol design. In practical applications, the size of Ω could be assumed to be as large as desired such that the states $x_i(t)$ stay inside Ω for all $t > 0$. In this sense, the analytic results given in Theorems 10.2 and 10.3 are semi-global. It is also worth noting that the results of this section will be global in the scenario where (10.7) and (10.8) hold for all $x_i(t) \in \mathbb{R}^m$, $i = 1, \ldots, N$.*

10.2.4 Numerical simulations

In this subsection, a numerical example is provided to verify the effectiveness of the analytical results.

In simulations, the leader is assumed to be a two-mass-spring system with a single force input, whose dynamics can be described by (10.3) with

$$
\theta(t) = \begin{bmatrix} \theta_1(t) \\ \theta_2(t) \\ \theta_3(t) \\ \theta_4(t) \end{bmatrix}, \quad \widehat{A} = \begin{bmatrix} 0 & 1 & 0 & 0 \\ \frac{-k_1-k_2}{m_1} & 0 & \frac{k_2}{m_2} & 0 \\ 0 & 0 & 0 & 1 \\ \frac{k_2}{m_2} & 0 & \frac{-k_2}{m_2} & 0 \end{bmatrix},
$$

$g(\theta(t), t) = [0, 0.2\sin(\theta_2(t)) + 2\cos(t), 0, 0]^T$, where $m_1 = 1.25$ and $m_2 = 1.2$ are two masses, $k_1 = 1.0$ and $k_2 = 1.5$ are spring constants. Take the output matrix

$$
\widehat{C} = \begin{bmatrix} 0 & 0 & 1 & 0 \\ 0 & 0 & 0 & 1 \end{bmatrix},
$$

it can be verified that $(\widehat{C}, \widehat{A})$ is detectable. The dynamics of each follower are governed by (10.6) with $A = \begin{bmatrix} 0 & 1 \\ 0 & 0 \end{bmatrix}$, $f_i(x_i(t)) = [x_{i1}(t)\sin(x_{i1}t), 2\cos(x_{i2}(t))]^T$, $h_i(t) = [h_{i1}(t), h_{i2}(t)]^T$, both $h_{i1}(t)$ and $h_{i2}(t)$ are taken as random and bounded by $|h_{ij}(t)| \leq 0.5$, $i = 1, \ldots, N$, $j = 1, 2$.

Consider a CNS with a leader given in (10.3) and followers given in (10.6), where the topology $\mathcal{G}^{\sigma(t)}$ switches between $\mathcal{G}^1$ and $\mathcal{G}^2$ every 0.9s with $\mathcal{G}^{\sigma(t)} = \mathcal{G}^1$ for $t \in [0, 0.9)$s. Topologies $\mathcal{G}^1$ and $\mathcal{G}^2$ are shown in Figure 10.2, where the communication weights are indicated on edges. By (10.42), one has that $\check{\lambda}_0 = 1.5858$. Set $\alpha = 25$,

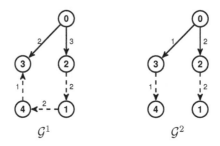

Figure 10.2 The communication graphs $\mathcal{G}^1$ and $\mathcal{G}^2$.

$\beta = 1$, $\check{c}_0 = 1.75$, and $\check{c}_2 = 0.1$, it can be yielded from Algorithm 10.2 that

$$F = - \begin{bmatrix} 2.3614 & 2.4691 & 0.1153 & 0.5780 \\ 33.6030 & 36.3046 & 0.5780 & 8.1780 \end{bmatrix}^T,$$

$$Q = \begin{bmatrix} 1.3354 & -0.0620 \\ -0.0620 & 1.2017 \end{bmatrix}.$$

In simulations, 4 neurons are employed for each NN. Sigmoid basis functions are adopted. The NN weight matrices $\hat{\omega}_i(t)$, $i = 1, 2, 3, 4$, are initialized as zero matrices. The initial state of the leader is set as $\theta(0) = [0, 0, 0, 0]^T$. Some calculations give that $\ln\mu = 1.5547$. Choose $\check{c}_1 = 0.001$ and $\Gamma_{\omega_i} = 1000$, for all $i = 1, 2, 3, 4$, it can be thus obtained from Theorem 10.3 that consensus tracking in the closed-loop CNS under the protocol designed by Algorithm 10.2 can be ensured. The evolutions of observer errors $\|e_i(t)\|$ are provided in Figure 10.3 which indicates that the states of observers asymptotically converge to those of the leader. Define $\delta_i^c(t) = [\delta_{i1}^c(t), \delta_{i2}^c(t)]^T = x_i(t) - y(t)$, $i = 1, 2, 3, 4$. The trajectories of consensus tracking errors $\delta_{i1}^c(t)$ and $\delta_{i2}^c(t)$ are shown in Figs. 10.4 and 10.5, respectively. These simulation results demonstrate the effectiveness of the theoretical results in Theorem 10.3.

10.3 ASYMPTOTIC CONSENSUS TRACKING OF CNSS WITH A HIGH DIMENSIONAL LEADER AND DIRECTED FIXED TOPOLOGY

10.3.1 Model formulation

Consider a CNS consisting of a leader and N followers, where the leader is labelled as agent 0 and the followers are, respectively, labelled as agents $1, \ldots, N$. As uncertainties and external disturbances exist everywhere in real applications, the dynamics of follower i are described by:

$$\dot{x}_i(t) = Ax_i(t) + f_i(x_i(t), t) + h_i + u_i(t), \; i = 1, \ldots, N, \quad (10.59)$$

where $x_i(t) \in \mathbb{R}^m$ is the state, $A \in \mathbb{R}^{m \times m}$ is the known system matrix, $f_i(\cdot, \cdot) : \mathbb{R}^m \times [0, +\infty) \mapsto \mathbb{R}^m$ represents the unknown uncertainties which is assumed to be continuous, h_i represents the external disturbances which satisfies $\|h_i\|_\infty \leq h_M$ for some $h_M > 0$, and $u_i(t)$ is the control input to be designed. As in the last section, the

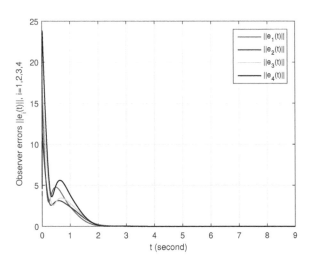

Figure 10.3 Trajectories of the observer errors $\|e_i(t)\|$, $i = 1, 2, 3, 4$.

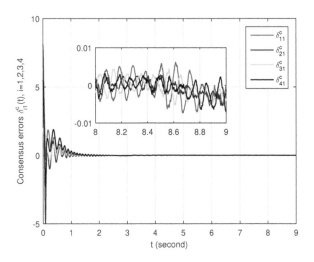

Figure 10.4 Trajectories of the consensus tracking errors $\delta_{i1}^c(t)$, $i = 1, 2, 3, 4$.

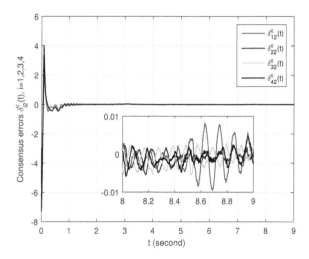

Figure 10.5 Trajectories of the consensus tracking errors $\delta_{i2}^c(t)$, $i = 1, 2, 3, 4$.

leader considered in this section has higher dimensional dynamics which are described by:

$$\dot{\theta}(t) = \widehat{A}\theta(t),$$
$$y(t) = \widehat{C}\theta(t), \tag{10.60}$$

where $\theta(t) \in \mathbb{R}^{nm}$, $y(t) \in \mathbb{R}^m$, $\widehat{A} \in \mathbb{R}^{nm \times nm}$, and $\widehat{C} \in \mathbb{R}^{m \times nm}$ are, respectively, the leader's states, the outputs, the system matrix, and the output matrix, $n \geq 1$.

By writing $\theta(t) = [\theta_1^T(t), \ldots, \theta_n^T(t)]^T$ with $\theta_i(t) \in \mathbb{R}^m$, $i = 1, \ldots, n$, we can divide $\widehat{A}$ as:

$$\widehat{A} = \begin{bmatrix} \widehat{A}_{11} & \widehat{A}_{12} & \cdots & \widehat{A}_{1n} \\ \widehat{A}_{21} & \widehat{A}_{22} & \cdots & \widehat{A}_{2n} \\ \vdots & \vdots & \ddots & \vdots \\ \widehat{A}_{n1} & \widehat{A}_{n2} & \cdots & \widehat{A}_{nn} \end{bmatrix}, \tag{10.61}$$

where $\widehat{A}_{ij} \in \mathbb{R}^{m \times m}$, $i, j = 1, \ldots, n$. Since our ultimate goal is to drive the followers to track the outputs of the leader, it is assumed that $\widehat{C} = [0_m, \ldots, 0_m, I_m]$. We should mention that this assumption will not cause any loss of generality in theoretical analysis.

As the leader is labelled as agent 0, then the Laplacian matrix $\mathcal{L}$ of topology $\mathcal{G}$ can be partitioned as

$$\mathcal{L} = \begin{bmatrix} 0 & 0_N^T \\ \mathbf{P} & \overline{\mathcal{L}} \end{bmatrix}, \quad \overline{\mathcal{L}} = \begin{bmatrix} \sum_{j \in \mathcal{N}_1} a_{1j} & -a_{12} & \cdots & -a_{1N} \\ -a_{21} & \sum_{j \in \mathcal{N}_2} a_{2j} & \cdots & -a_{2N} \\ \vdots & \vdots & \ddots & \vdots \\ -a_{N1} & -a_{N2} & \cdots & \sum_{j \in \mathcal{N}_N} a_{Nj} \end{bmatrix}, \tag{10.62}$$

where $\mathbf{P} = -[a_{10}, \ldots, a_{N0}]^T \in \mathbb{R}^N$.

Let $\Omega_i \subset \mathbb{R}^m$ be a compact set. According to the NN universal approximation theorem, for $x_i(t) \in \Omega_i$ and arbitrarily given $\epsilon_M > 0$, $f_i(x_i(t), t)$ given in (10.59) can be approximated on the compact set Ω_i as

$$f_i(x_i(t), t) = \omega_i^T \varphi_i(x_i(t)) + \epsilon_i, \tag{10.63}$$

where $\varphi_i(\cdot) : \mathbb{R}^m \to \mathbb{R}^p$ is the activation function, $\omega_i \in \mathbb{R}^{p \times m}$ is the unknown weight matrix, and $\epsilon_i \in \mathbb{R}^m$ represents the approximation error satisfying $\|\epsilon_i\|_\infty < \epsilon_M$.

In this section, the estimator embedded in follower i, $i = 1, \ldots, N$, is designed as:

$$\dot{z}_{ik}(t) = \sum_{j=1}^{n} \hat{A}_{kj} z_{ij}(t) + (\alpha_i(t) + \varrho_i(t)) F_k \zeta_{in}(t), \quad k = 1, \ldots, n, \tag{10.64}$$

where $\zeta_{ik}(t) = \sum_{j=1}^{N} a_{ij}(z_{ik}(t) - z_{jk}(t)) + a_{i0}(z_{ik}(t) - \theta_k(t))$, $\alpha_i(t) : [0, +\infty) \mapsto \mathbb{R}^+$ represents the time varying coupling strengths, $F = [F_1^T, \ldots, F_n^T]^T$ is the feedback gain matrix, and $\varrho_i(t)$ is a smooth function. Let $e_{ik}(t) = z_{ik}(t) - \theta_k(t)$. It follows from (10.64) and (10.60) that

$$\dot{e}_{ik}(t) = \sum_{j=1}^{n} \hat{A}_{kj} e_{ij}(t) + (\alpha_i(t) + \varrho_i(t)) F_k \sum_{j=1}^{N} \bar{l}_{ij} e_{in}(t),$$

where $\bar{\mathcal{L}} = [\bar{l}_{ij}]$ is defined in (10.62). Let $e_i(t) = [e_{i1}^T(t), \ldots, e_{in}^T(t)]^T$ and $e(t) = [e_1^T(t), \ldots, e_N^T(t)]^T$. We get

$$\dot{e}(t) = \left(I_N \otimes \hat{A} + (\Lambda + \varrho)\bar{\mathcal{L}} \otimes F\hat{C}\right) e(t),$$

where $\Lambda(t) = \text{diag}\{\alpha_1(t), \ldots, \alpha_N(t)\}$ and $\varrho(t) = \text{diag}\{\varrho_1(t), \ldots, \varrho_N(t)\}$. Let $\zeta(t) = [\zeta_1^T(t), \ldots, \zeta_N^T(t)]^T$, where $\zeta_i(t) = [\zeta_{i1}^T(t), \ldots, \zeta_{in}^T(t)]^T$ represents the relative states between estimator i and its neighbors. Noticed that $\zeta(t) = (\bar{\mathcal{L}} \otimes I_{nm})e(t)$, we have

$$\dot{\zeta}(t) = (I_N \otimes \hat{A} + \bar{\mathcal{L}}(\Lambda(t) + \varrho(t)) \otimes F\hat{C})\zeta(t). \tag{10.65}$$

To remove the effects of leader's unmeasurable states on consensus tracking, the following compensator is designed:

$$u_i^{ct}(t) = (\hat{A}_{nn} - A)x_i(t) + \sum_{j=1}^{n-1} \hat{A}_{nj} z_{ij}(t) + (\alpha_i(t) + \varrho_i(t)) F_n \zeta_{in}(t),$$

$$\dot{\alpha}_i(t) = \zeta_{in}^T(t) \zeta_{in}(t), \tag{10.66}$$

where the initial value $\alpha_i(t_0) > 0$.

The neuro-adaptive controller is designed as follows:

$$u_i^{ft}(t) = \beta(z_{in}(t) - x_i(t)) - \hat{\omega}_i^T(t)\varphi_i(x_i(t)) + \gamma \text{sgn}(P(z_{in}(t) - x_i(t))),$$

$$\dot{\hat{\omega}}_i(t) = -\nu_i \varphi_i(x_i(t))(z_{in}(t) - x_i(t))^T P, \tag{10.67}$$

where $\beta > 0$, $\gamma > 0$, and $P \in \mathbb{R}^{m \times m}$ are the parameters to be determined, $\nu_i > 0$ is a free parameter.

Then, the controller u_i in (10.59) is given by $u_i(t) = u_i^{ct}(t) + u_i^{ft}(t)$.

Let $\delta(t) = [\delta_1^T(t), \ldots, \delta_N^T(t)]^T$, where $\delta_i(t) = x_i(t) - z_{in}(t)$ represents the consensus tracking error. We conclude from (10.59), (10.63), (10.64), (10.66), and (10.67) that

$$
\begin{aligned}
\dot{\delta}(t) = &(I_N \otimes \hat{A}_{nn})\delta(t) - \beta\delta(t) - W^T(t)\Psi(x(t)) \\
&+ \epsilon + h(t) - \gamma\mathrm{sgn}((I_N \otimes P)\delta(t)),
\end{aligned}
\tag{10.68}
$$

where $W(t) = \mathrm{diag}\{\tilde{\omega}_1(t), \ldots, \tilde{\omega}_N(t)\}$, $\tilde{\omega}_i(t) = \hat{\omega}_i(t) - \omega_i$, $\Psi(x(t)) = [\varphi_1^T(t), \ldots, \varphi_N^T(t)]^T$, $\epsilon = [\epsilon_1^T, \ldots, \epsilon_N^T]^T$, $h(t) = [h_1^T(t), \ldots, h_N^T(t)]^T$, and $x(t) = [x_1^T(t), \ldots, x_N^T(t)]^T$.

10.3.2 Theoretical analysis

In this subsection, the main theorems and theoretical analyses are provided.

Algorithm 10.3 *Suppose $(\hat{C}, \hat{A})$ is detectable. The parameters β, P, F in (10.66) and (10.67) can be designed as follows:*

(1) Choose $\varsigma_1 > 0$. Solve the LMI

$$
\hat{A}^T Q + Q\hat{A} - \hat{C}^T\hat{C} + c_1 Q < 0
\tag{10.69}
$$

to get $Q > 0$. Then, we choose $F = -Q^{-1}\hat{C}^T$.

(2) Let $\chi = \max(\mathrm{Re}(\lambda(\hat{A}_{nn})))$. Choose $\beta > \chi + c_2/2$ with $c_2 > 0$. Solve the LMI

$$
\hat{A}_{nn}^T P + P\hat{A}_{nn} - (2\beta - c_2)P < 0
\tag{10.70}
$$

to get $P > 0$.

Before presenting the main theorem, the following assumption is made.

Assumption 10.5 *The graph $\mathcal{G}$ contains a directed spanning tree with the leader being the root.*

Under Assumption 10.5, it can be obtained from Lemma 2.15 that there exists a positive definite diagonal matrix $\Phi = \mathrm{diag}\{\phi_1, \ldots, \phi_N\}$ such that $\overline{\mathcal{L}}^T\Phi + \Phi\overline{\mathcal{L}} > 0$, where $\overline{\mathcal{L}}$ is defined in (10.62).

Theorem 10.4 *Suppose Assumption 10.5 holds. If $\gamma \geq \epsilon_M + h_M$, then consensus tracking of the CNS (10.59) and (10.60) under the neuro-adaptive controller (10.66), (10.67) with $\varrho_i(t) = \zeta_i^T(t)Q\zeta_i(t)$ and the control parameters determined by Algorithm 10.3 can be achieved. In addition, the NN approximation region Ω can be selected as:*

$$
\Omega \triangleq \left\{ x(t) : x(t) \in \mathbb{R}^{Nn}, \|x(t)\| \leq \sqrt{\frac{V_1(t_0)}{\lambda_{\min}(\overline{\mathcal{L}}^T\overline{\mathcal{L}})\phi_{\min}\alpha_{\min}(t_0)\lambda_{\min}(Q)}} \right.
$$
$$
\left. + \theta_M + \sqrt{\frac{V_2(t_0)}{\lambda_{\min}(P)}} \right\},
\tag{10.71}
$$

where $V_1(t_0) = \frac{1}{2}\sum_{i=1}^N \left(\phi_i(2\alpha_i(t_0) + \varrho_i(t_0))\varrho_i(t_0) + \phi_i(\alpha_i(t_0) - \alpha)^2\right)$, $V_2(t_0) = \delta^T(t_0)(I_N \otimes P)\delta(t_0) + \sum_{i=1}^N \text{tr}(\frac{1}{\nu_i}\tilde{\omega}_i^T(t_0)\tilde{\omega}_i(t_0))$, $\phi_{\min} = \min_{i=1,\dots,N} \phi_i$, $\alpha_{\min}(t_0) = \min_{i=1,\dots,N} \alpha_i(t_0)$, $\theta_M = \sup_t \|y(t)\|$, ν_i is a positive scalar given in (10.67), α is a positive scalar, Q and P are respectively the solutions of the LMIs (10.69) and (10.70).

Proof 10.3 *Let*

$$V_1(t) = \frac{1}{2}\sum_{i=1}^N \left(\phi_i(2\alpha_i(t) + \varrho_i(t))\varrho_i(t) + \phi_i(\alpha_i(t) - \alpha)^2\right), \qquad (10.72)$$

where α is a positive scalar to be given later. Since $\alpha_i(t_0) > 0$ and $\dot{\alpha}_i(t) \geq 0$, then $\alpha_i(t) > 0$ and thus $V_1(t)$ is positive definite. Taking time derivative on both sides of (10.72) gives

$$\dot{V}_1(t) = \sum_{i=1}^N \phi_i(\varrho_i(t) + \alpha_i(t) - \alpha)\dot{\alpha}_i(t) + \sum_{i=1}^N \phi_i\left(\varrho_i(t) + \alpha_i(t)\right)\dot{\varrho}_i(t). \qquad (10.73)$$

We obtain from (10.65), (10.66), and $F = -Q^{-1}\widehat{C}^T$ that

$$\sum_{i=1}^N \phi_i\left(\varrho_i(t) + \alpha_i(t)\right)\dot{\varrho}_i(t)$$

$$= \sum_{i=1}^N \phi_i\left(\varrho_i(t) + \alpha_i(t)\right)\left(\dot{\zeta}_i^T(t)Q\zeta_i(t) + \zeta_i^T(t)Q\dot{\zeta}_i(t)\right)$$

$$= \zeta^T(t)\left((\Lambda(t) + \varrho(t))\Phi \otimes (\widehat{A}^T Q + Q\widehat{A})\right.$$

$$\left. -(\Lambda(t) + \varrho(t))(\Phi\overline{\mathcal{L}} + \overline{\mathcal{L}}^T\Phi)(\Lambda(t) + \varrho(t)) \otimes \widehat{C}^T\widehat{C}\right)\zeta(t)$$

$$\leq \zeta^T(t)\left((\Lambda(t) + \varrho(t))\Phi \otimes (\widehat{A}^T Q + Q\widehat{A}) - \lambda_0(\Lambda(t) + \varrho(t))^2\Phi \otimes \widehat{C}^T\widehat{C}\right)\zeta(t), \quad (10.74)$$

where $\lambda_0 = \lambda_{\min}\left(\overline{\mathcal{L}} + \Phi^{-1}\overline{\mathcal{L}}^T\Phi\right)$. Noticed that $\zeta_{in}(t) = \widehat{C}\zeta_i(t)$, we get from (10.66) that

$$\sum_{i=1}^N \phi_i(\varrho_i(t) + \alpha_i(t) - \alpha)\dot{\alpha}_i(t)$$

$$= \sum_{i=1}^N \phi_i(\varrho_i(t) + \alpha_i(t) - \alpha)\zeta_i^T(t)\widehat{C}^T\widehat{C}\zeta_i(t)$$

$$= \zeta^T(t)\left((\varrho(t) + \Lambda(t) - \alpha I_N)\Phi \otimes \widehat{C}^T\widehat{C}\right)\zeta(t). \qquad (10.75)$$

Choose $\alpha > 1/\lambda_0$, it is not difficult to show that $\lambda_0(\Lambda(t)+\varrho(t))^2+\alpha I_N > 2(\Lambda(t)+\varrho(t))$. By substituting (10.74), (10.75) into (10.73), and by using (10.69), we have

$$\dot{V}_1(t) \leq \zeta^T(t)\left((\Lambda(t) + \varrho(t))\Phi \otimes \left(\widehat{A}^T Q + Q\widehat{A} - \widehat{C}^T\widehat{C}\right)\right)\zeta(t)$$

$$\leq -c_1\zeta^T(t)\left((\Lambda(t) + \varrho(t))\Phi \otimes Q\right)\zeta(t). \qquad (10.76)$$

Let

$$V_2(t) = \delta^T(t)(I_N \otimes P)\delta(t) + \sum_{i=1}^{N} \text{tr}\left(\frac{1}{\nu_i}\tilde{\omega}_i^T(t)\tilde{\omega}_i(t)\right), \tag{10.77}$$

where ν_i is a positive scalar given in (10.67). We obtain from (10.67) and (10.68) that

$$\dot{V}_2(t) = \delta^T(t)\left(I_N \otimes \left(\hat{A}_{nn}^T P + P\hat{A}_{nn} - 2\beta P\right)\right)\delta(t)$$
$$- 2\delta^T(t)(I_N \otimes P)\left(W^T(t)\Psi(x(t)) - \epsilon - h(t) + \gamma\text{sgn}((I_N \otimes P)\delta(t))\right)$$
$$+ 2\sum_{i=1}^{N} \text{tr}\left(\tilde{\omega}_i^T(t)\varphi_i(x_i(t))\delta_i^T(t)P\right). \tag{10.78}$$

By using the facts $a^T b = \text{tr}(ab^T)$, $\text{tr}(D) = \text{tr}(D^T)$, and $\text{tr}(DE) = \text{tr}(ED)$, we have

$$\delta^T(t)(I_N \otimes P)W^T(t)\Psi(x(t))$$
$$= \sum_{i=1}^{N} \delta_i^T(t)P\tilde{\omega}_i^T(t)\varphi_i(x_i(t)) = \sum_{i=1}^{N} \text{tr}\left(\delta_i(t)\varphi_i^T(x_i(t))\tilde{\omega}_i(t)P\right)$$
$$= \sum_{i=1}^{N} \text{tr}\left(P\tilde{\omega}_i^T(t)\varphi_i(x_i(t))\delta_i^T(t)\right) = \sum_{i=1}^{N} \text{tr}\left(\tilde{\omega}_i^T(t)\varphi_i(x_i(t))\delta_i^T(t)P\right). \tag{10.79}$$

Let $\tilde{\delta}_i(t) = P\delta_i(t)$. Then

$$\delta^T(t)(I_N \otimes P)(\epsilon + h(t)) \le (\epsilon_M + h_M)\sum_{i=1}^{N} \|\tilde{\delta}_i(t)\|_1. \tag{10.80}$$

Noticing that $a^T\text{sgn}(a) = \|a\|_1$, we have

$$\delta^T(t)(I_N \otimes P)\text{sgn}((I_N \otimes P)\delta(t)) = \sum_{i=1}^{N} \|\tilde{\delta}_i(t)\|_1. \tag{10.81}$$

Since $\gamma > \epsilon_M + h_M$, inserting (10.79)–(10.81) into (10.78) gives that

$$\dot{V}_2(t) \le \delta^T(t)\left(I_N \otimes \left(\hat{A}_{nn}^T P + P\hat{A}_{nn} - 2\beta P\right)\right)\delta(t)$$
$$< -c_2\delta^T(t)(I_N \otimes P)\delta(t), \tag{10.82}$$

where the second inequality follows from (10.70).

  Now, we construct the Lyapunov function $V(t) = V_1(t) + V_2(t)$. Combining (10.76) together with (10.82) gives that

$$\dot{V}(t) < -c_1\zeta^T(t)\left((\Lambda(t) + \varrho(t))\Phi \otimes Q\right)\zeta(t) - c_2\delta^T(t)(I_N \otimes P)\delta(t). \tag{10.83}$$

This together with $V(t) \ge 0$ implies $\lim_{t\to\infty} V(t)$ exists. Denote $\lim_{t\to\infty} V(t) = V(\infty)$. Since $V(t)$ is decreasing, so $0 \le V(\infty) \le V(t) \le V(t_0)$. Noticed that $\varrho(t) =$

$\zeta^T(t)(I_N \otimes Q)\zeta(t)$, we obtain from (10.72) and (10.77) that $\zeta(t), \delta(t), \tilde{\omega}_i(t) \in \mathbb{L}_\infty$, and $\alpha_i(t)$ is bounded. As $\Psi(x(t))$ is uniformly bounded, so we can get from (10.65) and (10.68) that $\dot{\zeta}(t), \dot{\delta}(t) \in \mathbb{L}_\infty$. On the other hand, we can show $\zeta(t), \delta(t) \in \mathbb{L}_2$ by integrating on both sides of (10.83). According to Lemma 2.13, both $\|\zeta(t)\|$ and $\|\delta(t)\|$ approach 0. Therefore, consensus tracking is achieved. Moreover, since $\dot{\alpha}_i(t) \geq 0$, each $\alpha_i(t)$ converges to a finite value.

Since $V_1(t) \leq V_1(t_0)$ and $V_2(t) \leq V_2(t_0)$, it follows from (10.72) and (10.77) that

$$\|e(t)\| \leq \sqrt{\frac{V_1(t_0)}{\lambda_{\min}(\overline{\mathcal{L}}^T \overline{\mathcal{L}})\phi_{\min}\alpha_{\min}(t_0)\lambda_{\min}(Q)}}, \quad \|\delta(t)\| \leq \sqrt{\frac{V_2(t_0)}{\lambda_{\min}(P)}}.$$

Combining this with $\delta_i(t) = x_i(t) - z_{in}(t)$ and $e_{ik}(t) = z_{ik}(t) - \theta_k(t)$ gives (10.71).

Remark 10.7 *In contrast to [192] in which both the agents' dynamics and $\lambda_{\min}(\overline{\mathcal{L}} + \Phi^{-1}\overline{\mathcal{L}}^T\Phi)$ were used for selecting the control parameters, the control parameters in the protocol (10.66), (10.67) depend only on the agents' dynamics. So the obtained consensus tracking is fully distributed. More importantly, both the estimation error and tracking error converge asymptotically to the zero vector, while the tracking error in [192] merely converges into a bounded set. A practical issue in implementing the discontinuous controller (10.67) is that it may cause the chattering phenomena. To avoid these undesired phenomena, one can use the boundary layer technique [196].*

Remark 10.8 *We learn from the proof of Theorem 10.4 that $\hat{\omega}_i$ will converge to some finite constants, but this does not imply $\hat{\omega}_i$ will converge to the ideal NN weight ω_i. Note also that the LMI (10.69) is feasible since $(\widehat{C}, \widehat{A})$ is detectable. According to Remark 5 of [192], (10.70) is always feasible since $\beta > \chi + \varsigma_2/2$.*

Remark 10.9 *As we noted above, the followers' dynamics are much simpler than those of the leader in some practical cases. So the identity matrix was used in (10.59) to be the input matrix. Based on this observation, the engineers can implement communication protocols for tracking a target with strong maneuverability by using simple yet low-cost agents.*

When the communication topology among the N followers is undirected, we design the following estimator and compensator:

$$\dot{z}_{ik}(t) = \sum_{j=1}^{n} \hat{A}_{kj} z_{ij}(t) + \alpha_i(t) F_k \zeta_{in}(t), \quad k = 1, \ldots, n, \tag{10.84}$$

$$u_i^{ct}(t) = (\hat{A}_{nn} - A)x_i(t) + \sum_{j=1}^{n-1} \hat{A}_{nj} z_{ij} + \alpha_i(t) F_n \zeta_{in}(t),$$

$$\dot{\alpha}_i(t) = \zeta_{in}^T(t)\zeta_{in}(t). \tag{10.85}$$

Theorem 10.5 *Suppose the communication topology among the N followers is undirected and Assumption 10.5 holds. If $\gamma \geq \epsilon_M + h_M$, then consensus tracking of the CNS (10.59) and (10.60) under the neuro-adaptive controller (10.85), (10.67) with control parameters determined by Algorithm 10.3 can be achieved. In addition, the NN approximation region Ω is selected as (10.86):*

$$\Omega \triangleq \left\{ x(t) : \ x(t) \in \mathbb{R}^{Nn}, \ \|x(t)\| \leq \sqrt{\frac{V_3(t_0)}{\lambda_{\min}(\overline{\mathcal{L}})\lambda_{\min}(Q)}} \right.$$
$$\left. +\theta_M + \sqrt{\frac{V_2(t_0)}{\lambda_{\min}(P)}} \right\}, \tag{10.86}$$

where $V_3(t_0) = e^T(t_0)(\overline{\mathcal{L}} \otimes Q)e(t_0) + \sum_{i=1}^{N}(\alpha_i(t_0) - \alpha)^2$, the other notations are the same as those in Theorem 10.4.

Proof 10.4 *We construct the Lyapunov function $V(t) = V_3(t) + V_2(t)$, where $V_2(t)$ is given by (10.77), and*

$$V_3(t) = e^T(t)(\overline{\mathcal{L}} \otimes Q)e(t) + \sum_{i=1}^{N}(\alpha_i(t) - \alpha)^2.$$

Then the theorem can be proven by using the same technique as that in Theorem 10.4.

Remark 10.10 *The 'linearity-in-parameters' assumption (i.e., $f_i(x_i(t)) = \omega_i^T \varphi_i(x_i(t))$) in [147] is removed in this section. To eliminate the effects of approximation error ϵ_i on asymptotical consensus, a discontinuous term $-\gamma\,\text{sgn}(P\delta_i(t))$ with $\gamma > \epsilon_M$ is added into the controller (10.67). Hence, our results are much convenient in engineering applications.*

10.3.3 Numerical simulations

To validate Theorem 10.4, we consider a CNS with communication topology depicted by Figure 10.6, where the dynamics of node 0 and node $1, \ldots, 5$ are, respectively, described by (10.60) and (10.59) with

$$\widehat{A} = \begin{bmatrix} 0 & 1 & 0 & 0 \\ -2 & 0 & 1.25 & 0 \\ 0 & 0 & 0 & 1 \\ 1.25 & 0 & -1.25 & 0 \end{bmatrix}, \quad \widehat{C} = \begin{bmatrix} 0 & 0 \\ 0 & 0 \\ 1 & 0 \\ 0 & 1 \end{bmatrix}^T,$$

$A = \begin{bmatrix} 0 & 1 \\ 0 & 0 \end{bmatrix}$, $f_i(x_i(t)) = [4x_{i1}(t)\cos(x_{i1}(t)), 2\sin(x_{i2}(t))]^T$ and $h_i(t) = [h_{i1}(t),$ $h_{i2}(t)]^T$, $|h_{ik}(t)| \leq 0.4$, $i = 1, \ldots, 5$, $k = 1, 2$. It is not difficult to show that $(\widehat{C}, \widehat{A})$ is detectable, Assumption 10.5 holds and $\chi = 0$. By choosing $c_1 = 0.3$, $c_2 = 0.25$, $\beta = 1 > 0.25/2$, we get from Algorithm 10.3 that

$$Q = \begin{bmatrix} 0.7878 & -0.2108 & -0.2832 & -0.5409 \\ -0.2108 & 0.5861 & 0.2511 & 0.3806 \\ -0.2832 & 0.2511 & 0.7331 & 0.1490 \\ -0.5409 & 0.3806 & 0.1490 & 0.9794 \end{bmatrix},$$

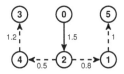

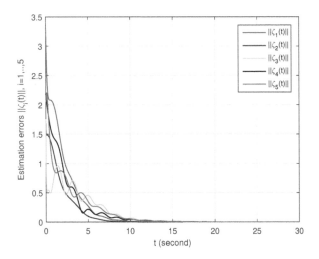

Figure 10.6 The communication graph $\mathcal{G}$.

Figure 10.7 Trajectories of the estimation errors $\|\zeta_i(t)\|$, $i = 1, \dots, 5$.

$$F = -\begin{bmatrix} 0.7554 & -0.8110 & 1.8417 & 0.4521 \\ 1.3136 & -1.0843 & 0.4521 & 2.0990 \end{bmatrix}^T, \quad P = I_2.$$

Assume the neuron layer has six neurons and each neuron is the sigmoid function. Suppose $\hat{\omega}_i(0) = 0_{6 \times 2}$ and $\epsilon_M < 0.1$. We choose $\gamma = 0.5$ and $\nu_1 = \dots = \nu_5 = 20000$. The evolutions of $\|\zeta_i(t)\|$ and $\delta_{ij}(t)$, $i = 1, \dots, 5$, $j = 1, 2$, are, respectively, depicted in Figs. 10.7–10.9 which indicate that the consensus tracking is achieved. Moreover, the evolutions of $\alpha_i(t)$ are plotted in Figure 10.10.

10.4 PRACTICAL AND ASYMPTOTIC CONTAINMENT TRACKING OF CNSS WITH MULTIPLE LEADERS

This section addresses the containment problem for uncertain CNSs with multiple leaders and detail balanced directed graph.

10.4.1 Model formulation

Suppose that there are N dynamic agents in the considered CNS, and M of them are designated as leaders, while the rest $N - M$ agents are designated as followers. Without loss of generality, one may assume that agents indexed by $1, \dots, M$ ($M \geq 1$) are the leaders, and the agents indexed by $M + 1, \dots, N$ are the followers. For illus-

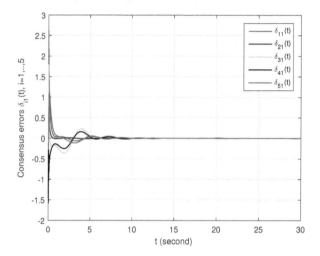

Figure 10.8 Trajectories of the consensus tracking errors $\delta_{i1}(t)$, $i = 1, \ldots, 5$.

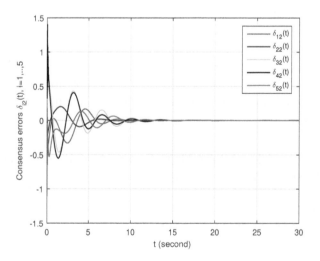

Figure 10.9 Trajectories of the consensus tracking errors $\delta_{i2}(t)$, $i = 1, \ldots, 5$.

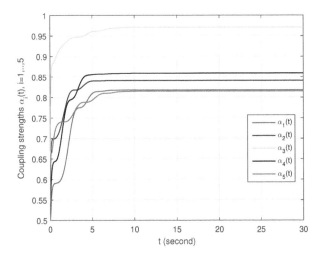

Figure 10.10 Trajectories of the coupling strengths $\alpha_i(t)$, $i = 1, \ldots, 5$.

tration convenience, denote $\mathbb{L} = \{1, \ldots, M\}$ and $\mathbb{F} = \{M + 1, \ldots, N\}$, respectively. Since none of the leaders can receive the information of the followers, so the Laplacian matrix $\mathcal{L}$ associated with the directed graph $\mathcal{G}$ can be written as

$$\mathcal{L} = \begin{bmatrix} 0_M & 0_{M \times (N-M)} \\ \mathcal{L}_1 & \mathcal{L}_2 \end{bmatrix}, \tag{10.87}$$

where $\mathcal{L}_1 \in \mathbb{R}^{(N-M) \times M}$ and $\mathcal{L}_2 \in \mathbb{R}^{(N-M) \times (N-M)}$.

Before moving forward, the following assumptions are provided which will be utilized in deriving the main results of this section.

Assumption 10.6 *For each follower $i \in \mathbb{F}$, there exists at least one leader $j \in \mathbb{L}$ that has a directed path to follower i.*

Assumption 10.7 *The induced subgraph with vertex set $\mathbb{F}$ is detailed balanced, that is, there exists a positive vector $\xi = [\xi_{M+1}, \ldots, \xi_N]^T$ such that $\Xi \mathcal{A}_2 = \mathcal{A}_2^T \Xi$, where $\mathcal{A}_2$ is the adjacency matrix of the induced subgraph with vertex set $\mathbb{F}$ in $\mathcal{G}$, and $\Xi = \text{diag}\{\xi_{M+1}, \ldots, \xi_N\}$.*

Lemma 10.1 *[106] Under Assumption 10.6, all the eigenvalues of $\mathcal{L}_2$ have positive real parts, each entry of $-\mathcal{L}_2^{-1}\mathcal{L}_1$ is nonnegative, and each row of $-\mathcal{L}_2^{-1}\mathcal{L}_1$ has sum equal to one.*

The dynamics of the leaders and followers are, respectively, given as

$$\dot{x}_i(t) = Ax_i(t) + Bu_i(t), \quad i \in \mathbb{L}, \tag{10.88}$$

and

$$\dot{x}_i(t) = Ax_i(t) + B\left[f_i(x_i(t)) + h_i(t) + u_i(t)\right], \quad i \in \mathbb{F}, \tag{10.89}$$

where $x_i(t) \in \mathbb{R}^n$ is the state of agent i, $A \in \mathbb{R}^{n \times n}$ represents the system matrix, $B \in \mathbb{R}^{n \times m}$ is the control input matrix with $n \geq m$, $u_i(t)$ is the control input acting on agent i, $f_i(x_i(t))$ represents the uncertainties which is assumed to be continuous, $h_i(t) \in \mathbb{R}^m$ describes the bounded matching disturbances such that

$$\|h_i(t)\|_\infty \leq h_M \tag{10.90}$$

for some given scalar $h_M > 0$.

According to the NN universal approximation theorem, the nonlinearity $f_i(x_i(t))$ can be approximated on a compact set $\Omega_i \subset \mathbb{R}^m$ to arbitrary accuracy by

$$f_i(x_i(t)) = \omega_i^T \varphi_i(x_i(t)) + \epsilon_i, \ \forall \ x_i(t) \in \Omega_i, \tag{10.91}$$

where $\varphi_i(\cdot) : \mathbb{R}^n \mapsto \mathbb{R}^p$ is a known basis function, $\omega_i \in \mathbb{R}^{p \times m}$ is the ideal NN weight matrix, $\epsilon_i \in \mathbb{R}^m$ is the NN approximation error vector such that

$$\|\epsilon_i\|_\infty \leq \epsilon_M$$

for all $i \in \mathbb{F}$. Suppose that $\max_{x_i(t) \in \Omega_i} \{\|\varphi_i(x_i(t))\|\}$ is finite, i.e., there exists a positive scalar $\varphi_{iM} > 0$ such that

$$\max_{x_i(t) \in \Omega_i} \|\varphi_i(x_i(t))\| \leq \varphi_{iM}, \tag{10.92}$$

for all $i \in \mathbb{F}$. For notational convenience, define $W = \text{diag}\{\omega_{M+1}, \ldots, \omega_N\}$. One then gets that there exists a positive scalar W_M such that $\|W\|_F \leq W_M$.

In this section, distributed practical containment control problem is firstly investigated where the control objective is to drive the states of followers to converge into the convex hull formed by those of the leaders with residual error $\varpi > 0$, by designing some containment controllers for followers. Then, distributed asymptotical containment control problem is studied where the control objective is to drive the states of followers to converge into the convex hull formed by those of the leaders asymptotically.

The considered practical containment and asymptotical containment problems are, respectively, stated below.

Definition 10.2 *Practical containment of CNSs with leaders given by (10.88) and followers given by (10.89) is said to be achieved, if there exist some nonnegative scalars $p_{ij} \geq 0$ with $\sum_{j=1}^M p_{ij} = 1$, such that*

$$\lim_{t \to \infty} \left\| x_i(t) - \sum_{j=1}^M p_{ij} x_j(t) \right\| \leq \varpi, \ i \in \mathbb{F}, \tag{10.93}$$

for some given positive scalars ϖ.

Definition 10.3 *Asymptotical containment of CNSs with leaders given by (10.88) and followers given by (10.89) is said to be achieved, if there exist some nonnegative scalars $p_{ij} \geq 0$ with $\sum_{j=1}^M p_{ij} = 1$, such that*

$$\lim_{t \to \infty} \left\| x_i(t) - \sum_{j=1}^M p_{ij} x_j(t) \right\| = 0, \ i \in \mathbb{F}. \tag{10.94}$$

Remark 10.11 *Notwithstanding our ability to design a well UUB cooperative controller for various CNSs [58, 122, 147, 197], we continue to lack an understanding of the design principles that govern the achievement of asymptotical cooperative behaviors. To solve this challenging problem, a new kind of distributed neuro-adaptive controllers will be designed and utilized. To the best of our knowledge, asymptotical containment problem for CNSs with unknown dynamics is successfully solved by using a distributed neuro-adaptive controller for the first time in this section.*

10.4.2   Practical containment of uncertain CNSs

The ideal approximating NN weight matrix ω_i given in (10.91) is generally unknown. This indicates that ω_i cannot be directly utilized in designing the containment controller. To effectively compensate for the unknown nonlinearities and the external disturbances, a kind of distributed NN-based containment controller is proposed as follows:

$$u_i(t) = -\alpha K\delta_i(t) - \beta\mathrm{sgn}(K\delta_i(t)) - \hat{\omega}_i^T(t)\varphi_i(x_i(t)), \ i \in \mathbb{F}, \tag{10.95}$$

where $\delta_i(t) = \sum_{j=1}^{N} a_{ij}(x_i(t) - x_j(t))$, $\alpha > 0$ and $\beta > 0$ are the coupling strengths to be selected, $K \in \mathbb{R}^{m \times n}$ is the feedback gain matrix to be designed, time-varying matrix $\hat{\omega}_i(t)$ is the estimation for the ideal weight for follower i at time t. It should be noted that the solutions of differential systems with non-smooth right-hand terms should be considered as those in the sense of Filippov [27] throughout this section.

Practical containment problem of the considered CNSs is studied in this subsection. To complete the goal of practical containment, the following neuro-adaptive evolution law for $\hat{\omega}_i(t)$ in (10.95) is proposed:

$$\dot{\hat{\omega}}_i(t) = \nu_i\left[\xi_i\varphi_i(x_i(t))\delta_i^T(t)(P^{-1}B) - c_i\hat{\omega}_i(t)\right], \ i \in \mathbb{F}, \tag{10.96}$$

of which ν_i and c_i are two positive scalars, ξ_i is provided in Assumption 10.7, P is a positive definite matrix to be designed later, $\hat{\omega}_i(0)$ is set as a real constant matrix with appropriate dimensions.

For the convenience of expression, let $\delta(t) = [\delta_{M+1}^T(t), \ldots, \delta_N^T(t)]^T$, $x_f(t) = [x_{M+1}^T(t), \ldots, x_N^T(t)]^T$, and $x_l(t) = [x_1^T(t), \ldots, x_M^T(t)]^T$. It is obviously that $\delta(t) = (\mathcal{L}_1 \otimes I_n)x_l(t) + (\mathcal{L}_2 \otimes I_n)x_f(t)$. Set $e(t) = x_f(t) - (-\mathcal{L}_2^{-1}\mathcal{L}_1 \otimes I_n)x_l(t)$ as the containment error vector of the considered CNS. It can be derived from the above analysis that $e(t) = (\mathcal{L}_2^{-1} \otimes I_n)\delta(t)$. This means that

$$\|e(t)\| \le \varrho\|\delta(t)\| \tag{10.97}$$

with ϱ being the largest singular value of $\mathcal{L}_2^{-1}$, i.e., $\varrho = \sqrt{\lambda_{\max}\left((\mathcal{L}_2^{-1})^T\mathcal{L}_2^{-1}\right)}$. Then, combining (10.95) together with (10.88)–(10.91), we have

$$\begin{aligned}
\dot{\delta}(t) =& [(I_{N-M} \otimes A) - \alpha(\mathcal{L}_2 \otimes BK)]\delta(t) + (\mathcal{L}_1 \otimes B)u_l(t) \\
&+ (\mathcal{L}_2 \otimes B)g(t) - (\mathcal{L}_2 \otimes B)(\widetilde{W}^T\Psi - \epsilon) \\
&- \beta(\mathcal{L}_2 \otimes B)\mathrm{sgn}((I_{N-M} \otimes K)\delta(t)),
\end{aligned} \tag{10.98}$$

where $u_l(t) = [u_1^T(t), \ldots, u_M^T(t)]^T$, $g(t) = [g_{M+1}^T(t), \ldots, g_N^T(t)]^T$, $\widetilde{W} = \text{diag}\{\hat{\omega}_{M+1}(t) - \omega_{M+1}, \ldots, \hat{\omega}_N(t) - \omega_N\}$, $\epsilon = [\epsilon_{M+1}^T(t), \ldots, \epsilon_N^T(t)]^T \in \mathbb{R}^{(N-M)m}$, $\Psi = [\varphi_{M+1}^T(t), \ldots, \varphi_N^T(t)]^T \in \mathbb{R}^{(N-M)p}$. To facilitate the analysis in the next section, the following assumption is made.

Assumption 10.8 *For any given $x_l(0) \in \mathbb{R}^{Mn}$, there exist two positive scalars $\eta(x_l(0))$ and $\hat{\eta}$ such that*

$$\|x_l(t)\|_\infty \le \eta(x_l(0)), \quad \|u_l(t)\|_\infty \le \hat{\eta}, \tag{10.99}$$

for all $t \ge 0$.

Remark 10.12 *Note that Assumption 10.8 is very mild. For example, Assumption 10.8 holds if the system matrix A is marginally stable and $u_i(t)$ is set as $\mathbf{0}_m$ for $i \in \mathbb{L}$. On the other hand, under the condition that (A, B) is stabilizable, Assumption 10.8 holds if the control input $u_i(t)$ for each $i \in \mathbb{L}$ is designed such that $u_i(t) = Fx_i(t)$, $i \in \mathbb{L}$, and $A + BF$ is marginally stable or Hurwitz stable.*

For notational brevity, let

$$\xi_{\min} = \min_{i \in \{M+1,\ldots,N\}} \{\xi_i\}. \tag{10.100}$$

One may then get the following theorem which summarizes the main analytical results of this subsection.

Theorem 10.6 *Suppose that Assumptions 10.6–10.8 hold and the matrix pair (A, B) is stabilizable. Then, for arbitrarily given $x_i(0) \in \mathbb{R}^n$, $i = 1, \ldots, N$, distributed practical containment for CNSs with leaders given by (10.88) and followers given by (10.89) under controller (10.95) associated with adaptive law (10.96) will be achieved if the control parameters are appropriately designed such that $\beta > \epsilon_M + \hat{\eta} + h_M$, $\alpha > (\chi_0 \lambda_{\max}(\Xi \mathcal{L}_2^{-1}))/(2\xi_{\min})$ for some given $\chi_0 > 0$ and $K = B^T P^{-1}$, where P is a positive definite solution of the LMI:*

$$AP + PA^T - \chi_0 BB^T + \theta_1 P < 0, \tag{10.101}$$

h_M is given in (10.90), θ_1 is a positive scalar. Particularly, the NN approximation region Ω can be selected as

$$\Omega = \left\{ z : z \in \mathbb{R}^{(N-M)n}, \ \|z\| \le \frac{\varrho(\sqrt{V_1(0)} + \bar{c}_0/c_0)}{\sqrt{\lambda_{\min}\left(\Xi \mathcal{L}_2^{-1} \otimes P^{-1}\right)}} \right.$$

$$\left. + \sqrt{M} \|\mathcal{L}_2^{-1} \mathcal{L}_1 \otimes I_n\|_F \eta(x_l(0)) \right\}, \tag{10.102}$$

where ϱ is the largest singular value of $\mathcal{L}_2^{-1}$, $c_0 = \min\{\theta_1, 2c_m\nu_m\}$, $\bar{c}_0 = 2c_M W_M \sqrt{\nu_M}$, $c_M = \max_{i \in \{M+1,\ldots,N\}}\{c_i\}$, $c_m = \min_{i \in \{M+1,\ldots,N\}}\{c_i\}$, $\nu_m = \min_{i \in \{M+1,\ldots,N\}}\{\nu_i\}$, $\eta(x_l(0))$ is given in Assumption 10.8 and

$$V_1(0) = \delta^T(0)(\Xi \mathcal{L}_2^{-1} \otimes P^{-1})\delta(0) + \sum_{i=M+1}^N \text{tr}\left(\frac{1}{\nu_i}\tilde{\omega}_i^T(0)\tilde{\omega}_i(0)\right).$$

Proof 10.5 *Under Assumption 10.7, one knows that there exists a positive vector* $\xi = [\xi_{M+1}, \ldots, \xi_N]^T$ *such that* $\Xi \mathcal{L}_2 = \mathcal{L}_2^T \Xi$, *of which* $\Xi = \text{diag}\{\xi_{M+1}, \ldots, \xi_N\} > 0$. *On the other hand, it can be obtained from Assumption 10.6 and Lemma 10.1 that* $\mathcal{L}_2$ *is nonsingular. Since* $\Xi > 0$, *we obtain that* $\Xi \mathcal{L}_2$ *is also a nonsingular matrix. The above analysis indicates that* $\Xi \mathcal{L}_2$ *is a nonsingular and symmetric real matrix. Thus,* $\Xi \mathcal{L}_2$ *is positive definite. As* $\mathcal{L}_2 \Xi^{-1} = \Xi^{-1} (\Xi \mathcal{L}_2) \Xi^{-1}$, *one knows* $\mathcal{L}_2 \Xi^{-1}$ *is also positive definite. Based upon the above analysis, we may choose the following Lyapunov function candidate for system (10.98):*

$$V_1(t) = \delta^T(t)(\Xi \mathcal{L}_2^{-1} \otimes P^{-1})\delta(t) + \sum_{i=M+1}^{N} \text{tr}\Big(\frac{1}{\nu_i}\tilde{\omega}_i^T(t)\tilde{\omega}_i(t)\Big), \qquad (10.103)$$

where $\tilde{\omega}_i(t) = \hat{\omega}_i(t) - \omega_i$, $i = M+1, \ldots, N$, $P > 0$ *is a solution of LMI (10.101). Noticeably,* $V_1(t)$ *depending on* $\delta(t)$ *and* $\tilde{\omega}_i(t)$, $i \in \mathbb{F}$, *is regular and locally Lipschitz. According to the properties of the Filippov set-valued map* $\mathcal{F}[\cdot]$ *[27], one may get the set-valued Lie derivative of* $V_1(t)$ *along the solution of (10.98) as follows:*

$$
\begin{aligned}
\dot{V}_1(t) \in\ & \delta^T(t)\big[\Xi \mathcal{L}_2^{-1} \otimes (P^{-1}A + A^T P^{-1}) - \alpha\Xi \otimes (P^{-1}BK + K^T B^T P^{-1})\big]\delta(t) \\
& + 2\delta^T(t)(\Xi \mathcal{L}_2^{-1}\mathcal{L}_1 \otimes P^{-1}B)u_l(t) \\
& + 2\delta^T(t)(\Xi \otimes P^{-1}B)\Big[h(t) - \widetilde{W}^T\Psi + \epsilon\Big] \\
& - 2\beta\mathcal{F}\Big[\delta^T(t)(\Xi \otimes P^{-1}B)\text{sgn}((I_{N-M} \otimes K)\delta(t))\Big] \\
& + 2\sum_{i=M+1}^{N} \text{tr}\Big(\frac{1}{\nu_i}\tilde{\omega}_i^T(t)\dot{\hat{\omega}}_i(t)\Big). \qquad (10.104)
\end{aligned}
$$

Since the diagonal matrix Ξ *is positive definite, we can get that* $\text{sgn}((I_{N-M} \otimes K)\delta(t)) = \text{sgn}((\Xi \otimes K)\delta(t))$. *Furthermore, the equality* $\eta\text{sgn}(\eta) = \|\eta\|_1$ *holds for an arbitrarily given real column vector* η. *Based on the above analysis and the fact* $\mathcal{F}[f] = \{f\}$ *for an arbitrarily given continuous function* f, *one has that the set-valued Lie derivative* $\dot{V}_1(t)$ *is actually a singleton. Substituting* $K = B^T P^{-1}$ *into (10.104) gives*

$$
\begin{aligned}
\dot{V}_1(t) =\ & \delta^T(t)\big[\Xi \mathcal{L}_2^{-1} \otimes (P^{-1}A + A^T P^{-1}) - 2\alpha\Xi \otimes (P^{-1}BB^T P^{-1})\big]\delta(t) \\
& + 2\delta^T(t)(\Xi \mathcal{L}_2^{-1}\mathcal{L}_1 \otimes P^{-1}B)u_l(t) \\
& + 2\delta^T(t)(\Xi \otimes P^{-1}B)\Big[h(t) - \widetilde{W}^T\Psi + \epsilon\Big] \\
& - 2\beta\|(\Xi \otimes B^T P^{-1})\delta(t)\|_1 - 2\sum_{i=M+1}^{N} \text{tr}\Big(c_i\tilde{\omega}_i^T(t)\hat{\omega}_i(t)\Big) \\
& + 2\sum_{i=M+1}^{N} \text{tr}\Big(\tilde{\omega}_i^T(t)\xi_i\varphi_i(t)\delta_i^T(t)(P^{-1}B)\Big). \qquad (10.105)
\end{aligned}
$$

Since $\text{tr}(CD) = \text{tr}(DC)$ *holds for any matrices* C, D *with compatible dimensions, some calculations give that*

$$\delta^T(t)(\Xi \otimes P^{-1}B)(\widetilde{W}^T\Psi) = \sum_{i=M+1}^{N} \text{tr}\Big(\tilde{\omega}_i^T(t)\xi_i\varphi_i(t)\delta_i^T(t)(P^{-1}B)\Big). \qquad (10.106)$$

Furthermore, by Hölder's inequality, one obtains

$$
\begin{aligned}
&2\delta^T(t)(\Xi\mathcal{L}_2^{-1}\mathcal{L}_1\otimes P^{-1}B)u_l(t)\\
&\leq 2\|\delta^T(t)(\Xi\otimes P^{-1}B)(\mathcal{L}_2^{-1}\mathcal{L}_1\otimes I_n)u_l(t)\|_1\\
&\leq 2\|(\mathcal{L}_2^{-1}\mathcal{L}_1\otimes I_n)u_l(t)\|_\infty \cdot \|(\Xi\otimes B^T P^{-1})\delta(t)\|_1.
\end{aligned}
\tag{10.107}
$$

Based on the fact $\|(\mathcal{L}_2^{-1}\mathcal{L}_1\otimes I_n)u_l(t)\|_\infty \leq \|(\mathcal{L}_2^{-1}\mathcal{L}_1\otimes I_n)\|_\infty\|u_l(t)\|_\infty \leq \widehat{\eta}$, *it can be got from (10.107) that*

$$
2\delta^T(t)(\Xi\mathcal{L}_2^{-1}\mathcal{L}_1\otimes P^{-1}B)u_l(t) \leq 2\widehat{\eta}\|(\Xi\otimes B^T P^{-1})\delta(t)\|_1.
\tag{10.108}
$$

Similarly, one gets

$$
\begin{aligned}
2\delta^T(t)(\Xi\otimes P^{-1}B)h(t) &\leq 2\|h(t)\|_\infty \cdot \|(\Xi\otimes B^T P^{-1})\delta(t)\|_1\\
&\leq 2h_M\|(\Xi\otimes B^T P^{-1})\delta(t)\|_1.
\end{aligned}
\tag{10.109}
$$

It can be thus obtained from (10.105) to (10.109) that

$$
\begin{aligned}
\dot{V}_1(t) \leq{}& \delta^T(t)\left[\Xi\mathcal{L}_2^{-1}\otimes\left(P^{-1}A+A^T P^{-1}-\frac{2\alpha\xi_{min}}{\lambda_{max}(\Xi\mathcal{L}_2^{-1})}P^{-1}BB^T P^{-1}\right)\right]\delta(t)\\
&-2\left(\beta-\epsilon_M-\widehat{\eta}-h_M\right)\|(\Xi\otimes B^T P^{-1})\delta(t)\|_1\\
&-2\sum_{i=M+1}^{N}\mathrm{tr}\left(c_i\tilde{\omega}_i^T(t)\tilde{\omega}_i(t)\right)+2\sum_{i=M+1}^{N}\mathrm{tr}\left(c_i\tilde{\omega}_i^T(t)\omega_i\right)\\
\leq{}& -\theta_1\delta^T(t)\left(\Xi\mathcal{L}_2^{-1}\otimes P^{-1}\right)\delta(t)-2\sum_{i=M+1}^{N}\mathrm{tr}\left(c_i\tilde{\omega}_i^T(t)\tilde{\omega}_i(t)\right)\\
&+2\sum_{i=M+1}^{N}\mathrm{tr}\left(c_i\tilde{\omega}_i^T(t)\omega_i\right),
\end{aligned}
\tag{10.110}
$$

where the last inequality is derived by using LMI (10.101), the condition $\beta > \epsilon_M + \widehat{\eta}+h_M$ *and* $-2\sum_{i=M+1}^{N}\mathrm{tr}\left(c_i\tilde{\omega}_i^T(t)\tilde{\omega}_i(t)\right) \leq 0$. *Furthermore, the following inequality can be derived:*

$$
\begin{aligned}
2\sum_{i=M+1}^{N}\mathrm{tr}\left(c_i\tilde{\omega}_i^T(t)\omega_i\right) &\leq 2c_M W_M\|\widetilde{W}\|_F,\\
-2\sum_{i=M+1}^{N}\mathrm{tr}\left(c_i\tilde{\omega}_i^T(t)\tilde{\omega}_i(t)\right) &\leq -2c_m\nu_m\sum_{i=M+1}^{N}\mathrm{tr}\left(\frac{1}{\nu_i}\tilde{\omega}_i^T(t)\tilde{\omega}_i(t)\right),
\end{aligned}
\tag{10.111}
$$

where $c_M = \max_{i\in\{M+1,\dots,N\}}\{c_i\}$, $c_m = \min_{i\in\{M+1,\dots,N\}}\{c_i\}$, *and* $\nu_m = \min_{i\in\{M+1,\dots,N\}}\{\nu_i\}$. *Combining (10.110) and (10.111) give that*

$$
\dot{V}_1(t) \leq -c_0 V_1(t)+2c_M W_M\|\widetilde{W}\|_F,
\tag{10.112}
$$

where $c_0 = \min\{\theta_1, 2c_m\nu_m\}$. Some mathematical calculations give that

$$2c_M W_M \|\widetilde{W}\|_F \leq 2c_M W_M \sqrt{\nu_M} \sqrt{V_1(t)}, \tag{10.113}$$

where $\nu_M = \max_{i \in \{M+1,\dots,N\}}\{\nu_i\}$. According to the above analysis, we get

$$\dot{V}_1(t) \leq -c_0 V_1(t) + \bar{c}_0 \sqrt{V_1(t)},$$

i.e.,

$$\frac{d}{dt}\left(\sqrt{V_1(t)}\right) \leq -(c_0/2)\sqrt{V_1(t)} + \bar{c}_0/2, \tag{10.114}$$

where $\bar{c}_0 = 2c_M W_M \sqrt{\nu_M}$. Integrating both sides of (10.114) from 0 to t gives

$$\sqrt{V_1(t)} \leq \sqrt{V_1(0)} \exp\left(-\frac{c_0}{2}t\right) + \frac{\bar{c}_0}{c_0}\left(1 - \exp\left(-\frac{c_0}{2}t\right)\right)$$

$$\leq \sqrt{V_1(0)} + \frac{\bar{c}_0}{c_0}. \tag{10.115}$$

This means that $\|\delta(t)\|$ is uniformly bounded for any given $V_1(0)$. On the other hand, recall that $\lambda_{\min}\left(\Xi\mathcal{L}_2^{-1} \otimes P^{-1}\right)\|\delta(t)\|^2 \leq V_1(t)$, we then conclude that

$$\|\delta(t)\| \leq \frac{\sqrt{V_1(0)} + \frac{\bar{c}_0}{c_0}}{\sqrt{\lambda_{\min}\left(\Xi\mathcal{L}_2^{-1} \otimes P^{-1}\right)}}. \tag{10.116}$$

Combining (10.97) and (10.116) give that

$$\|e(t)\| \leq \frac{\varrho\left(\sqrt{V_1(0)} + \frac{\bar{c}_0}{c_0}\right)}{\sqrt{\lambda_{\min}\left(\Xi\mathcal{L}_2^{-1} \otimes P^{-1}\right)}},$$

where $\varrho = \sqrt{\lambda_{\max}\left((\mathcal{L}_2^{-1})^T \mathcal{L}_2^{-1}\right)}$. According to the definition of e(t), one may further get

$$\|x_f(t)\| \leq \frac{\varrho\left(\sqrt{V_1(0)} + \frac{\bar{c}_0}{c_0}\right)}{\sqrt{\lambda_{\min}\left(\Xi\mathcal{L}_2^{-1} \otimes P^{-1}\right)}} + \left\|\mathcal{L}_2^{-1}\mathcal{L}_1 \otimes I_n\right\|_F \|x_l(t)\|$$

$$\leq \frac{\varrho\left(\sqrt{V_1(0)} + \frac{\bar{c}_0}{c_0}\right)}{\sqrt{\lambda_{min}\left(\Xi\mathcal{L}_2^{-1} \otimes P^{-1}\right)}} + \sqrt{M}\left\|\mathcal{L}_2^{-1}\mathcal{L}_1 \otimes I_n\right\|_F \|x_l(t)\|_\infty$$

$$\leq \frac{\varrho\left(\sqrt{V_1(0)} + \frac{\bar{c}_0}{c_0}\right)}{\sqrt{\lambda_{\min}\left(\Xi\mathcal{L}_2^{-1} \otimes P^{-1}\right)}} + \sqrt{M}\left\|\mathcal{L}_2^{-1}\mathcal{L}_1 \otimes I_n\right\|_F \eta(x_l(0)),$$

where $\eta(x_l(0))$ is given in Assumption 10.8. Based on the above analysis, one has that $x_f(t) \in \Omega$ for $t \geq 0$, where Ω is given by (10.102).

Noticeably, Assumption 10.7 is satisfied if the subgraph describing the communication topology among the $N - M$ followers is an undirected graph. In this case, the following corollary can be established.

Corollary 10.1 *Suppose that Assumptions 10.6, 10.8 hold, the subgraph describing the interaction topology among $N - M$ followers is an undirected graph, and the matrix pair (A, B) is stabilizable. Then, for arbitrarily given $x_i(0) \in \mathbb{R}^n$, $i = 1, \ldots, N$, distributed practical containment for CNSs with leaders given by (10.88) and followers given by (10.89) under controller (10.95) associated with adaptive law (10.96) will be achieved if the control parameters are appropriately designed such that $\beta > \epsilon_M + \widehat{\eta} + h_M$, $\alpha > (\widehat{\chi}_0 \lambda_{\max}(\mathcal{L}_2^{-1}))/2$ for some given $\widehat{\chi}_0 > 0$ and $K = B^T P^{-1}$, where $P > 0$ is a solution of the following LMI:*

$$AP + PA^T - \widehat{\chi}_0 BB^T + \widehat{\theta}_1 P < 0, \tag{10.117}$$

h_M is defined in (10.90), $\widehat{\theta}_1$ is a positive scalar. Particularly, the NN approximation region Ω can be selected as

$$\Omega = \left\{ z : z \in \mathbb{R}^{(M-N)n}, \|z\| \leq \frac{\varrho\left(\sqrt{\widehat{V}_1(0)} + \bar{c}_0/c_0\right)}{\sqrt{\lambda_{\min}\left(\mathcal{L}_2^{-1} \otimes P^{-1}\right)}} \right.$$
$$\left. + \sqrt{M} \left\| \mathcal{L}_2^{-1}\mathcal{L}_1 \otimes I_n \right\|_F \eta(x_l(0)) \right\},$$

where $\widehat{V}_1(0) = \delta^T(0)(\mathcal{L}_2^{-1} \otimes P^{-1})\delta(0) + \sum_{i=M+1}^{N} \operatorname{tr}\left(\frac{1}{\nu_i}\widetilde{\omega}_i^T(0)\widetilde{\omega}_i(0)\right)$, and the other parameters are defined the same as those in Theorem 10.6.

Proof 10.6 *Since the subgraph describing the interaction topology among $N - M$ followers is an undirected graph, it can be obtained from Assumption 10.6 that $\mathcal{L}_2$ is positive definite. We may choose the following Lyapunov function candidate for (10.98):*

$$\widehat{V}_1(t) = \delta^T(t)(\mathcal{L}_2^{-1} \otimes P^{-1})\delta(t) + \sum_{i=M+1}^{N} \operatorname{tr}\left(\frac{1}{\nu_i}\widetilde{\omega}_i^T(t)\widetilde{\omega}_i(t)\right), \tag{10.118}$$

where $P > 0$ is a solution of LMI (10.117), and the other notations are defined the same as those in (10.103). This corollary can be then proven by using some similar analysis as that in the proof of Theorem 10.6.

Remark 10.13 *The feasibility analysis of LMI (10.101) is provided as follows. According to Theorem 10.6, one knows that practical containment can be achieved if the LMI (10.101) is solvable for some given positive scalar θ_1 and the control parameters of controller (10.95) are appropriately designed. Note that the specific selection of θ_1 does not influence the qualitative results of Theorem 10.6. Thus, in feasibility analysis, one may choose the positive scalar θ_1 as small as possible. According to the continuity, one gets that LMI (10.101) is feasible for some $\theta_1 > 0$ if LMI*

$AP + PA^T - \chi_0 BB^T < 0$ *is feasible for some* $\chi_0 > 0$ *and* $P > 0$. *Note that LMI* $AP + PA^T - \chi_0 BB^T < 0$ *is feasible for some* $\chi_0 > 0$ *and* $P > 0$ *if and only if the following LMI*

$$AP + PA^T - BB^T < 0 \tag{10.119}$$

is feasible for some $P > 0$. *According to Finsler's lemma, one knows that LMI (10.119) is feasible if and only if the matrix pair* (A, B) *is stabilizable. One may further get that, for an arbitrarily given* $\theta_1 > 0$, *there always exists a positive definite matrix* $P > 0$ *and a positive scalar* χ_0, *such that LMI (10.101) holds if* (A, B) *is completely controllable. The feasibility analysis of the LMI (10.117) can be similarly obtained and thus omitted.*

10.4.3 Asymptotical containment of uncertain CNSs

Asymptotical containment for the considered CNSs is studied in this subsection where the control objective is to drive the states of followers to converge into a convex hull spanned by those of the leaders asymptotically.

To achieve the goal of asymptotical containment, the designed controller for each follower should be able to exactly compensate for the external matching disturbances and the unknown nonlinearities of that follower. Based on the analysis given in the last subsection and partly motivated by adaptive controllers given in [107, 147, 214], the following neuro-adaptive evolution law for $\hat{\omega}_i(t)$ in (10.95) is proposed:

$$\dot{\hat{\omega}}_i(t) = \nu_i \left[\xi_i \varphi_i(t) \delta_i^T(t) (P^{-1} B) - c_i \left(\hat{\omega}_i(t) - \overline{\omega}_i(t) \right) \right],$$
$$\dot{\overline{\omega}}_i(t) = c_i d_0 \left(\hat{\omega}_i(t) - \overline{\omega}_i(t) \right), \ i \in \mathbb{F}, \tag{10.120}$$

where ν_i, c_i, and d_0 are positive scalars, ξ_i is provided in Assumption 10.7, $\overline{\omega}_i(t)$ is the pseudo ideal approximating weight matrix, P is a positive definite matrix to be designed later, $\hat{\omega}_i(0)$ and $\overline{\omega}_i(0)$ are set as constant real matrices with suitable dimensions. We may then obtain the following theorem in which the main analytical results of this subsection are summarized.

Theorem 10.7 *Suppose that Assumptions 10.6–10.8 hold and the matrix pair* (A, B) *is stabilizable. Then asymptotical containment for CNSs with leaders given by (10.88) and followers given by (10.89) under controller (10.95) associated with adaptive law (10.120) can be achieved if the control parameters are appropriately designed such that* $\alpha > (\chi_1 \lambda_{\max}(\Xi \mathcal{L}_2^{-1}))/(2\xi_{min})$ *for some given* $\chi_1 > 0$, $\beta > \epsilon_M + \hat{\eta} + h_M$, *and* $K = B^T P^{-1}$, *where* $P > 0$ *is a positive definite solution of the LMI*

$$AP + PA^T - \chi_1 BB^T + \theta_2 P < 0, \tag{10.121}$$

h_M *is given in (10.90) and* θ_2 *is a positive scalar. Moreover,* $\lim_{t \to +\infty} \|\hat{\omega}_i(t) - \overline{\omega}_i(t)\| = 0$, $\forall i \in \mathbb{F}$. *Particularly, the NN approximation region* Ω *can be selected as*

$$\Omega = \left\{ z : z \in \mathbb{R}^{(N-M)n}, \|z\| \leq \frac{\varrho \sqrt{V_2(0)}}{\sqrt{\lambda_{min} \left(\Xi \mathcal{L}_2^{-1} \otimes P^{-1} \right)}} \right.$$

$$+ \sqrt{M} \left\| \mathcal{L}_2^{-1} \mathcal{L}_1 \otimes I_n \right\|_F \eta(x_l(0)) \}, \qquad (10.122)$$

where $\eta(x_l(0))$ is given in Assumption 10.8, ϱ is the largest singular value of $\mathcal{L}_2^{-1}$, and

$$V_2(0) = \delta^T(0)(\Xi \mathcal{L}_2^{-1} \otimes P^{-1})\delta(0) + \sum_{i=M+1}^{N} \text{tr}\left(\frac{1}{\nu_i}\tilde{\omega}_i^T(0)\tilde{\omega}_i(0)\right)$$

$$+ \frac{1}{d_0} \sum_{i=M+1}^{N} \text{tr}\left(\tilde{\bar{\omega}}_i^T(0)\tilde{\bar{\omega}}_i(0)\right).$$

Proof 10.7 *Under Assumptions 10.6, 10.7, it can be obtained from the analysis given in the proof of Theorem 10.6 that $\Xi \mathcal{L}_2^{-1}$ is a positive definite matrix. We may then choose the following Lyapunov function candidate for (10.98):*

$$V_2(t) = \delta^T(t)(\Xi \mathcal{L}_2^{-1} \otimes P^{-1})\delta(t) + \sum_{i=M+1}^{N} \text{tr}\left(\frac{1}{\nu_i}\tilde{\omega}_i^T(t)\tilde{\omega}_i(t)\right)$$

$$+ \frac{1}{d_0} \sum_{i=M+1}^{N} \text{tr}\left(\tilde{\bar{\omega}}_i^T(t)\tilde{\bar{\omega}}_i(t)\right), \qquad (10.123)$$

where $\tilde{\omega}_i(t) = \hat{\omega}_i(t) - \omega_i$ and $\tilde{\bar{\omega}}_i(t) = \bar{\omega}_i(t) - \omega_i$ represent the error matrices of $\hat{\omega}_i(t)$ and $\bar{\omega}_i(t)$, respectively, $P > 0$ is a solution of LMI (10.121). Calculating the set-valued Lie derivative of $V_2(t)$ along the solution of (10.98) under controller (10.95) associated with adaptive law (10.120) gives

$$\dot{V}_2(t) \in \delta^T(t)[\Xi \mathcal{L}_2^{-1} \otimes (P^{-1}A + A^T P^{-1}) - 2\alpha\Xi \otimes (P^{-1}BB^T P^{-1})]\delta(t)$$

$$+ 2\delta^T(t)(\Xi \mathcal{L}_2^{-1}\mathcal{L}_1 \otimes P^{-1}B)u_l(t)$$

$$+ 2\delta^T(t)(\Xi \otimes P^{-1}B)\left[h(t) - \tilde{W}^T\Psi + \epsilon\right]$$

$$- 2\beta\mathcal{F}\left[\delta^T(t)(\Xi \otimes P^{-1}B) \cdot \text{sgn}((I_{N-M} \otimes K)\delta(t))\right]$$

$$+ 2\sum_{i=M+1}^{N} \text{tr}\left(\tilde{\omega}_i^T(t)\xi_i\varphi_i(t)\delta_i^T(t)(P^{-1}B)\right)$$

$$- 2\sum_{i=M+1}^{N} \text{tr}\left(c_i\tilde{\omega}_i^T(t)(\hat{\omega}_i(t) - \bar{\omega}_i(t))\right)$$

$$+ 2\sum_{i=M+1}^{N} \text{tr}\left(c_i\tilde{\bar{\omega}}_i^T(t)(\hat{\omega}_i(t) - \bar{\omega}_i(t))\right). \qquad (10.124)$$

By employing some similar analysis as that employed in (10.110) and using the fact $\tilde{\bar{\omega}}_i(t) - \tilde{\omega}_i(t) = -(\hat{\omega}_i(t) - \bar{\omega}_i(t))$, it can be got from (10.124) that

$$\dot{V}_2(t) \leq -\theta_2\delta^T(t)\left(\mathcal{L}_2^{-1} \otimes P^{-1}\right)\delta(t)$$

$$- 2\sum_{i=M+1}^{N} \text{tr}\left(c_i(\hat{\omega}_i(t) - \bar{\omega}_i(t))^T(\hat{\omega}_i(t) - \bar{\omega}_i(t))\right)$$

$$\leq -\theta_2 \delta^T(t)\left(\Xi\mathcal{L}_2^{-1}\otimes P^{-1}\right)\delta(t). \tag{10.125}$$

It can be obtained from (10.125) that $V_2(t)$ is non-increasing, i.e., $V_2(t)\leq V_2(0)$ for all $t\geq 0$. Since $\lambda_{\min}(\Xi\mathcal{L}_2^{-1}\otimes P^{-1})\delta^T(t)\delta(t)\leq V_2(t)$ for all $t\geq 0$, one may thus get that $\delta(t)$ is uniformly bounded over time t. Since $V_2(t)$ is non-increasing, one may also get from (10.123) that all the elements of $\tilde{w}_i(t)$ and $\tilde{\overline{w}}_i(t)$ are uniformly bounded. Noticing the fact that ω_i is a constant matrix for each $i\in\mathbb{F}$, one gets that the elements of $\hat{w}_i(t)$ and $\overline{w}_i(t)$ are uniformly bounded over time t. According to (10.98) and the assumption that $u_l(t)$ is uniformly bounded, it can be got from the above analysis that $\dot{\delta}(t)$ is uniformly bounded over time t. Since $V_2(t)\leq V_2(0)$ and $V_2(t)$ is non-increasing, it thus has a finite limit V_2^∞ as $t\to+\infty$. Integrating both sides of (10.125) yields

$$\int_0^{+\infty}\theta_2\delta^T(t)\left(\Xi\mathcal{L}_2^{-1}\otimes P^{-1}\right)\delta(t)dt\leq V_2(0)-V_2^\infty. \tag{10.126}$$

As $\dot{\delta}(t)$ is uniformly bounded over time t, one knows that $\theta_2\delta^T(t)\left(\Xi\mathcal{L}_2^{-1}\otimes P^{-1}\right)\delta(t)$ is uniformly continuous. By using Lemma 2.13, one has that $\theta_2\delta^T(t)\left(\Xi\mathcal{L}_2^{-1}\otimes P^{-1}\right)\delta(t)\to 0$ as $t\to+\infty$, i.e., $\|\delta(t)\|\to 0$ as $t\to+\infty$.

On the other hand, it can be obtained from (10.120) and by letting $W_i^e(t)=\widehat{W}_i(t)-\overline{W}_i(t)$ that

$$\dot{W}_i^e(t)=-(\nu_i c_i+c_i)W_i^e(t)+\nu_i\xi_i\varphi_i(t)\delta_i^T(t)(P^{-1}B),\quad i\in\mathbb{F}. \tag{10.127}$$

Since ν_i, c_i, ξ_i are given positive scalars, $\varphi_i(t)$ is uniformly bounded, and $\|\delta(t)\|\to 0$ as $t\to+\infty$, it can be got by taking the term $\nu_i\xi_i\varphi_i(t)\delta_i^T(t)(P^{-1}B)$ as the control input of system (10.127) that the asymptotic gain property holds for (10.127) [139]. The above analysis indicates that (10.127) is input to state stable. One may then conclude that $\lim_{t\to+\infty}\|\widehat{W}_i(t)-\overline{W}_i(t)\|=0,\ \forall i\in\mathbb{F}$.

Furthermore, according to the fact $V_2(t)\leq V_2(0)$ for all $t\geq 0$, we may get that

$$\|\delta(t)\|\leq\frac{\sqrt{V_2(0)}}{\sqrt{\lambda_{\min}\left(\Xi\mathcal{L}_2^{-1}\otimes P^{-1}\right)}},\ \forall\,t\geq 0. \tag{10.128}$$

Combining (10.97) and (10.128) give that

$$\|e(t)\|\leq\frac{\varrho\sqrt{V_2(0)}}{\sqrt{\lambda_{\min}\left(\Xi\mathcal{L}_2^{-1}\otimes P^{-1}\right)}},\ \forall\,t\geq 0,$$

where $\varrho=\sqrt{\lambda_{\max}\left((\mathcal{L}_2^{-1})^T\mathcal{L}_2^{-1}\right)}$. According to the definition of $e(t)$, one may further get

$$\|x_f(t)\|\leq\frac{\varrho\sqrt{V_2(0)}}{\sqrt{\lambda_{min}\left(\Xi\mathcal{L}_2^{-1}\otimes P^{-1}\right)}}+\left\|\mathcal{L}_2^{-1}\mathcal{L}_1\otimes I_n\right\|_F\|x_l(t)\|$$

$$\leq \frac{\varrho\sqrt{V_2(0)}}{\sqrt{\lambda_{\min}\left(\Xi\mathcal{L}_2^{-1}\otimes P^{-1}\right)}} + \sqrt{M}\left\|\mathcal{L}_2^{-1}\mathcal{L}_1\otimes I_n\right\|_F \|x_l(t)\|_\infty$$

$$\leq \frac{\varrho\sqrt{V_2(0)}}{\sqrt{\lambda_{\min}\left(\Xi\mathcal{L}_2^{-1}\otimes P^{-1}\right)}} + \sqrt{M}\left\|\mathcal{L}_2^{-1}\mathcal{L}_1\otimes I_n\right\|_F \eta(x_l(0)),$$

where $\eta(x_l(0))$ is given in Assumption 10.8. Based on the above analysis, one has that $x_f(t) \in \Omega$ for each $t \geq 0$, where Ω is given by (10.122).

With the condition that the subgraph describing the interaction topology among $N - M$ followers is undirected, the following corollary can be obtained from Theorem 10.7 where the detailed proof is omitted for brevity.

Corollary 10.2 *Suppose that Assumptions 10.6, 10.8 hold, the subgraph describing the interaction topology among $N - M$ followers is undirected, and the matrix pair (A, B) is stabilizable. Then asymptotical containment for CNSs with leaders given by (10.88) and followers given by (10.89) under controller (10.95) associated with adaptive law (10.120) can be achieved if the control parameters are appropriately designed such that $\alpha > (\chi_1\lambda_{\max}(\mathcal{L}_2^{-1}))/2$ for some given $\widehat{\chi}_1 > 0$, $\beta > \epsilon_M + \widehat{\eta} + h_M$, and $K = B^T P^{-1}$, where $P > 0$ is a solution of the LMI*

$$AP + PA^T - \widehat{\chi}_1 BB^T + \widehat{\theta}_2 P < 0, \tag{10.129}$$

h_M is given in (10.90) and $\widehat{\theta}_2$ is a positive scalar. Moreover, $\lim_{t\to+\infty}\|\widehat{W}_i(t) - \overline{W}_i(t)\| = 0$, $\forall i \in \mathbb{F}$. Particularly, the NN approximation region Ω can be selected as

$$\Omega = \left\{z : z \in \mathbb{R}^n, \|z\| \leq \frac{\varrho\sqrt{\widehat{V}_2(0)}}{\sqrt{\lambda_{min}\left(\mathcal{L}_2^{-1}\otimes P^{-1}\right)}} \right.$$
$$\left. + \sqrt{M}\left\|\mathcal{L}_2^{-1}\mathcal{L}_1\otimes I_n\right\|_F \eta(x_l(0))\right\},$$

where $\eta(x_l(0))$ is given in Assumption 10.8, ϱ is the largest singular value of $\mathcal{L}_2^{-1}$, $\eta(x_l(0))$ is given in Assumption 10.8 and

$$\widehat{V}_2(0) = \delta^T(0)\left(\mathcal{L}_2^{-1}\otimes P^{-1}\right)\delta(0) + \sum_{i=M+1}^N \text{tr}\left(\frac{1}{\nu_i}\widetilde{\omega}_i^T(0)\widetilde{\omega}_i(0)\right)$$

$$+ \frac{1}{d_0}\sum_{i=M+1}^N \text{tr}\left(\widetilde{\omega}_i^T(0)\widetilde{\omega}_i(0)\right).$$

Remark 10.14 *Compared with the results provided in Theorem 10.6 on achieving practical containment by employing the NN-based containment controller (10.95) associated with adaptive law (10.96), asymptotic containment can be ensured by employing the NN-based containment controller (10.95) associated with adaptive law*

(10.120). Obviously, the adaptive law for NN approximating weight matrix $\hat{\omega}_i(t)$ designed in (10.120) is more complex than that designed in (10.96) as the pseudo ideal approximating weight matrix has been involved in the adaptive law designed in (10.120). This indicates that a relatively higher computing ability is required for executing the NN-based containment controller (10.95) associated with adaptive law (10.120) compared with executing the NN-based containment controller (10.95) associated with adaptive law (10.96). This indicates that an in-depth study on how to achieve practical containment under various environments is particularly meaningful for CNSs with limited computing ability. Nevertheless, the achievement of asymptotic containment in CNSs implies that the Euclidean norm of the containment error vector will converge to zero as time approaches infinity.

Remark 10.15 *It can be seen from Theorems 10.6, 10.7, and Corollaries 10.1, 10.2 that, for arbitrarily given $x_i(0) \in \mathbb{R}^n$, $i = 1, \ldots, N$, the feasible NN approximation regions can be determined under Assumptions 10.6–10.8. Hence, the theoretical results given in Theorems 10.6, 10.7, and Corollaries 10.1, 10.2 are semi-global. Furthermore, it can be seen from the proofs of Theorems 10.6 and 10.7 that the non-smooth term in (10.95) is utilized to reject the effect of bounded uncertain term $h_i(t)$ and the bounded unknown input $u_l(t)$ acting on the leaders. However, the bounds of $h_i(t)$ and $u_l(t)$ should be explicitly known and will be involved in selecting the parameters of the non-smooth term in (10.95). To achieve containment in the considered CNS, the NN adaptive term is designed in (10.95) to compensate for the effect of totally unknown term $f_i(x_i(t))$ with an unknown bound. It is still an open issue whether practical containment or asymptotic containment can be ensured in CNSs described by (10.88) and (10.89) by designing a distributed controller without employing the NN adaptive term.*

Remark 10.16 *Though it has been shown in both Theorem 10.7 and Corollary 10.2 that $\lim_{t \to +\infty} \|\hat{\omega}_i(t) - \overline{\omega}_i(t)\| = 0$, $\forall i \in \mathbb{F}$, we still can not determine whether the estimation weight matrix $\hat{\omega}_i(t)$ converges to the ideal weight matrix ω_i for $i \in \mathbb{F}$. It is also worth noting that the introductions of pseudo ideal weight matrix $\overline{\omega}_i(t)$ and the non-smooth feedback term in the containment protocol provide us with the ability to derive the states of followers to asymptotically converge into the convex hull spanned by those of the multiple leaders. Furthermore, as the dynamic evolution of each follower given by (10.89) is influenced by unknown nonlinear dynamics and external bounded disturbances, then the achievement of containment given in Theorem 10.7 and Corollary 10.2 implies semi-globally robust convergence of the states of followers, with different $f_i(x_i(t))$ and $h_i(t)$ for followers, to the convex hull formed by the states of multiple leaders.*

Remark 10.17 *A practical issue in implementing the proposed non-smooth coupling laws is that the chattering phenomenon may occur in the evolution of the closed-loop CNSs. Nevertheless, in real applications, the boundary layer technique can be employed to avoid coupling discontinuities and fast switchings of the actuators. Particularly, the signum function in the controller (10.95) can be replaced by the saturation function $\text{sat}(\cdot)$ [134] so as to reduce the chattering effects. It should also be noted that*

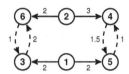

Figure 10.11 The communication graph $\mathcal{G}$.

only globally bounded containment can be achieved if the boundary layer technique is utilized in designing the coupling law for CNSs.

10.4.4 Numerical simulations

Two numerical examples are provided in this section to verify the effectiveness of the analytical results derived in this section.

Example 1: The analytical results provided in Theorem 10.6 are illustrated in this numerical example. Suppose that there are six agents in the considered CNSs with two leaders. The agents labeled as 1 and 2 are designated as the leaders while the rest are designated as the followers. The interaction topology is depicted in Figure 10.11. According to (10.87), one gets that

$$
\mathcal{L}_2 = \begin{bmatrix} 3 & 0 & 0 & -1 \\ 0 & 4 & -1 & 0 \\ 0 & -1.5 & 3.5 & 0 \\ -2 & 0 & 0 & 4 \end{bmatrix}.
$$

It can be verified that Assumptions 10.6 and 10.7 hold for the considered CNS associated with $\xi = [2, 1.5, 1, 1]^T$. The inherent dynamics of leaders are described by linearized model of the longitudinal dynamics of an aircraft [55], described by (10.88) with

$$
A = \begin{bmatrix} -0.2770 & 1.0000 & -0.0002 \\ -17.1000 & -0.1780 & -12.2000 \\ 0 & 0 & -6.6700 \end{bmatrix}, \quad B = \begin{bmatrix} 0 \\ 0 \\ 6.67 \end{bmatrix}.
$$

The state vector $x_i(t) = [x_{i1}(t), x_{i2}(t), x_{i3}(t)]^T \in \mathbb{R}^3$, where $x_{i1}(t)$ represents the angle of attack, $x_{i2}(t)$ represents the pitch rate, and $x_{i3}(t)$ is the elevator angle, for each $i \in \mathbb{L}$. Set $u_1(t) = -1$, $u_2(t) = 0$, $x_1(0) = [-4, -2, 5]^T$, and $x_2(0) = [2, 1, -3.5]^T$. One may thus choose $\hat{\eta} = 1$. According the NN approximation theory, the positive scalar ϵ_M can be arbitrarily selected. In simulations, set $f_i(x_i(t)) = 4x_{i1}(t)\sin(x_{i1}(t)) + 2\cos(x_{i2}(t))$, $h_i(t) = 0.5\sin(it)$, $\epsilon_M = 0.01$, $\alpha = 0.25$, $\beta = 1.52$, $\nu_i = 500$, and $c_i = 0.5$ for $i \in \mathbb{F}$. Furthermore, three neurons are utilized in simulations for each NN approximator. Sigmoid basis functions are employed and the estimation matrices $\hat{\omega}_i(t)$, $i \in \mathbb{F}$, are initialized to be zero matrices. One thus gets that the conditions given in Theorem 10.6 are satisfied. This indicates that distributed practical containment in the closed-loop CNS can be guaranteed. The state trajectories of agents are respectively provided in Figs. 10.12–10.14.

Example 2: In this numerical example, the theoretical results provided in Theorem 10.7 are firstly verified. Then, some comparison numerical simulations between

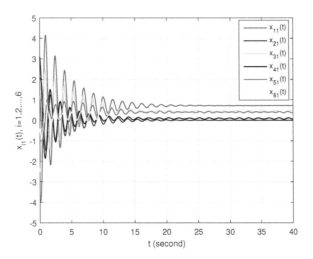

Figure 10.12 Trajectories of $x_{i1}(t)$, $i = 1, 2, \ldots, 6$, in Example 1.

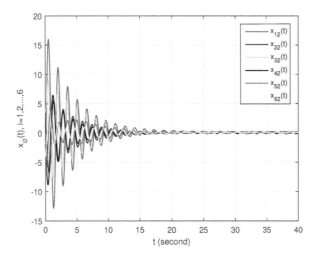

Figure 10.13 Trajectories of $x_{i2}(t)$, $i = 1, 2, \ldots, 6$, in Example 1.

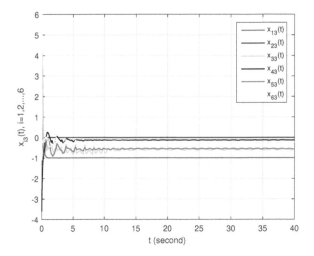

Figure 10.14 Trajectories of $x_{i3}(t)$, $i = 1, 2, \ldots, 6$, in Example 1.

practical containment and asymptotic containment are performed. The inherent dynamics of the agents, the interaction topology, and the control inputs acting on the leaders are taken respectively the same as those in Example 1. In simulations, let $d_0 = 25$. Furthermore, three neurons are utilized in simulations for each NN approximator. Sigmoid basis functions are employed and the estimation matrices $\hat{\omega}_i(t)$, $i = 3, 4, 5, 6$, are initialized to be zero matrices. One then gets that the conditions given in Theorem 10.7 are satisfied. This indicates that asymptotic containment in the considered CNS can be guaranteed. The state trajectories of agents are respectively provided in Figs. 10.15–10.17. Moreover, the profiles of Euclidean norm of $\hat{\omega}_i(t) - \overline{\omega}_i(t)$, $i = 3, 4, 5, 6$, are plotted in Figure 10.18. Moreover, the profiles of $\|e(t)\|$ for practical containment and asymptotic containment are respectively plotted in Figure 10.19 which indicates that a smaller steady containment error can be yielded for the case of asymptotic containment in comparison with the case of practical containment. The simulation results verify the analytical results given in Theorem 10.7 very well.

10.5 CONCLUSIONS

In this chapter, the consensus tracking problems for CNSs with uncertain dynamics and directed topology were studied based on the NN universal approximation theory. In Section 10.2, we have shown the consensus tracking error of the uncertain CNSs with a high-dimensional leader and directed switching topologies was UUB under the designed neuro-adaptive controllers. In Section 10.3, we have successfully solved the asymptotical neuro-adaptive consensus tracking problem for uncertain CNSs with a high dimensional leader and directed fixed topology by designing a novel discontinuous controller which can be implemented in a fully distributed way. In Section 10.4, we have successfully solved the asymptotical neuro-adaptive containment problem for

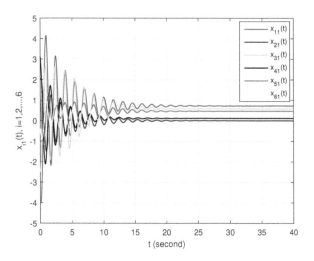

Figure 10.15 Trajectories of $x_{i1}(t)$, $i = 1, 2, \ldots, 6$, in Example 2.

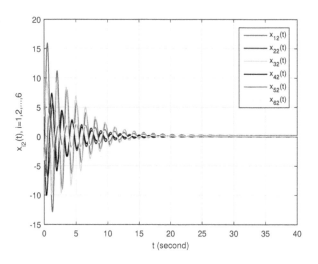

Figure 10.16 Trajectories of $x_{i2}(t)$, $i = 1, 2, \ldots, 6$, in Example 2.

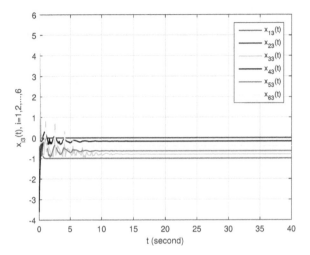

Figure 10.17 Trajectories of $x_{i3}(t)$, $i = 1, 2, \ldots, 6$, in Example 2.

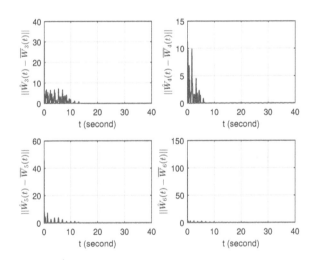

Figure 10.18 Trajectories of $||\hat{\omega}_i(t) - \overline{\omega}_i(t)||$, $i = 3, 4, 5, 6$, in Example 2.

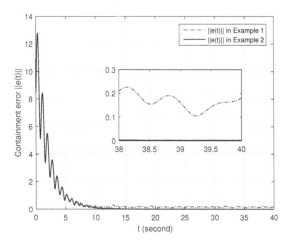

Figure 10.19 Trajectories of the Euclidean norm of containment errors in Example 1 and Example 2.

uncertain CNSs with multiple leaders and detail balanced directed communication topology by designing a feedback controller together with second order neuro-adaptive coupling updating laws. Although some works have been done in this chapter, the asymptotical neuro-adaptive consensus problem for CNSs with uncertain dynamics under directed topology remains unsolved. We leave this problem as an open problem and hope it can be successfully solved in the near future.

CHAPTER 11

Resilient consensus of CNSs with input saturation and malicious attack under switching topologies

This chapter studies the resilient consensus of CNSs with switching topologies in the presence of input saturation or malicious attack. Section 11.2 firstly proposes an edge-based distributed adaptive anti-windup protocol with relative output information by designing distributed observer and anti-windup compensator separately. As this design approach may suffer from heavy calculation burden as it doubles the order of observers, the case with absolute output information is then studied. Section 11.3 studies the robustness of the switching topologies to present necessary and sufficient condition for resilient consensus of complex network systems under malicious attack. This section proposes a novel concept of jointly (r, s)-robust topology to put forward a sufficient and necessary condition for resilient consensus achievement of Weighted-Mean-Subsequence-Reduced algorithm.

11.1 INTRODUCTION

Under most circumstances, the evolution of CNSs always suffers from systematic faults or external attacks. One commonly investigated fault is the input saturation, which is caused by physical limit as the actuation power cannot be arbitrarily large in practical applications. The consensus of first-order CNSs with input saturation is studied in [40,75]. [46] considers the consensus of different input saturation levels with first-order CNSs. A distributed saturation controller is presented in [207] for consensus of second-order CNSs, where the sign function is introduced to ensure the control input of each agent essentially being no larger than 1. The time-delayed consensus of second-order CNSs with input saturation is studied in [210]. Output feedback consensus algorithms under input saturation constraint are designed in [1] for second-order CNSs. Distributed anti-windup approach is proposed in [44] for second-order CNSs with input saturation. The containment problem of saturated second-order CNSs with multiple leaders is studied in [45]. [132] studies the consensus of second-order nonlinear CNSs with saturation and dead-zone, where the neural network approximation and back-stepping methods are used to generate distributed adaptive controllers. By using a low gain method, saturation consensus of linear CNSs is solved in [142], and the saturated bipartite consensus or saturated output consensus problem with switching topologies are studied in [125] and [144], respectively. Distributed event-triggered protocol is presented in [212] to achieve consensus of linear CNSs with input saturation. An LMI-based approach is presented in [30] to achieve robust consensus of CNSs with input saturation.

On the other hand, the attacker may make some nodes in CNSs be malicious nodes, which apply the control signal different from the preset controller, or even send wrong information to normal nodes. How to ensure the normal nodes realize consensus despite of malicious attack is quite important in practice. [72,223] propose some novel notions to capture the robustness of the graphs, and the resilient consensus of F-total malicious model under Weighted-Mean-Subsequence-Reduced (W-MSR) algorithm is realized if and only if the graph is $(F + 1, F + 1)$-robust. The resilient consensus problems of second-order and high-order CNSs are further studied in [33] and [71], respectively. To reduce the amount of information interactions in each time instant, the switching topologies are introduced. It is presented in [71,72] that a sufficient condition for resilient consensus is that there are infinite time instants at which the time-varying graph satisfies the robust condition. [131] proposes the Sliding Weighted Mean-Subsequence-Reduced (SW-MSR) algorithm, where each agent uses all the neighbors' values within T steps to update its value, and the resilient consensus can be achieved if the dynamic graph is $(T, 2F + 1)$-robust. The resilient leader-follower consensus is further studied with the SW-MSR algorithm, where the strongly $(T, 2F + 1)$-robust condition is proposed [164].

This chapter concerns on the resilient consensus of CNSs with switching topologies in the presence of input saturation or malicious attack. This chapter is organized as follows. Section 11.2 investigates the fully distributed adaptive output feedback controller design for consensus of linear CNSs with input saturation under switching topologies. Section 11.3 studies the resilient consensus of CNSs with malicious attacks under switching topologies.

11.2　CONSENSUS OF LINEAR CNSS WITH INPUT SATURATION UNDER SWITCHING TOPOLOGIES

In this section, we discuss the fully distributed adaptive protocol design for linear CNSs with input saturation under switching topologies.

11.2.1 Problem formulation

Consider the linear CNS consisting of N agents with input saturation. The dynamics of the agents are described as

$$\begin{aligned} \dot{x}_i(t) &= Ax_i(t) + B\mathrm{sat}_\iota(u_i(t)), \\ y_i(t) &= Cx_i(t), \ i = 1, \ldots, N, \end{aligned} \tag{11.1}$$

where $x_i(t) \in \mathbb{R}^n$, $u_i(t) \in \mathbb{R}^p$, and $y_i(t) \in \mathbb{R}^m$ are the state, input, and output of the agent i, the matrices A, B, and C are the system matrix, input matrix, and output matrix with corresponding dimensions, and $\mathrm{sat}_\iota(u_i(t)) = [\mathrm{sat}_\iota(u_{i1}(t)), \ldots, \mathrm{sat}_\iota(u_{ip}(t))]^T$ is the saturation vector defined as

$$\mathrm{sat}_\iota(s) = \begin{cases} s, & \text{if } |s| < \iota, \\ \mathrm{sgn}(s)\iota, & \text{if } |s| \geq \iota, \end{cases}$$

with $\iota > 0$ as the upper bound of control input.

The following assumption is made on the system matrices.

Assumption 11.1 *All the eigenvalues of the dynamic matrix A are in closed left-hand plane, i.e., no eigenvalue of A has positive real part. Moreover, (A, B) is stabilizable and (A, C) is detectable.*

Remark 11.1 *Assumption 11.1 is a sufficient condition for the semi-global stabilization via linear dynamic output feedback [158], consisting of the well-known asymptotically null controllable with bounded control (ANCBC) condition and the detectability of (A, C).*

The communication topology of the CNS under consideration is assumed to be dynamically switching over a graph set $\hat{\mathcal{G}} = \{\mathcal{G}^1, \ldots, \mathcal{G}^\kappa\}, \kappa \geq 1$, where each $\mathcal{G}^k$ denotes an undirected topology, i.e., $\mathcal{G}(t) \in \hat{\mathcal{G}}$ for all t. We assume that

Assumption 11.2 *For each $k \in \{1, \ldots, \kappa\}$, the graph $\mathcal{G}^k$ is undirected and connected.*

Under Assumption 11.2, we have the following property.

Lemma 11.1 *The Laplacian matrix $\mathcal{L}^{\sigma(t)}$ is positive semi-definite containing a simple zero eigenvalue with $\mathbf{1}$ as the eigenvector.*

The fully distributed consensus problem for the CNS (11.1) under input saturation is defined as follows:

Definition 11.1 *Design appropriate internal state $\varrho_i(t)$ and distributed controller $u_i(t)$ for each agent $i = 1, \ldots, N$ in the form*

$$\dot{\varrho}_i(t) = h_i\left(\varrho_i(t), \sum_{j=1}^N a_{ij}(\varrho_i(t) - \varrho_j(t)), y_i(t), \sum_{j=1}^N a_{ij}(y_i(t) - y_j(t))\right),$$

$$u_i(t) = k_i\left(\varrho_i(t), \sum_{j=1}^N a_{ij}(\varrho_i(t) - \varrho_j(t)), y_i(t), \sum_{j=1}^N a_{ij}(y_i(t) - y_j(t))\right),$$

such that for any initial condition $x_i(t_0)$, it holds $\|x_i(t) - x_j(t)\| \to 0$ as $t \to \infty$, where $h_i(\cdot)$ and $k_i(\cdot)$ are nonlinear functions independent of global graph connectivity information.

Before moving forward, we have to introduce the following lemma, which is of vital importance to our protocol design.

Lemma 11.2 ([160]) *For the following system satisfying ANCBC:*

$$\dot{z}(t) = Az(t) + B\left[\operatorname{sat}_\iota(u(t) + h(t)) - h(t)\right],$$

there is a globally Lipschitz feedback controller $u(t) = f(z(t))$ such that if $h(t) \in \mathbb{L}_2$, we have $z(t) \in \mathbb{L}_2$. Furthermore, the controller $u(t)$ can be designed by the multi-level saturation feedback algorithm 11.1.

Algorithm 11.1 ([159]) *Multi-level saturation feedback controller design.*

(1) Make the nonsingular linear transformation $\bar{z}(t) = Tz(t)$ with transformation matrix T such that $\bar{A} = TAT^{-1} = \operatorname{diag}(J_1, \ldots, J_q, J_{q+1})$ is the Jordan normal form of A, where J_{q+1} includes all the Jordan blocks of negative real part eigenvalues, and $J_i \in \mathbb{R}^{m_i \times m_i}$ represents the Jordan block of a zero real part eigenvalue λ_i with $m_1 \geq \ldots \geq m_q$.

(2) Reformulate the system $\bar{z}(t) = [\bar{z}_1^T(t), \ldots, \bar{z}_{q+1}^T(t)]^T$ into $\hat{z}(t) = [\hat{z}_1^T(t), \ldots, \hat{z}_{m_1}^T(t)]^T$ with $\hat{z}_{m_1}(t) = [\bar{z}_{1,m_1}(t), \ldots, \bar{z}_{q,m_q}(t), \bar{z}_{q+1}^T(t)]^T$ and $\hat{z}_{m_1-i}(t) = [\bar{z}_{1,m_1-i}(t), \ldots, \bar{z}_{q,m_q-i}(t)]^T$ for $i = 1, \ldots, m_1 - 1$, where $\bar{z}_{j,m_j-i}(t)$ is null if $m_j - i \leq 0$.

(3) Let $\hat{Z}_i(t) = [\hat{z}_i^T(t), \ldots, \hat{z}_{m_1}^T(t)]^T$, $i = 1, \ldots, m_1$, and we have $\dot{\hat{Z}}_i(t) = \hat{A}_i \hat{Z}_i(t) + \hat{B}_i \left[\operatorname{sat}_\iota(u(t) + h(t)) - h(t)\right]$.

(4) Let $g_{m_1}(t) = -\hat{B}_{m_1}^T P_{m_1} \hat{Z}_{m_1}(t)$ with P_{m_1} being the positive definite solution of $\hat{A}_{m_1}^T P_{m_1} + P_{m_1} \hat{A}_{m_1} \leq 0$.

(5) Let $g_i(t) = -\hat{B}_i^T P_i \hat{Z}_i(t)$ with P_i being the positive definite solution of $\tilde{A}_i^T P_i + P_i \tilde{A}_i \leq 0$, where $\tilde{A}_i = \hat{A}_i + \frac{\partial}{\partial \hat{Z}_i(t)}(\hat{B}_i \sum_{j=i+1}^{m_1} g_j(t))$, for $i = m_1 - 1, \ldots, 1$.

(6) Choose $f_i(t) = \mu_i \operatorname{sat}_\iota\left(\frac{g_i(t) + f_{i-1}(t)}{\mu_i}\right)$ with μ_i being sufficiently small constant and $f_0(t) = 0$, for $i = 1, \ldots, m_1$. Then, $f(z(t)) = f_{m_1}(t)$.

Remark 11.2 *Algorithm 11.1 constructs an m_1-level saturation feedback controller, whose computational complexity depends on the dimension of the largest Jordan block associated with zero real part eigenvalues, i.e., m_1. For the case that A is stable as $m_1 \in \{0, 1\}$, the controller can be designed as $f(z(t)) = -B^T \bar{P} z(t)$ with $\bar{P}$ being the positive definite solution of $A^T \bar{P} + \bar{P} A \leq 0$.*

11.2.2 CNSs with relative output information

In this subsection, the edge-based distributed adaptive anti-windup protocol with relative output information is proposed by designing distributed observer and anti-windup compensator separately.

Based on relative output information among neighboring agents, we can formulate the following distributed adaptive anti-windup observers:

$$\dot{v}_i(t) = (A + BK)v_i(t) + F \sum_{j=1}^{N} a_{ij}c_{ij}(t)[C(v_i(t) - v_j(t))$$

$$+ C(w_i(t) - w_j(t)) - (y_i(t) - y_j(t))],$$

$$\dot{w}_i(t) = Aw_i(t) + B[\text{sat}_\iota(u_i(t)) - Kv_i(t)],$$

$$\dot{c}_{ij}(t) = \mu_{ij}a_{ij}\|C(v_i(t) - v_j(t)) + C(w_i(t) - w_j(t)) - (y_i(t) - y_j(t))\|^2, \quad (11.2)$$

where $v_i(t)$ and $w_i(t)$ are respectively the distributed observer and the anti-windup compensator for agent i, $c_{ij}(t)$ is the adaptive coupling weight acting on the edge (i, j) with initial value satisfying $c_{ij}(t_0) = c_{ji}(t_0) > 0$, μ_{ij} is a positive constant satisfying $\mu_{ij} = \mu_{ji}$, K and F are feedback gain matrices.

Define $\eta_i(t) = x_i(t) - v_i(t) - w_i(t)$ as the potential state estimation. And we have the following equation to describe the dynamics of $\eta_i(t)$:

$$\dot{\eta}_i(t) = A\eta_i(t) + FC \sum_{j=1}^{N} a_{ij}c_{ij}(t)(\eta_i(t) - \eta_j(t)),$$

$$\dot{c}_{ij}(t) = \mu_{ij}a_{ij}\|C(\eta_i(t) - \eta_j(t))\|^2. \quad (11.3)$$

The following lemma shows how to design feedback gain matrix F to achieve consensus of $\eta_i(t)$.

Lemma 11.3 ([80]) *Suppose that Assumptions 11.1 and 11.2 hold. Then, the potential state estimation $\eta_i(t)$ can reach consensus if $F = -Q^{-1}C^T$ with $Q > 0$ being the positive definite solution of the LMI:*

$$QA + A^T Q - 2C^T C < 0. \quad (11.4)$$

Besides, each adaptive coupling weight $c_{ij}(t)$ converges to finite constant.

Proof 11.1 *Let $e_i(t) = \eta_i(t) - \frac{1}{N}\sum_{j=1}^{N} \eta_j(t)$ be the consensus error of $\eta_i(t)$, and $e(t) = [e_1^T(t), \ldots, e_N^T(t)]^T$. Consider the Lyapunov function*

$$V_1(t) = \frac{1}{2} \sum_{i=1}^{N} e_i^T(t)Qe_i(t) + \sum_{i=1}^{N} \sum_{j=1,j\neq i}^{N} \frac{(c_{ij}(t) - \alpha)^2}{4\mu_{ij}}, \quad (11.5)$$

where $\alpha \geq \frac{1}{\lambda_2}$ with λ_2 being the smallest nonzero eigenvalue of $\mathcal{L}^k, k = 1, \ldots, \kappa$. It can be easily verified that $V_1(t)$ is positive definite about the variables $e_i(t)$ and $c_{ij}(t) - \alpha$.

The time derivative of V_1 is given by

$$
\dot{V}_1(t) = \sum_{i=1}^{N} [e_i(t)^T Q A e_i(t) - e_i^T(t) \sum_{j=1}^{N} a_{ij} c_{ij}(t) C^T C (e_i(t) - e_j(t))]
$$

$$
+ \frac{1}{2} \sum_{i=1}^{N} \sum_{\substack{j=1 \\ j \neq i}}^{N} (c_{ij}(t) - \alpha) a_{ij} (e_i(t) - e_j(t))^T C^T C (e_i(t) - e_j(t))
$$

$$
\tag{11.6}
$$

$$
= \sum_{i=1}^{N} [e_i^T(t) Q A e_i(t) - \alpha e_i^T(t) C^T C \sum_{j=1}^{N} a_{ij}(e_i(t) - e_j(t))]
$$

$$
\leq \frac{1}{2} e^T(t)[I_N \otimes (QA + A^T Q - 2C^T C)]e(t)
$$

$$
\leq 0,
$$

where the second equality is obtained by using symmetric property of a_{ij} and $c_{ij}(t)$ to derive

$$
\frac{1}{2} \sum_{i=1}^{N} \sum_{\substack{j=1 \\ j \neq i}}^{N} (c_{ij}(t) - \alpha) a_{ij} (e_i(t) - e_j(t))^T C^T C (e_i(t) - e_j(t))
$$

$$
= \sum_{i=1}^{N} e_i^T(t) C^T C \sum_{j=1}^{N} a_{ij}(c_{ij}(t) - \alpha)(e_i(t) - e_j(t)).
$$

Therefore, $V_1(t)$ is bounded, and so are $e_i(t)$ and $c_{ij}(t)$. Since $\dot{V}_1(t) \equiv 0$ can derive $e(t) \equiv 0$, by LaSalle's invariance principle, we can conclude $e(t) \to 0$. That is, $\eta_i(t)$ reaches consensus. Notice that the derivative of $c_{ij}(t)$ is nonnegative, meaning that $c_{ij}(t)$ is nondecreasing. On the other hand, $c_{ij}(t)$ is bounded, which in turn indicates that each adaptive coupling weight $c_{ij}(t)$ converges to a finite constant.

Then, we have the following theorem to design the distributed adaptive anti-windup controller $u_i(t)$.

Theorem 11.1 *Suppose Assumptions 11.1 and 11.2 hold. Then the consensus of the N agents in (11.1) with input saturation can be realized by designing the controller*

$$
u_i(t) = K v_i(t) + f(w_i(t)), \tag{11.7}
$$

where K is chosen such that $A + BK$ is Hurwitz, and $f(w_i(t))$ is the controller designed by Algorithm 11.1 in Lemma 11.2.

Proof 11.2 *By (11.6), we can obtain that $V_1(t)$ is bounded and nonincreasing, and it has finite limit V_1^∞ as $t \to \infty$. Integrating the third inequality of (11.6) yields*

$$
- \int_{t_0}^{\infty} \frac{1}{2} e^T(t)[I_N \otimes (QA + A^T Q - 2C^T C)]e(t)dt \leq V_1(t_0) - V_1(\infty).
$$

Therefore, $e(t) \in \mathbb{L}_2$ and so is $e_i(t)$.

Next, we will show $v_i(t) \in \mathbb{L}_2$ and $w_i(t) \in \mathbb{L}_2$. Since $e_i(t) \in \mathbb{L}_2$, we can conclude from the dynamics of $c_{ij}(t)$ in (11.3) that each $c_{ij}(t)$ is bounded. Thus, it is not difficult to obtain $F \sum_{j=1}^{N} a_{ij} c_{ij}(t) C(e_i(t) - e_j(t)) \in \mathbb{L}_2$. Since $A + BK$ is Hurwitz, we have $v_i(t) \in \mathbb{L}_2$ by the dynamics of $v_i(t)$ in (11.2). In light of Lemma 11.2 and the fact $Kv_i(t) \in \mathbb{L}_2$, we can conclude that $w_i(t) \in \mathbb{L}_2$.

Define $\xi_i(t) = x_i(t) - \frac{1}{N} \sum_{j=1}^{N} x_j(t)$, and we have $\xi_i(t) = e_i(t) + v_i(t) + w_i(t) + \frac{1}{N} \sum_{j=1}^{N} (v_j(t) + w_j(t)) \in \mathbb{L}_2$. That is, the consensus is achieved.

Remark 11.3 *By the proof of Theorem 11.1, $v_i(t)$ and $w_i(t)$ converge to zero, making $\eta_i(t)$ converge to the state $x_i(t)$. That is why we named $\eta_i(t)$ the potential state estimation. It can be revealed from the construction of distributed observer $v_i(t)$ and the adaptive coupling weight $c_{ij}(t)$ that the output information of the state, the distributed observer and the anti-windup compensator, i.e., $y_i(t)$, $Cv_i(t)$, and $Cw_i(t)$, should be transmitted to neighboring agents via communication channel.*

Remark 11.4 *It should be noted that the controller (11.7) designed by Algorithm 11.1 is utilized to ensure the convergence of $w_i(t)$, where only agents' dynamics and the observer information $v_i(t)$ and $w_i(t)$ are involved. Therefore, the protocol designed in (11.7) can be implemented by each agent in a fully distributed manner.*

Remark 11.5 *Note that simply combining the techniques of designing adaptive protocol presented in [80] and the saturation compensator proposed in [159, 160] cannot yield the distributed adaptive saturated protocol (11.7). The benefit of designing controller (11.7) lies in the fact that it could decouple the nonlinearities caused by adaptive control and the input saturation by proposing distributed adaptive observer $v_i(t)$ and the anti-windup compensator $w_i(t)$ with coupled dynamics of $v_i(t)$ and $w_i(t)$ in (11.2). In other words, the dynamics of $v_i(t)$ and $w_i(t)$ are coupled with each other, however, such a design structure makes the consensus analysis much simplified since the unpleasant nonlinearities are skillfully decoupled as shown in Lemma 11.3 and Theorem 11.1.*

Since both the distributed observer $v_i(t)$ and the anti-windup compensator $w_i(t)$ would converge to zero, it may be meaningful to discuss the dynamics of $\zeta_i(t) = v_i(t) + w_i(t)$, which will also converge to zero. The dynamics of $\zeta_i(t)$ is given by

$$\dot{\zeta}_i(t) = A\zeta_i(t) + B\mathrm{sat}_\iota(u_i(t)) + F \sum_{j=1}^{N} a_{ij} c_{ij}(t)[C(\zeta_i(t) - \zeta_j(t)) - (y_i(t) - y_j(t))],$$

$$\dot{c}_{ij}(t) = \mu_{ij} a_{ij} \| C(\zeta_i(t) - \zeta_j(t)) - (y_i(t) - y_j(t)) \|^2.$$

$$(11.8)$$

And the control input $u_i(t)$ can be written as

$$u_i(t) = f(w_i(t)) + K(\zeta_i(t) - w_i(t)),$$
$$\dot{w}_i(t) = Aw_i(t) + B[\mathrm{sat}_\iota(u_i(t)) - K(\zeta_i(t) - w_i(t))].$$
$$(11.9)$$

We can conclude the following corollary to show the effectiveness of consensus protocol (11.9).

Corollary 11.1 *Suppose Assumptions 11.1 and 11.2 hold. Then, the consensus of the N agents in (11.1) with input saturation can be achieved under distributed adaptive anti-windup protocol (11.9).*

We can also write the control input $u_i(t)$ into the following alternative way:

$$u_i(t) = f(\zeta_i(t) - v_i(t)) + Kv_i(t),$$
$$\dot{v}_i(t) = (A + BK)v_i(t) + F\sum_{j=1}^{N} a_{ij}c_{ij}(t)[C(\zeta_i(t) - \zeta_j(t)) - (y_i(t) - y_j(t))]. \tag{11.10}$$

Corollary 11.2 *Suppose Assumptions 11.1 and 11.2 hold. Then, the consensus of the N agents in (11.1) with input saturation can be achieved under distributed adaptive anti-windup protocol (11.10).*

Remark 11.6 *The equivalence of these three protocols (11.7), (11.9), and (11.10) can be easily verified, as any one of the internal states $v_i(t)$, $w_i(t)$, and $\zeta_i(t)$ can be derived from the other two. And all of the three protocols are of $2n + l_{ii}$ order for agent i. However, compared with the protocol (11.7), the protocols (11.9) and (11.10) take the advantage of saving communication burden, since only the output of the state and internal state $\zeta_i(t)$, i.e., $y_i(t)$ and $C\zeta_i(t)$ should be transmitted to neighboring agents via communication channel.*

11.2.3 CNSs with absolute output information

In the previous subsection, we have presented the edge-based distributed adaptive anti-windup protocol by designing a distributed observer $v_i(t)$ and an anti-windup compensator $w_i(t)$ for each agent with relative output information among neighboring agents. Such design approach may suffer from heavy calculation burden as it doubles the order of observers. Thus, it is natural to ask if the order of the protocol can be decreased. Note that the key to achieve consensus is the second term of the right hand in the dynamics of the distributed observer $v_i(t)$ depending on the output matrix C, and the reason anti-windup compensator $w_i(t)$ can handle input saturation mainly locates on the input matrix B. Therefore, it may not be possible to design integrated distributed anti-windup observer under the architecture of protocol (11.7) with relative output information. This forces us to seek for novel designing structure.

For the case the absolute output information is available, we can reconstruct the distributed adaptive anti-windup protocol (11.7) into

$$\dot{\hat{x}}_i(t) = A\hat{x}_i(t) + B\text{sat}_\iota(u_i(t)) + F(C\hat{x}_i(t) - y_i(t)),$$
$$\dot{\tilde{v}}_i(t) = (A + BK)\tilde{v}_i(t) + BK\sum_{j=1}^{N} a_{ij}\tilde{c}_{ij}(t)[(\tilde{v}_i(t) - \tilde{v}_j(t))$$
$$+ (\tilde{w}_i(t) - \tilde{w}_j(t)) - (\hat{x}_i(t) - \hat{x}_j(t))] + F(C\hat{x}_i - y_i(t)),$$
$$\dot{\tilde{w}}_i(t) = A\tilde{w}_i(t) + B[\text{sat}_\iota(u_i(t)) - K\tilde{v}_i(t)],$$
$$\dot{\tilde{c}}_{ij}(t) = \mu_{ij}a_{ij}\|K[(\tilde{v}_i(t) - \tilde{v}_j(t)) + (\tilde{w}_i(t) - \tilde{w}_j(t)) - (\hat{x}_i(t) - \hat{x}_j(t))]\|^2,$$

$$u_i(t) = K\tilde{v}_i(t) + f(\tilde{w}_i(t)), \tag{11.11}$$

where $\hat{x}_i(t)$ is the local observer to estimate the state $x_i(t)$, $\tilde{v}_i(t)$ is the distributed observer to provide certain consensus variable, $\tilde{w}_i(t)$ is the anti-windup compensator to tackle input saturation constraint, $\tilde{c}_{ij}(t)$ is the adaptive coupling gain acting on the edge (i,j) with initial value satisfying $\tilde{c}_{ij}(t_0) = \tilde{c}_{ji}(t_0) > 0$, F, K, and P are the feedback gain matrices.

Let $\tilde{\eta}_i(t) = \hat{x}_i(t) - \tilde{v}_i(t) - \tilde{w}_i(t)$, whose dynamics are given as

$$\dot{\tilde{\eta}}_i(t) = A\tilde{\eta}_i(t) + BK \sum_{j=1}^{N} a_{ij}\tilde{c}_{ij}(t)(\tilde{\eta}_i(t) - \tilde{\eta}_j(t)), \tag{11.12}$$

$$\dot{\tilde{c}}_{ij}(t) = \mu_{ij} a_{ij} \| K(\tilde{\eta}_i(t) - \tilde{\eta}_j(t)) \|^2.$$

The following lemma shows how to design feedback gain matric K and P to achieve consensus of $\tilde{\eta}_i$.

Lemma 11.4 ([80]) *Suppose that Assumptions 11.1 and 11.2 hold. Then, $\tilde{\eta}_i(t)$ can reach consensus if $K = -B^T P$ with $\bar{P} = P^{-1} > 0$ being the positive definite solution of the LMI:*

$$A\bar{P} + \bar{P}A^T - 2BB^T < 0. \tag{11.13}$$

Besides, each adaptive coupling weight $\tilde{c}_{ij}(t)$ converges to a finite constant.

Proof 11.3 *Let $\tilde{e}_i(t) = \tilde{\eta}_i(t) - \frac{1}{N}\sum_{j=1}^{N}\tilde{\eta}_j(t)$ be the consensus error of $\tilde{\eta}_i(t)$, and $\tilde{e}(t) = [\tilde{e}_1^T(t), \ldots, \tilde{e}_N^T(t)]^T$. Consider the Lyapunov function*

$$V_2(t) = \frac{1}{2}\sum_{i=1}^{N} \tilde{e}_i^T(t) P\tilde{e}_i(t) + \sum_{i=1}^{N}\sum_{j=1}^{N} \frac{(\tilde{c}_{ij}(t) - \alpha)^2}{4\mu_{ij}}. \tag{11.14}$$

The time derivative of $V_2(t)$ is given by

$$\dot{V}_2(t) = \sum_{i=1}^{N}[\tilde{e}_i^T(t)PA\tilde{e}_i(t) - \alpha\tilde{e}_i^T(t)PBB^T P \sum_{j=1}^{N} a_{ij}(\tilde{e}_i(t) - \tilde{e}_j(t))]$$

$$\leq \frac{1}{2}\tilde{e}^T(t)[I_N \otimes (PA + A^T P - 2PBB^T P)]\tilde{e}(t) \tag{11.15}$$

$$\leq 0.$$

The rest of the proof is similar to that in Lemma 11.3, which is omitted here for brevity.

Theorem 11.2 *Suppose Assumptions 11.1 and 11.2 hold. Then the consensus of the N agents in (11.1) with input saturation can be realized under distributed adaptive anti-windup protocol (11.11) by choosing F such that $A + FC$ is Hurwitz.*

Proof 11.4 *Since $A + FC$ is Hurwitz, it is not difficult to obtain $(\hat{x}_i(t) - x_i(t)) \in \mathbb{L}_2$ and $F(C\hat{x}_i(t) - y_i(t)) \in \mathbb{L}_2$. Similar to the proof of Theorem 11.1, we have $\tilde{e}_i(t) \in \mathbb{L}_2$*

and $BK \sum_{j=1}^{N} a_{ij} \tilde{c}_{ij}(t)(\tilde{\eta}_i(t) - \tilde{\eta}_j(t)) \in \mathbb{L}_2$. Thus, $\tilde{v}_i(t) \in \mathbb{L}_2$. In light of Lemma 11.2, we can obtain $\tilde{w}_i(t) \in \mathbb{L}_2$, which further implies $\hat{\xi}_i(t) \triangleq \hat{x}_i(t) - \frac{1}{N}\sum_{j=1}^{N} \hat{x}_j(t) = \tilde{e}_i + \tilde{v}_i(t) + \tilde{w}_i(t) + \sum_{j=1}^{N}(\tilde{v}_j(t) + \tilde{w}_j(t)) \in \mathbb{L}_2$. Therefore, we arrive at $\xi_i(t) = \hat{\xi}_i(t) - (\hat{x}_i(t) - x_i(t)) + \frac{1}{N}\sum_{j=1}^{N}(\hat{x}_j(t) - x_j(t)) \in \mathbb{L}_2$. That is, the consensus is achieved.

Remark 11.7 *Note that the protocol (11.11) with absolute output information is of $3n + l_{ii}$ order for agent i, where local observer $\hat{x}_i(t)$, distributed observer $\tilde{v}_i(t)$, anti-windup compensator $\tilde{w}_i(t)$ as well as the adaptive coupling weight $\tilde{c}_{ij}(t)$ are designed. From the construction of distributed observer $\tilde{v}_i(t)$ and the adaptive coupling weight $\tilde{c}_{ij}(t)$, the local observer, the distributed observer and the anti-windup compensator, i.e., $\hat{x}_i(t)$, $\tilde{v}_i(t)$ and $\tilde{w}_i(t)$, should be transmitted to neighboring agents via communication channel. Though the protocol (11.11) takes the drawback of higher order and heavier communication burden than the relative output feedback protocol (11.7), it provides opportunity to propose integrated distributed anti-windup observer, as the input matrix B now is the crux of both consensus achievement and input saturation handling.*

By choosing $\tilde{\zeta}_i(t) = \tilde{v}_i(t) + \tilde{w}_i(t)$ as the integrated distributed anti-windup observer, and making certain modification on its dynamics, the following distributed adaptive anti-windup protocol is presented:

$$
\begin{aligned}
\dot{\hat{x}}_i(t) =& A\hat{x}_i(t) + B\mathrm{sat}_\iota(u_i(t)) + F(C\hat{x}_i(t) - y_i(t)), \\
\dot{\tilde{\zeta}}_i(t) =& A\tilde{\zeta}_i(t) + B\mathrm{sat}_\iota(u_i(t)) + BK\sum_{j=1}^{N} a_{ij}\tilde{c}_{ij}(t)[(\tilde{\zeta}_i(t) - \tilde{\zeta}_j(t)) - (\hat{x}_i(t) - \hat{x}_j(t))], \\
\dot{\tilde{c}}_{ij}(t) =& \mu_{ij}a_{ij}\|K[(\tilde{\zeta}_i(t) - \tilde{\zeta}_j(t)) - (\hat{x}_i(t) - \hat{x}_j(t))]\|^2, \\
u_i(t) =& f(\tilde{\zeta}_i(t)) - K\sum_{j=1}^{N} a_{ij}\tilde{c}_{ij}(t)[(\tilde{\zeta}_i(t) - \tilde{\zeta}_j(t)) - (\hat{x}_i(t) - \hat{x}_j(t))],
\end{aligned}
\tag{11.16}
$$

where the other variables are the same as in (11.11).

Theorem 11.3 *Suppose Assumptions 11.1 and 11.2 hold. Then the consensus of the N agents in (11.1) with input saturation can be realized under distributed adaptive anti-windup protocol (11.16).*

Proof 11.5 *Under protocol (11.16), the dynamics of $\tilde{\eta}_i(t) = \hat{x}_i(t) - \tilde{\zeta}_i(t)$ would be*

$$
\begin{aligned}
\dot{\tilde{\eta}}_i(t) =& A\tilde{\eta}_i(t) + BK\sum_{j=1}^{N} a_{ij}\tilde{c}_{ij}(t)(\tilde{\eta}_i(t) - \tilde{\eta}_j(t)) + FC(\hat{x}_i(t) - x_i(t)), \\
\dot{\tilde{c}}_{ij}(t) =& \mu_{ij}a_{ij}\|K(\tilde{\eta}_i(t) - \tilde{\eta}_j(t))\|^2.
\end{aligned}
\tag{11.17}
$$

To show the consensus of $\tilde{\eta}_i(t)$, the following Lyapunov function is constructed:

$$
V_3(t) = V_2(t) + \gamma \sum_{i=1}^{N} \tilde{\xi}_i^T(t)\bar{Q}\tilde{\xi}_i(t),
\tag{11.18}
$$

where $\tilde{\xi}_i(t) = \hat{\xi}_i(t) - \xi_i(t)$, $\bar{Q}$ is a positive definite matrix such that $W = \bar{Q}(A + FC) + (A + FC)^T\bar{Q} < 0$, and $\gamma = \frac{\bar{\gamma}}{\lambda_{\min}(-W)}$, $\bar{\gamma} \geq \frac{\lambda_{\max}((PFC)^T(PFC))}{\lambda_{\min}(-\bar{W})}$ with $\bar{W} = PA + A^TP - 2PBB^TP$.

The time derivative of V_3 is given by

$$\dot{V}_3(t) = \sum_{i=1}^N [\tilde{e}_i^T(t)PA\tilde{e}_i(t) - \alpha\tilde{e}_i^T(t)PBB^TP\sum_{j=1}^N a_{ij}(\tilde{e}_i(t) - \tilde{e}_j(t))$$
$$+ \tilde{e}_i^T(t)PFC\tilde{\xi}_i(t) + \gamma\tilde{\xi}_i^T(t)W\tilde{\xi}_i(t)]$$
$$\leq \frac{1}{4}\tilde{e}^T(t)[I_N \otimes \bar{W}]\tilde{e}(t) - \left(\bar{\gamma} - \frac{\lambda_{\max}((PFC)^T(PFC))}{\lambda_{\min}(-\bar{W})}\right)\tilde{\xi}^T(t)\tilde{\xi}(t)$$
$$\leq 0,$$

(11.19)

where the second inequality is obtained by the fact

$$\tilde{e}_i^T(t)PFC\tilde{\xi}_i(t) \leq -\frac{1}{4}\tilde{e}_i^T(t)\bar{W}\tilde{e}_i(t) + \frac{\lambda_{\max}((PFC)^T(PFC))}{\lambda_{\min}(-\bar{W})}\tilde{\xi}_i^T(t)\tilde{\xi}_i(t).$$

Similar to the discussions in Theorem 11.1, we can get $\tilde{e}(t) \in \mathbb{L}_2$. The rest of the proof is similar to that in Theorem 11.2, which is omitted here for brevity.

Remark 11.8 The integrated distributed anti-windup observer $\tilde{\zeta}_i(t)$ not only generates the consensus variable $\tilde{\eta}_i(t)$, but also results in the distributed adaptive anti-windup protocol $u_i(t)$ in (11.16) tackling the input saturation constraint based on Lemma 11.2. The order of the protocol (11.16) is $2n + l_{ii}$ for agent i, and the information that should be transmitted to neighboring agents via communication channel includes the local observer $\hat{x}_i(t)$ and the integrated distributed anti-windup observer $\tilde{\zeta}_i(t)$. Note that the local observer $\hat{x}_i(t)$ is designed with full-order. Thus, we can obtain reduced-order distributed adaptive anti-windup protocol by designing reduced-order local observer.

Based on the discussion above, we can introduce the following reduce-order distributed adaptive anti-windup protocol

$$\dot{\chi}_i(t) = G\chi_i(t) + Hy_i(t) + TB\text{sat}_\iota(u_i(t)),$$
$$\dot{\tilde{\zeta}}_i(t) = A\tilde{\zeta}_i(t) + B\text{sat}_\iota(u_i(t)) + BK\sum_{j=1}^N a_{ij}\tilde{c}_{ij}(t)[(\tilde{\zeta}_i(t) - \tilde{\zeta}_j(t)) - (\hat{\chi}_i(t) - \hat{\chi}_j(t))],$$
$$\dot{\tilde{c}}_{ij}(t) = \mu_{ij}a_{ij}\|K[(\tilde{\zeta}_i(t) - \tilde{\zeta}_j(t)) - (\hat{\chi}_i(t) - \hat{\chi}_j(t))]\|^2,$$
$$u_i(t) = f(\tilde{\zeta}_i(t)) - K\sum_{j=1}^N a_{ij}\tilde{c}_{ij}(t)[(\tilde{\zeta}_i(t) - \tilde{\zeta}_j(t)) - (\hat{\chi}_i(t) - \hat{\chi}_j(t))],$$

(11.20)

where $G \in \mathbb{R}^{(n-m)\times(n-m)}$ is a Hurwitz matrix sharing no common eigenvalue with A, $H \in \mathbb{R}^{(n-m)\times m}$ is a matrix such that (G, H) is controllable, $T \in \mathbb{R}^{(n-m)\times n}$ is the solution of the Sylvester equation

$$TA - GT = HC,$$

(11.21)

satisfying that $\begin{bmatrix} C \\ T \end{bmatrix}$ is nonsingular with $\begin{bmatrix} C \\ T \end{bmatrix}^{-1} = \begin{bmatrix} S_1 & S_2 \end{bmatrix}$, K is designed the same as this in Theorem 11.1, $\hat{\chi}_i(t) = S_1 y_i(t) + S_2 \chi_i(t)$ is the estimation of $x_i(t)$.

Define $\tilde{\chi}_i(t) = \chi_i(t) - Tx_i(t)$, and it is not difficult to obtain

$$\dot{\tilde{\chi}}_i(t) = G\tilde{\chi}_i(t),$$

which implies that $\chi_i(t)$ can estimate $Tx_i(t)$, and $\hat{\chi}_i(t) = x_i(t) + S_2\tilde{\chi}_i(t)$ can estimate the state $x_i(t)$. Then the dynamics of $\tilde{\eta}_i(t) = \hat{\chi}_i(t) - \tilde{\zeta}_i(t)$ would be

$$\dot{\tilde{\eta}}_i(t) = A\tilde{\eta}_i(t) + BK \sum_{j=1}^{N} a_{ij}\tilde{c}_{ij}(t)(\tilde{\eta}_i(t) - \tilde{\eta}_j(t)) + S_2 G\tilde{\chi}_i(t), \tag{11.22}$$

$$\dot{\tilde{c}}_{ij}(t) = \mu_{ij}a_{ij}\|K(\tilde{\eta}_i(t) - \tilde{\eta}_j(t))\|^2.$$

Theorem 11.4 *Suppose Assumptions 11.1 and 11.2 hold, and (A, C) is observable. Then the consensus of the N agents in (11.1) with input saturation can be realized under distributed adaptive anti-windup protocol (11.20).*

Proof 11.6 *To show the consensus of $\tilde{\eta}_i(t)$, the following Lyapunov function is constructed:*

$$V_4(t) = V_2(t) + \delta \sum_{i=1}^{N} \psi_i^T(t)\tilde{Q}\psi_i(t), \tag{11.23}$$

where $\psi_i(t) = \tilde{\chi}_i(t) - \frac{1}{N}\sum_{j=1}^{N}\tilde{\chi}_j(t)$, $\tilde{Q}$ is a positive definite matrix such that $\tilde{W} = \tilde{Q}G + G^T\tilde{Q} < 0$, and $\delta = \frac{\bar{\delta}}{\lambda_{\min}(-\tilde{W})}$, $\bar{\delta} \geq \frac{\lambda_{\max}((PS_2G)^T(PS_2G))}{\lambda_{\min}(-\tilde{W})}$. Let $\psi(t) = [\psi_1^T(t), \ldots, \psi_N^T(t)]^T$. The time derivative of $V_4(t)$ is given by

$$\begin{aligned}\dot{V}_4(t) &= \sum_{i=1}^{N}[\tilde{e}_i^T(t)PA\tilde{e}_i(t) - \alpha\tilde{e}_i^T(t)PBB^TP\sum_{j=1}^{N}a_{ij}(\tilde{e}_i(t) - \tilde{e}_j(t)) \\ &\quad + \tilde{e}_i^T(t)PS_2G\psi_i(t) + \delta\psi_i^T(t)\tilde{W}\psi_i(t)] \\ &\leq \frac{1}{4}\tilde{e}^T(t)[I_N \otimes \bar{W}]\tilde{e}(t) - \left(\bar{\delta} - \frac{\lambda_{\max}((PS_2G)^T(PS_2G))}{\lambda_{\min}(-\bar{W})}\right)\psi^T(t)\psi(t) \\ &\leq 0.\end{aligned} \tag{11.24}$$

Following the similar discussions on Theorem 11.3, it is not difficult to illustrate the consensus of the N agents in (11.1).

Remark 11.9 *Compared with the edge-based adaptive anti-windup protocols (11.7), (11.9), and (11.10), the advantage of the integrated distributed anti-windup observer-based protocol (11.20) is that it reduces the calculation burden for each agent. Specifically, the order of protocol (11.20) is $2n - m + l_{ii}$ for each agent, which is smaller than $2n + l_{ii}$ for protocols (11.7), (11.9), and (11.10). However, the integrated distributed anti-windup observer-based protocol (11.20) raises higher demands on the system dynamics as well as the access of information. That is, to present integrated distributed*

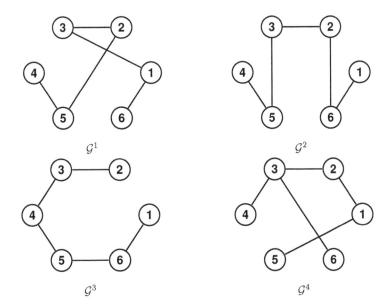

Figure 11.1 The communication graphs $\mathcal{G}^i$, $i = 1, \ldots, 4$.

anti-windup observer $\tilde{\zeta}_i(t)$, the local reduced-order observer $\chi_i(t)$ is necessary, which requires absolute output information of each agent, and the observability of (A, C) is also needed to ensure the existence of feedback gain matrix T satisfying the invertibility of $M = \begin{bmatrix} C \\ T \end{bmatrix}$. The information needed to be transmitted via communication channel is the distributed observer $\tilde{\zeta}_i(t)$ and the local observer $\hat{\chi}_i(t)$. It should be noted that the observability of (A, C) is the necessary condition for the invertibility of M. That is, after the determination of G, M may be singular if H is chosen inappropriate. However, as is mentioned in [19], the probability for M to be nonsingular is 1 with H randomly selected.

11.2.4 Numerical simulation

In this subsection, numerical examples are provided to verify the effectiveness of the analytical results.

Consider a CNS with 6 agents whose dynamics given as in (11.1) with

$$A = \begin{bmatrix} 0 & 1 & 0 \\ 0 & 0 & 1 \\ 0 & 0 & 0 \end{bmatrix}, \quad B = \begin{bmatrix} 0 \\ 0 \\ 1 \end{bmatrix}, \quad C = \begin{bmatrix} 1 & 0 & 0 \end{bmatrix}, \quad \iota = 2.$$

Let $\mathcal{G}(t)$ switch randomly among the graphs $\mathcal{G}^1$–$\mathcal{G}^4$ given in Fig. 11.1.

First, we will show how to design $f(z(t))$ constructed by Algorithm 11.1. Since in this case $m_1 = 3$ and $\hat{A}_3 = 0$, we can choose $P_3 = 1$ and $g_3(t) = -z_3(t)$. Then $\tilde{A}_2 = \hat{A}_2 + \frac{\partial}{\partial Z_2(t)}(\hat{B}_2 g_3(t)) = \begin{bmatrix} 0 & 1 \\ 0 & -1 \end{bmatrix}$, and choose $P_2 = \begin{bmatrix} 1 & 1 \\ 1 & 2 \end{bmatrix}$, $g_2(t) =$

$-z_2(t) - 2z_3(t)$. We have $\tilde{A}_1 = \hat{A}_1 + \frac{\partial}{\partial Z_1(t)}[\hat{B}_1(g_2(t) + g_3(t))] = \begin{bmatrix} 0 & 1 & 0 \\ 0 & 0 & 1 \\ 0 & -1 & -3 \end{bmatrix}$, and

choose $P_3 = \begin{bmatrix} 1 & 3 & 1 \\ 3 & 10 & 3 \\ 1 & 3 & 2 \end{bmatrix}$, $g_1(t) = -z_1(t) - 3z_2(t) - 2z_3(t)$. Let $\mu_i = 1, i = 1, 2, 3$,

and $f_1(z(t)) = \text{sat}_\iota(-[1\ 3\ 2]z(t))$, $f_2(z(t)) = \text{sat}_\iota(f_1(z(t)) - [0\ 1\ 2]z(t))$, $f(z(t)) = f_3(z(t)) = \text{sat}_\iota(f_2(z(t)) - [0\ 0\ 1]z(t))$.

Case 1: The protocol with relative output information. Solving the LMI (11.4) gives $Q = \begin{bmatrix} 0.8849 & -0.4741 & -0.3395 \\ -0.4741 & 0.8822 & -0.5373 \\ -0.3395 & -0.5373 & 2.0383 \end{bmatrix}$. The control parameters are chosen as

$F = -Q^{-1}C^T = \begin{bmatrix} -2.5039 \\ -1.9056 \\ -0.9194 \end{bmatrix}$, and $K = \begin{bmatrix} -1 & -3 & -2 \end{bmatrix}$, $\mu_{ij} = 1$. The initial values

$c_{ij}(t_0) = 1$ and the state and observer initial values are randomly chosen. The state $x_i(t)$, consensus error $e_i(t)$, distributed observer $v_i(t)$ and anti-windup compensator $w_i(t)$ of the agents are depicted in Figs. 11.2–11.5, demonstrating that the consensus is indeed reached. The adaptive gains $c_{ij}(t)$ are given in Fig. 11.6, which converge to finite values, and the control inputs with saturations are presented in Fig. 11.7 with the solid lines, which satisfy the input saturation constraints with $\iota = 2$ shown with dash lines.

Case 2: The protocol with absolute output information. We choose $G = \begin{bmatrix} 0 & 1 \\ -1 & -1 \end{bmatrix}$ and $H = \begin{bmatrix} 0 \\ 1 \end{bmatrix}$. Solving the Sylvester equation (11.21) gives $T = \begin{bmatrix} 1 & -1 & 0 \\ 0 & 1 & -1 \end{bmatrix}$. And we have $S_1 = \begin{bmatrix} 1 \\ 1 \\ 1 \end{bmatrix}$ and $S_2 = \begin{bmatrix} 0 & 0 \\ -1 & 0 \\ -1 & -1 \end{bmatrix}$. Solving the LMI (11.13)

gives $\bar{P} = \begin{bmatrix} 2.0383 & -0.5373 & -0.3395 \\ -0.5373 & 0.8822 & -0.4741 \\ -0.3395 & -0.4741 & 0.8849 \end{bmatrix}$, and $K = \begin{bmatrix} -0.9194 & -1.9056 & -2.5039 \end{bmatrix}$.

The state $x_i(t)$, local observer $\chi_i(t)$ and the consensus error $\tilde{e}_i(t)$, integrated distributed anti-windup observer $\zeta_i(t)$ of the agents are depicted in Figs. 11.8–11.11, while the adaptive gains $\tilde{c}_{ij}(t)$ are given in Fig. 11.12, which indicate that the consensus is achieved.

11.3 RESILIENT CONSENSUS OF CNSS WITH MALICIOUS ATTACK UNDER SWITCHING TOPOLOGIES

In this section, we study the robustness of the switching topologies to present necessary and sufficient condition for resilient consensus of CNSs under malicious attack.

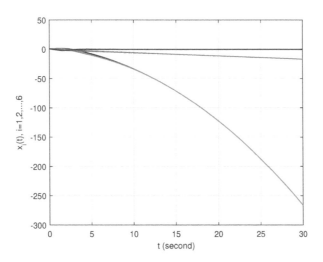

Figure 11.2 Trajectories of $x_i(t)$, $i = 1, \ldots, 6$, under distributed adaptive anti-windup protocols (11.2) and (11.7).

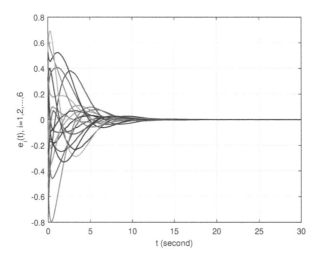

Figure 11.3 Trajectories of the consensus error $e_i(t)$, $i = 1, \ldots, 6$, under distributed adaptive anti-windup protocols (11.2) and (11.7).

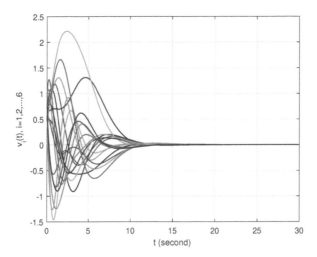

Figure 11.4 Trajectories of the distributed observer $v_i(t)$, $i = 1,\ldots,6$, under distributed adaptive anti-windup protocols (11.2) and (11.7).

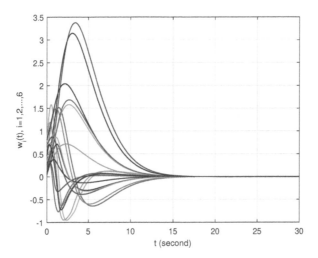

Figure 11.5 Trajectories of the anti-windup compensator $w_i(t)$, $i = 1,\ldots,6$, under distributed adaptive anti-windup protocols (11.2) and (11.7).

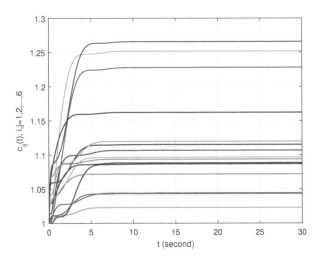

Figure 11.6 Trajectories of the adaptive gains $c_{ij}(t)$, $i, j = 1, \ldots, 6$, under distributed adaptive anti-windup protocols (11.2) and (11.7).

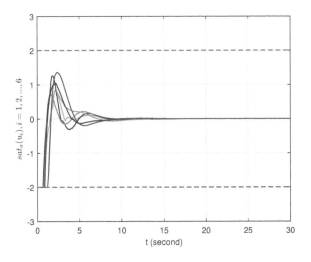

Figure 11.7 Trajectories of the control input $\mathrm{sat}_\iota(u_i(t))$, $i = 1, \ldots, 6$, under distributed adaptive anti-windup protocols (11.2) and (11.7).

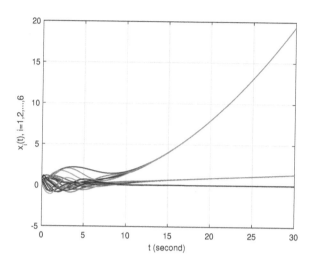

Figure 11.8 Trajectories of $x_i(t)$, $i = 1, \ldots, 6$, under distributed adaptive anti-windup protocol (11.20).

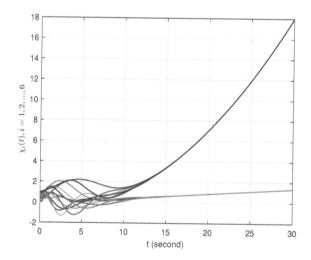

Figure 11.9 Trajectories of the local observer $\chi_i(t)$, $i = 1, \ldots, 6$, under distributed adaptive anti-windup protocol (11.20).

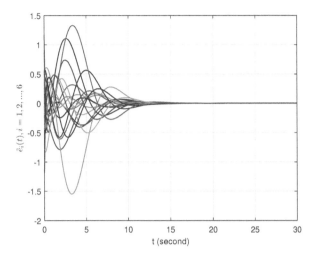

Figure 11.10 Trajectories of the consensus error $\tilde{e}_i(t)$, $i = 1, \ldots, 6$, under distributed adaptive anti-windup protocol (11.20).

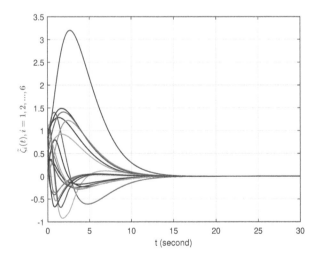

Figure 11.11 Trajectories of the integrated distributed anti-windup observer $\tilde{\zeta}_i(t)$, $i = 1, \ldots, 6$, under distributed adaptive anti-windup protocol (11.20).

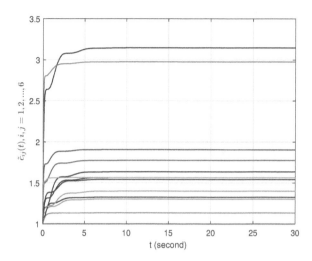

Figure 11.12 Trajectories of the adaptive gains $\tilde{c}_{ij}(t)$, $i, j = 1, \ldots, 6$, under distributed adaptive anti-windup protocol (11.20).

11.3.1 Problem formulation

Consider a discrete-time first-order CNS, whose dynamics are given by

$$x_i[k + 1] = x_i[k] + u_i[k], \quad i = 1, \ldots, N, \tag{11.25}$$

where $x_i[k]$ and $u_i[k]$ are the state and control input of the agent i at step k.

The node set $\mathcal{V}$ contains normal nodes and malicious nodes.

Definition 11.2 ([72]) *A node i is said to be normal if it follows the preset controller to make updates and sends its value to all neighbors at each step; while it is malicious if it sends its value to all neighbors at each step, but does not follow the preset controller to make updates at some steps.*

Under the Definition 11.2, the node set $\mathcal{V}$ is divided into two disjoint subsets, the normal node set $\mathcal{N}$ and the malicious node set $\mathcal{M}$, i.e., $\mathcal{N} \bigcap \mathcal{M} = \emptyset$ and $|\mathcal{N}| + |\mathcal{M}| = N$.

Definition 11.3 ([72]) *The network is F-total malicious if $\mathcal{M}$ contains at most F nodes, i.e., $|\mathcal{M}| \leq F$.*

The following resilient consensus problem is studied.

Definition 11.4 ([33]) *The CNS (11.25) realizes resilient consensus if for any initial values of nodes, any possible malicious node set,*

(1) all the normal nodes reach consensus in the sense that $\lim_{t \to \infty}(x_i[k] - x_j[k]) = 0$, $\forall i, j \in \mathcal{N}$;

(2) *the states of the normal nodes remain in a bounded interval* Υ, *i.e.*, $x_i[k] \in \Upsilon, \forall i \in \mathcal{N}, \forall k$.

Let $\mathcal{N}_i[k] = \{j | (i, j) \in \mathcal{E}[k]\}$ be the neighboring set of node i. The W-MSR algorithm is introduced [72]:

(1) At step k, each normal agent i sorts the relative values $x_j[k] - x_i[k]$ of its neighbors in a decreasing order.

(2) Each normal agent removes the largest F positive values $x_j[k] - x_i[k]$ and the smallest F negative values $x_j[k] - x_i[k]$. If the number of positive (negative) values is less than F then remove all the positive (negative) values. Denote $\mathcal{R}_i[k]$ the set of the removed neighbors of agent i.

(3) Each normal agent generates its control input by

$$u_i[k] = \sum_{j=1}^{N} w_{ij}[k](x_j[k] - x_i[k]), \tag{11.26}$$

where $w_{ij}[k] = 0$ if $j \notin \mathcal{N}_i[k] \backslash \mathcal{R}_i[k]$ and $w_{ij}[k] > 0$ if $j \in \mathcal{N}_i[k] \backslash \mathcal{R}_i[k]$ with $\sum_{j=1}^{N} w_{ij}[k] < 1$.

The significant feature of the W-MSR algorithm is that each agent removes enough extreme values compared with its own value to resist the effect of the misbehaving agents.

The concept of (r, s)-robustness was firstly introduced in [72] for the resilient consensus achievement of first-order integrators under fixed graph.

Definition 11.5 ([72]) *The graph* $\mathcal{G}$ *is* (r, s)-*robust if for every pair of nonempty disjoint subsets* $\mathcal{S}_1$ *and* $\mathcal{S}_2$, *at least one of the following conditions holds:*

- $|\mathcal{X}_{\mathcal{S}_1}^r| = |\mathcal{S}_1|$;

- $|\mathcal{X}_{\mathcal{S}_2}^r| = |\mathcal{S}_2|$;

- $|\mathcal{X}_{\mathcal{S}_1}^r| + |\mathcal{X}_{\mathcal{S}_2}^r| \geq s$,

where $\mathcal{X}_{\mathcal{S}_j}^r, j = 1, 2$, *is the set of nodes in* $\mathcal{S}_j$ *with at least* r *neighbors outside of* $\mathcal{S}_j$, *i.e.,* $\mathcal{X}_{\mathcal{S}_j}^r = \{i \in \mathcal{S}_j : |\mathcal{N}_i \backslash \mathcal{S}_j| \geq r\}$.

With the definition of (r, s)-robustness, [72] has revealed that the sufficient and necessary condition for resilient consensus of time-invariant F-total malicious network under W-MSR algorithm is that the topology $\mathcal{G}$ is $(F + 1, F + 1)$-robust. And the sufficient condition for resilient consensus of time-varying F-total malicious network under W-MSR algorithm is that there exists an infinite step sequence $\{k_j\}$ such that for each step k_j the topology $\mathcal{G}[k_j]$ is $(F + 1, F + 1)$-robust.

It should be noticed that the above condition for resilient consensus of time-varying F-total malicious network is not necessary. And this section intends to figure out the sufficient and necessary condition on resilient consensus of switching network under W-MSR algorithm.

11.3.2 Joint (r, s)-robustness

In this subsection, we intend to introduce the following definition of the joint robustness of switching topologies.

Definition 11.6 (joint (r, s)-robustness) *The switching network $\mathcal{G}[k] = (\mathcal{V}, \mathcal{E}[k])$ is jointly (r, s)-robust, if for every pair of nonempty, disjoint subsets $\mathcal{S}_1$ and $\mathcal{S}_2$ of $\mathcal{V}$, there exists an infinite sequence of bounded step intervals $[k_j, k_{j+1})$ such that in each step interval, at least one of the following conditions holds:*

(1) $|\mathcal{X}^r_{\mathcal{S}_1}[k_j, k_{j+1}]| = |\mathcal{S}_1|$;

(2) $|\mathcal{X}^r_{\mathcal{S}_2}[k_j, k_{j+1}]| = |\mathcal{S}_2|$;

(3) $|\mathcal{X}^r_{\mathcal{S}_1}[k_j, k_{j+1}]| + |\mathcal{X}^r_{\mathcal{S}_2}[k_j, k_{j+1}]| \geq s$,

where $\mathcal{X}^r_{\mathcal{S}_l}[k_j, k_{j+1}), l = 1, 2$, is the set of nodes in $\mathcal{S}_l$ with at least r neighbors outside of $\mathcal{S}_l$ for at least one step, i.e., $\mathcal{X}^r_{\mathcal{S}_l}[k_j, k_{j+1}) = \{i \in \mathcal{S}_l : \exists k^i_{T_j} \in [k_j, k_{j+1}) \ s.t. \ |\mathcal{N}_i[k^i_{T_j}]\backslash\mathcal{S}_l| \geq r\}$.

Compared with the (r, s)-robustness of fixed graph, the joint (r, s)-robustness captures the robust connectivity of switching topologies. It is clear that the joint (r, s)-robustness is degenerated into (r, s)-robustness for the time-invariant graph.

We have the following result to show the relationship between joint spanning tree and the joint (r, s)-robustness of switching topologies.

Lemma 11.5 *The switching network $\mathcal{G}[k] = (\mathcal{V}, \mathcal{E}[k])$ jointly contains a directed spanning tree, if and only if $\mathcal{G}[k]$ is jointly $(1, 1)$-robust.*

Proof 11.7 (Necessity) *According to [128], $\mathcal{G}$ jointly contains a directed spanning tree if and only if there exists an infinite sequence of bounded step intervals $[k_j, k_{j+1})$ such that in each step interval, the union of the graphs contains a directed spanning tree. Then for every pair of nonempty, disjoint subsets $\mathcal{S}_1$ and $\mathcal{S}_2$ of $\mathcal{V}$, there must be a node i in one subset having at least one neighbor outside its subset for at least one step in each step interval $[k_j, k_{j+1})$, or else the union of the graphs cannot contain a directed spanning tree.*

(Sufficiency) On the other hand, if $\mathcal{G}$ does not jointly contain a directed spanning tree, after finite step $\bar{k}$, the node set $\mathcal{V}$ can be divided into two nonempty, disjoint subsets, where any node in one subset has no neighbor in another subset at any step. Clearly, the network $\mathcal{G}[k]$ is not jointly $(1, 1)$-robust in this case. This completes the proof.

Remark 11.10 *It is pointed out in [128] that the consensus can be realized for first-order integrators if and only if the switching network jointly contains a directed spanning tree. In other word, the resilient consensus of 0-total malicious model can be achieved under W-MSR algorithm if and only if the graph is jointly $(1, 1)$-robust. This motivates us to present the sufficient and necessary condition for resilient consensus of general F-total malicious model.*

11.3.3 Resilient consensus of switching topologies

Let $\bar{x}[k]$ and $\underline{x}[k]$ be the maximum and minimum values of the normal nodes in step k. We have the following result.

Theorem 11.5 *Under the F-total malicious model, the agents in (11.25) under W-MSR algorithm can realize resilient consensus if and only if the switching graph $\mathcal{G}[k]$ is jointly $(F+1, F+1)$-robust. Moreover, the safety interval $\Upsilon = [\underline{x}[0], \bar{x}[0]]$.*

Proof 11.8 (Necessity) *If $\mathcal{G}[k]$ is not jointly $(F+1, F+1)$-robust, then there are nonempty, disjoint subsets $\mathcal{S}_1, \mathcal{S}_2$ of $\mathcal{V}$ such that after some finite step $\bar{k}$, the following three conditions hold.*

$1°$ $|\mathcal{X}_{\mathcal{S}_1}^{F+1}[\bar{k}, \infty)| < |\mathcal{S}_1|;$

$2°$ $|\mathcal{X}_{\mathcal{S}_2}^{F+1}[\bar{k}, \infty)| < |\mathcal{S}_2|;$

$3°$ $|\mathcal{X}_{\mathcal{S}_1}^{F+1}[\bar{k}, \infty)| + |\mathcal{X}_{\mathcal{S}_2}^{F+1}[\bar{k}, \infty)| \leq F.$

Let the values of all the agents in $\mathcal{S}_1$ and $\mathcal{S}_2$ at step $\bar{k}$ be a and b, respectively, where $a < b$. By condition $3°$, we can choose all the agents in $\mathcal{X}_{\mathcal{S}_1}^{F+1}[\bar{k}, \infty) \cup \mathcal{X}_{\mathcal{S}_2}^{F+1}[\bar{k}, \infty)$ as the malicious nodes, which keeps the values unchanged. Let the agents in $\mathcal{S}_1 \backslash \mathcal{X}_{\mathcal{S}_1}^{F+1}[\bar{k}, \infty)$ and $\mathcal{S}_2 \backslash \mathcal{X}_{\mathcal{S}_2}^{F+1}[\bar{k}, \infty)$ be normal nodes. By conditions $1°$ and $2°$, there are normal nodes in both subsets $\mathcal{S}_1$ and $\mathcal{S}_2$. For any normal node $i \in \mathcal{S}_1$, it has at most F neighbors with value different from a, implying that all the nodes in $\{i\} \cup \mathcal{N}_i[k] \backslash \mathcal{R}_i[k]$ have the same value a at any step k, and $x_i[k]$ keeps the value a unchanged. Similarly, we have that the values of normal nodes in $\mathcal{S}_2$ remain b. Consequently, the resilient consensus can never be realized.

(Sufficiency) We first show the safety condition. Substituting (11.26) into (11.25) yields

$$x_i[k+1] = \sum_{j=1}^{N} \bar{w}_{ij}[k] x_j[k], \quad (11.27)$$

where $\bar{w}_{ij}[k] = w_{ij}[k]$ if $i \neq j$ and $\bar{w}_{ii}[k] = 1 - \sum_{j=1}^{N} w_{ij}[k]$, meaning that the value of each normal agent in step $k+1$ is a convex combination of the values of itself and its neighbors in the set $\mathcal{N}_i[k] \backslash \mathcal{R}_i[k]$. It is not difficult to verify that $x_j[k] \in [\underline{x}[k], \bar{x}[k]]$ if $j \in \{i\} \cup \mathcal{N}_i[k] \backslash \mathcal{R}_i[k]$ for all normal agent i. Therefore, we have $x_i[t+1] \in [\underline{x}[k], \bar{x}[k]]$ for all normal agent i, implying that $[\underline{x}[0], \bar{x}[0]] \supset [\underline{x}[1], \bar{x}[1]] \supset \ldots \supset [\underline{x}[k], \bar{x}[k]] \supset \ldots$. Thus, the values of all normal agents remain in the safety interval $\Upsilon = [\underline{x}[0], \bar{x}[0]]$.

Next, we present the achievement of resilient consensus. Since both $\bar{x}[k]$ and $\underline{x}[k]$ are monotone and bounded, the limits of $\bar{x}[k]$ and $\underline{x}[k]$ exist, denoted by $\overline{A}$ and $\underline{A}$, respectively. And the resilient consensus is achieved if and only if $\overline{A} = \underline{A}$.

Suppose that $\overline{A} > \underline{A}$. Then there exists a constant $\epsilon_0 > 0$ such that $\overline{A} - \epsilon_0 > \underline{A} + \epsilon_0$. Since $\mathcal{G}[k]$ is jointly $(F+1, F+1)$-robust, for every pair of nonempty, disjoint subsets $\mathcal{S}_1$ and $\mathcal{S}_2$ of $\mathcal{V}$, there exists an infinite sequence of bounded step intervals $[k_j, k_{j+1})$ such that at least one of the following conditions holds:

(1) $|\mathcal{X}_{\mathcal{S}_1}^{F+1}[k_j, k_{j+1})| = |\mathcal{S}_1|;$

(2) $|\mathcal{X}_{\mathcal{S}_2}^{F+1}[k_j, k_{j+1}]| = |\mathcal{S}_2|$;

(3) $|\mathcal{X}_{\mathcal{S}_1}^{F+1}[k_j, k_{j+1}]| + |\mathcal{X}_{\mathcal{S}_2}^{F+1}[k_j, k_{j+1}]| \geq F + 1$.

Let M be the number of normal nodes, and $\alpha \in (0, \frac{1}{2})$ be the lower bound of the nonzero weights $\bar{w}_{ij}[k]$ in the control input, i.e., $\bar{w}_{ij}[k] \geq \alpha, \forall k \geq 0, \forall i, \forall j \in \{i\} \cup \mathcal{N}_i[k] \backslash \mathcal{R}_i[k]$. Further let T be the maximum length of time interval $[k_j, k_{j+1})$, and choose $\epsilon = \frac{\alpha^{MT+1}}{1-\alpha^{MT+1}} \epsilon_0$. Define k_p as the finite step such that $\overline{x}[k] < \overline{A} + \epsilon$ and $\underline{x}[k] > \underline{A} - \epsilon, \forall k \geq k_p$. We have $0 < \epsilon < \frac{\alpha^{k_p+M-k_p}}{1-\alpha^{k_p+M-k_p}} \epsilon_0 < \epsilon_0$.

Define the sequence $\{\epsilon_l\}$ by

$$\epsilon_{l+1} = \alpha \epsilon_l - (1-\alpha)\epsilon, \ l = 0, \ldots, k_{p+M} - k_p - 1.$$

It is easy to verify that the sequence $\{\epsilon_l\}$ is strictly monotone decreasing. Noting that

$$\epsilon_{k_p+M-k_p} = \alpha^{k_p+M-k_p}\epsilon_0 - \sum_{l=0}^{k_p+M-k_p-1} \alpha^l (1-\alpha)\epsilon$$
$$= \alpha^{k_p+M-k_p}\epsilon_0 - (1 - \alpha^{k_p+M-k_p})\epsilon > 0,$$

we can conclude that each $\epsilon_l > 0$ and $\overline{A} - \epsilon_l > \underline{A} + \epsilon_l$. Let $\mathcal{Y}_1(k_p+l, \epsilon_l)$ and $\mathcal{Y}_2(k_p+l, \epsilon_l)$ be the subsets of $\mathcal{V}$ with nodes of value larger than $\overline{A} - \epsilon_l$ and smaller than $\underline{A} + \epsilon_l$ at step $k_p + l$, respectively, i.e.,

$$\mathcal{Y}_1(k_p + l, \epsilon_l) = \{i \in \mathcal{V}: \ x_i[k_p + l] > \overline{A} - \epsilon_l\},$$
$$\mathcal{Y}_2(k_p + l, \epsilon_l) = \{i \in \mathcal{V}: \ x_i[k_p + l] < \underline{A} + \epsilon_l\}. \tag{11.28}$$

From the above definition, we can easily get that $\mathcal{Y}_1(k_p + l, \epsilon_l)$ and $\mathcal{Y}_2(k_p + l, \epsilon_l)$ are disjoint for all l.

In the following, we shall show that

$$|(\mathcal{Y}_1(k_j, \epsilon_{k_j-k_p}) \cup \mathcal{Y}_2(k_j, \epsilon_{k_j-k_p})) \cap \mathcal{N}|$$
$$> |(\mathcal{Y}_1(k_{j+1}, \epsilon_{k_{j+1}-k_p}) \cup \mathcal{Y}_2(k_{j+1}, \epsilon_{k_{j+1}-k_p})) \cap \mathcal{N}|. \tag{11.29}$$

To do this, we first demonstrate that $\{\mathcal{Y}_1(k_p+l, \epsilon_l) \cap \mathcal{N}\} \supset \{\mathcal{Y}_1(k_p+l+1, \epsilon_{l+1}) \cap \mathcal{N}\}$, and $\{\mathcal{Y}_2(k_p+l, \epsilon_l) \cap \mathcal{N}\} \supset \{\mathcal{Y}_2(k_p+l+1, \epsilon_{l+1}) \cap \mathcal{N}\}$ hold for all l. The normal agents at step $k_p + l$ are divided into five disjoint subsets:

$$\mathcal{Z}_1(k_p + l, \epsilon_l) = \{i \in \mathcal{Y}_1(k_p + l, \epsilon_l) \cap \mathcal{N}: |\mathcal{N}_i[k_p + l] \backslash \mathcal{Y}_1(k_p + l, \epsilon_l)| \geq F + 1\},$$
$$\mathcal{Z}_2(k_p + l, \epsilon_l) = \{\mathcal{Y}_1(k_p + l, \epsilon_l) \cap \mathcal{N}\} \backslash \mathcal{Z}_1(k_p + l, \epsilon_l),$$
$$\mathcal{Z}_3(k_p + l, \epsilon_l) = \{i \in \mathcal{Y}_2(k_p + l, \epsilon_l) \cap \mathcal{N}: |\mathcal{N}_i[k_p + l] \backslash \mathcal{Y}_2(k_p + l, \epsilon_l)| \geq F + 1\},$$
$$\mathcal{Z}_4(k_p + l, \epsilon_l) = \{\mathcal{Y}_2(k_p + l, \epsilon_l) \cap \mathcal{N}\} \backslash \mathcal{Z}_3(k_p + l, \epsilon_l),$$
$$\mathcal{Z}_5(k_p + l, \epsilon_l) = \mathcal{N} \backslash \{(\mathcal{Y}_1(k_p + l, \epsilon_l) \cup \mathcal{Y}_2(k_p + l, \epsilon_l)) \cap \mathcal{N}\}.$$

For any agent $i \in \mathcal{Z}_1(k_p + l, \epsilon_l)$, at least one neighbor with value no more than $\overline{A} - \epsilon_l$ would be used in the control algorithm and we have

$$x_i[k_p + l + 1] \leq \alpha(\overline{A} - \epsilon_l) + (1-\alpha)\overline{x}[k_p + l]$$
$$< \alpha(\overline{A} - \epsilon_l) + (1-\alpha)(\overline{A} + \epsilon)$$
$$= \overline{A} - \epsilon_{l+1},$$

meaning that $\mathcal{Z}_1(k_p + l, \epsilon_l) \cap \mathcal{Y}_1(k_p + l + 1, \epsilon_{l+1}) = \emptyset$. *On the other hand, the value of itself would be used in the control algorithm and we have*

$$\begin{aligned} x_i[k_p + l + 1] &> \alpha(\overline{A} - \epsilon_l) + (1 - \alpha)\underline{x}[k_p + l] \\ &> \alpha(\underline{A} + \epsilon_l) + (1 - \alpha)(\underline{A} - \epsilon) \\ &= \underline{A} + \epsilon_{l+1}, \end{aligned}$$

meaning that $\mathcal{Z}_1(k_p + l, \epsilon_l) \cap \mathcal{Y}_2(k_p + l + 1, \epsilon_{l+1}) = \emptyset$. *Thus,* $\mathcal{Z}_1(k_p + l, \epsilon_l) \subset \mathcal{Z}_5(k_p + l + 1, \epsilon_{l+1})$.

For any agent $i \in \mathcal{Z}_2(k_p + l, \epsilon_l)$, *all the neighbors with value no more than* $\overline{A} - \epsilon_l$ *are removed, and we have* $x_i[k_p + l + 1] > \overline{A} - \epsilon_l > \underline{A} + \epsilon_{l+1}$. *Thus,* $\mathcal{Z}_2(k_p + l, \epsilon_l) \cap \mathcal{Y}_2(k_p + l + 1, \epsilon_{l+1}) = \emptyset$.

Similarly, we can obtain $\mathcal{Z}_3(k_p + l, \epsilon_l) \subset \mathcal{Z}_5(k_p + l + 1, \epsilon_{l+1})$ *and* $\mathcal{Z}_4(k_p + l, \epsilon_l) \cap \mathcal{Y}_1(k_p + l + 1, \epsilon_{l+1}) = \emptyset$.

For any agent $i \in \mathcal{Z}_5(k_p + l, \epsilon_l)$, *the value of itself would be used in the control algorithm. We have*

$$\begin{aligned} x_i[k_p + l + 1] &\leq \alpha(\overline{A} - \epsilon_l) + (1 - \alpha)\overline{x}[k_p + l] \\ &< \underline{A} - \epsilon_{l+1}, \end{aligned}$$

and

$$\begin{aligned} x_i[k_p + l + 1] &\geq \alpha(\underline{A} + \epsilon_l) + (1 - \alpha)\underline{x}[k_p + l] \\ &> \underline{A} + \epsilon_{l+1}. \end{aligned}$$

Thus, $\mathcal{Z}_5(k_p + l, \epsilon_l) \subset \mathcal{Z}_5(k_p + l + 1, \epsilon_{l+1})$.

From the above analysis, we can derive that the normal nodes in $\mathcal{Y}_1(k_p + l, \epsilon_l)$ *would be in* $\mathcal{V} \backslash \mathcal{Y}_2(k_p + l + 1, \epsilon_{l+1})$, *the normal nodes in* $\mathcal{Y}_2(k_p + l, \epsilon_l)$ *would be in* $\mathcal{V} \backslash \mathcal{Y}_1(k_p + l + 1, \epsilon_{l+1})$, *and the normal nodes in* $\mathcal{V} \backslash \{\mathcal{Y}_1(k_p + l, \epsilon_l) \cup \mathcal{Y}_2(k_p + l, \epsilon_l)\}$ *would still be in* $\mathcal{V} \backslash \{\mathcal{Y}_1(k_p + l + 1, \epsilon_{l+1}) \cup \mathcal{Y}_2(k_p + l + 1, \epsilon_{l+1})\}$. *Therefore, we can conclude that* $\{\mathcal{Y}_1(k_p + l, \epsilon_l) \cap \mathcal{N}\} \supset \{\mathcal{Y}_1(k_p + l + 1, \epsilon_{l+1}) \cap \mathcal{N}\}$, $\{\mathcal{Y}_2(k_p + l, \epsilon_l) \cap \mathcal{N}\} \supset \{\mathcal{Y}_2(k_p + l + 1, \epsilon_{l+1}) \cap \mathcal{N}\}$, *which immediately results in* $|(\mathcal{Y}_1(k_j, \epsilon_{k_j - k_p}) \cup \mathcal{Y}_2(k_j, \epsilon_{k_j - k_p})) \cap \mathcal{N}| \geq |(\mathcal{Y}_1(k_{j+1}, \epsilon_{k_{j+1} - k_p}) \cup \mathcal{Y}_2(k_{j+1}, \epsilon_{k_{j+1} - k_p})) \cap \mathcal{N}|$.

To verify that the equality does not hold in the above inequality, we consider the pair of nonempty, disjoint subsets $\mathcal{Y}_1(k_j, \epsilon_{k_j - k_p})$ *and* $\mathcal{Y}_2(k_j, \epsilon_{k_j - k_p})$. *There exists a normal node* $i_j \in \mathcal{Y}_1(k_j, \epsilon_{k_j - k_p}) \cup \mathcal{Y}_2(k_j, \epsilon_{k_j - k_p})$ *and a step* $\bar{k}_j \in [k_j, k_{j+1})$ *such that the agent* i_j *has at least* $F + 1$ *neighbors outside from the set it belongs to in step* $\bar{k}_j$. *Without generality, we assume that* $i_j \in \mathcal{Y}_1(k_j, \epsilon_{k_j - k_p})$. *If* $i_j \notin \mathcal{Y}_1(\bar{k}_j, \epsilon_{\bar{k}_j - k_p})$, *we have*

$$\begin{aligned} &|(\mathcal{Y}_1(k_j, \epsilon_{k_j - k_p}) \cup \mathcal{Y}_2(k_j, \epsilon_{k_j - k_p})) \cap \mathcal{N}| \\ &> |(\mathcal{Y}_1(\bar{k}_j, \epsilon_{\bar{k}_j - k_p}) \cup \mathcal{Y}_2(\bar{k}_j, \epsilon_{\bar{k}_j - k_p})) \cap \mathcal{N}| \\ &\geq |(\mathcal{Y}_1(k_{j+1}, \epsilon_{k_{j+1} - k_p}) \cup \mathcal{Y}_2(k_{j+1}, \epsilon_{k_{j+1} - k_p})) \cap \mathcal{N}|. \end{aligned}$$

For the case $i_j \in \mathcal{Y}_1(\bar{k}_j, \epsilon_{\bar{k}_j - k_p})$, *we have* $i_j \notin \mathcal{Y}_1(\bar{k}_j + 1, \epsilon_{\bar{k}_j + 1 - k_p})$ *and thereby*

$$\begin{aligned} &|(\mathcal{Y}_1(k_j, \epsilon_{k_j - k_p}) \cup \mathcal{Y}_2(k_j, \epsilon_{k_j - k_p})) \cap \mathcal{N}| \\ &\geq |(\mathcal{Y}_1(\bar{k}_j, \epsilon_{\bar{k}_j - k_p}) \cup \mathcal{Y}_2(\bar{k}_j, \epsilon_{\bar{k}_j - k_p})) \cap \mathcal{N}| \\ &> |(\mathcal{Y}_1(k_{j+1}, \epsilon_{k_{j+1} - k_p}) \cup \mathcal{Y}_2(k_{j+1}, \epsilon_{k_{j+1} - k_p})) \cap \mathcal{N}|. \end{aligned}$$

Up till now, we have obtained (11.29), which immediately leads to $|(\mathcal{Y}_1(k_{p+M},$ $\epsilon_{k_{p+M}-k_p}) \cup \mathcal{Y}_2(k_{p+M}, \epsilon_{k_{p+M}-k_p})) \cap \mathcal{N}| = 0$ *by noticing*

$$M \geq |(\mathcal{Y}_1(k_p, \epsilon_0) \cup \mathcal{Y}_2(k_p, \epsilon_0)) \cap \mathcal{N}|$$
$$> |(\mathcal{Y}_1(k_{p+1}, \epsilon_{k_{p+1}-k_p}) \cup \mathcal{Y}_2(k_{p+1}, \epsilon_{k_{p+1}-k_p}) \cap \mathcal{N}|$$
$$> \dots$$
$$> |(\mathcal{Y}_1(k_{p+M}, \epsilon_{k_{p+M}-k_p}) \cup \mathcal{Y}_2(k_{p+M}, \epsilon_{k_{p+M}-k_p})) \cap \mathcal{N}|.$$

This contradict with $|(\mathcal{Y}_1(k_j, \epsilon_{k_j-k_p}) \cup \mathcal{Y}_2(k_j, \epsilon_{k_j-k_p})) \cap \mathcal{N}| > 0$. *Thus, we have* $\overline{A} = \underline{A}$, *i.e., the resilient consensus is realized.*

Remark 11.11 *Theorem 11.5 shows that the resilient consensus for F-total malicious model can be achieved if and only if the switching graph $\mathcal{G}[k]$ is jointly $(F+1, F+1)$-robust. Compared with [72], the proposed sufficient and necessary condition of joint $(F+1, F+1)$-robustness captures the more general case of switching topologies, where the graph $\mathcal{G}[k]$ in every step k may not be $(F+1, F+1)$-robust. Such condition reduces the connectivity requirement and releases the communication burden in each step, making the robust graph theory much more practical for real applications. Compared with the results using SW-MSR algorithm in [131, 164], which requires to know the step interval T with the union of the graphs satisfying certain robust condition and store all the information of neighbors within T steps, the result in this section removes such global graph information, and the information storage is also not needed. While the price it takes is a little bit tight requirement of the switching topologies.*

11.3.4 Numerical simulation

The numerical example is presented to illustrate the result in previous subsection. The switching topologies among six agents is jointly $(2,2)$-robust, which switches periodically among the graphs in Fig. 11.13, i.e., the topology would be Fig. 11.13(1) at step $6k$, Fig. 11.13(2) at step $6k+1$, Fig. 11.13(3) at step $6k+2$, Fig. 11.13(4) at step $6k+3$, Fig. 11.13(5) at step $6k+4$, Fig. 11.13(6) at step $6k+5$, respectively. To verify the joint $(2,2)$-robustness of the switching network, every nonempty disjoint pair of nodes' subsets should be checked. Take the pair of subsets $\{1,6\}$ and $\{2,3,4,5\}$ as an instant, which satisfies the condition since there exists infinite step interval $[6k, 6k+6)$, the node 1 has two neighbors outside its set at step $6k$, node 5 has two neighbors outside its set at step $6k+4$, and node 6 has two neighbors outside its set at step $6k+5$.

Let the node 1 be the malicious node, which is decreasing with a constant speed $u_1[k] = -0.1$, i.e., $x_1[k+1] = x_1[k] - 0.1$. And the normal nodes implement the W-MSR algorithm, where the control input of node i is chosen as the mean of the values of itself and the neighbors in $\mathcal{N}_i[k] \backslash \mathcal{R}_i[k]$, i.e., $\bar{w}_{ij}[k] = \frac{1}{1+|\mathcal{N}_i[k] \backslash \mathcal{R}_i[k]|}$. The initial values of each agent are randomly chosen in the interval $[-3, 3]$. The trajectories of the six agents are shown in Fig. 11.14, where the dash line represents the trajectory of malicious node 1, and the solid lines represent the trajectories of normal nodes.

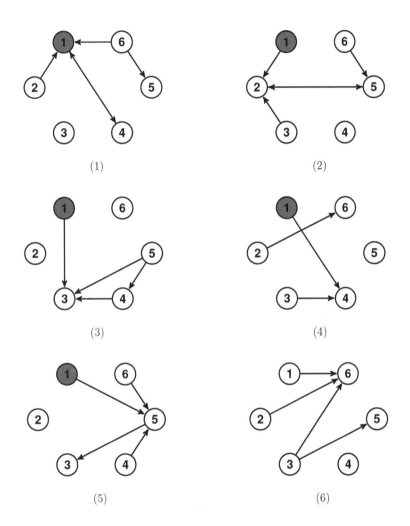

Figure 11.13 The jointly $(2, 2)$-robust graph.

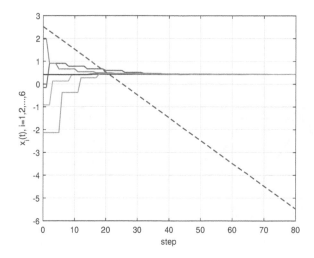

Figure 11.14 Resilient consensus among normal agents with the jointly $(2, 2)$-robust topology is achieved under W-MSR algorithm.

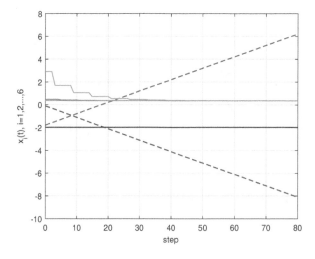

Figure 11.15 Resilient consensus fails under W-MSR algorithm as the network is not jointly $(3, 3)$-robust.

It is clearly that the consensus among normal agents is achieved, and the states of all normal agents remain in the interval $[-3, 3]$ though the value of malicious node would be out of $[-3, 3]$.

If node 2 also turns into malicious node, i.e., there are two malicious nodes in the network, the resilient consensus would not be realized according to Theorem 11.5 as the topology is not jointly $(3, 3)$-robust. Let the node 2 be the malicious node, which is increasing with a constant speed $u_2[k] = 0.1$, i.e., $x_2[k + 1] = x_2[k] + 0.1$. The trajectories of the agents are illustrated in Fig. 11.15, which is coincident with the theoretical analysis.

11.4 CONCLUSIONS

This chapter has investigated the resilient consensus problem for CNSs with input saturation or malicious attack under switching topologies. For the consensus of linear CNSs with input saturation, fully distributed adaptive anti-windup protocols have been proposed by introducing the edge-based adaptive gains, which can be implemented for each agent in a fully distributed manner, without using any global information. For the resilient consensus of CNSs under malicious attack, the novel concept of jointly (r, s)-robust topology has been presented to put forward sufficient and necessary condition for resilient consensus achievement of W-MSR algorithm. Notice that the fully distributed adaptive anti-windup protocols are only applicable to the undirected switching topologies. It is much difficult to design fully distributed adaptive anti-windup protocols for directed switching topologies, which still remains unsolved.

Bibliography

[1] A. Abdessameud and A. Tayebi. On consensus algorithms for double-integrator dynamics without velocity measurement and with input constraints. *Systems & Control Letters*, 59(12):812–821, 2010.

[2] R. Agaev and P. Chebotarev. The matrix of maximum out forests of a digraph and its applications. *Automation and Remote Control*, 61(9):1424–1450, 2000.

[3] R. Albert and A. L. Barabási. Statistical mechanics of complex networks. *Reviews of Modern Physics*, 74(1):47–97, 2002.

[4] M. Amin. National Infrastructures as Complex Interactive Networks. In *Automation, Control, and Complexity: An Integrated Approach*, Samad & Weyrauch (Eds.), John Wiley and Sons, 263–286, 2000.

[5] P. Amster and M. C. Mariani. Some results on the forced pendulum equation. *Nonlinear Analysis: Theory, Methods & Applications*, 68(7):1874–1880, 2008.

[6] R. B. Bapat. *Linear Algebra and Linear Models*. London: Springer, 2012.

[7] I. V. Belykh, V. N. Belykh, and M. Hasler. Blinking model and synchronization in small-world networks with a time-varying coupling. *Physica D: Nonlinear Phenomena*, 195(1–2):188–206, 2004.

[8] D. S. Bernstein. *Matrix Mathematics: Theory, Facts, and Formulas*. Princeton: Princeton University Press, 2009.

[9] D. Biles and P. Binding. On Carathéodory's conditions for the initial value problem. *Proceedings of the American Mathematical Society*, 125(5):1371–1376, 1997.

[10] S. Boccaletti, V. Latora, Y. Moreno, M. Chavez, and D. U. Hwang. Complex networks: Structure and dynamics. *Physics Reports*, 424(4–5):175–308, 2006.

[11] S. Boccaletti, D. U. Hwang, M. Chavez, A. Amann, J. Kurths, and L. M. Pecora. Synchronization in dynamical networks: Evolution along commutative graphs. *Physical Review E*, 74(1):016102-1–016102-5, 2006.

[12] S. Boyd and L. Vandenberghe. *Convex Optimization*. Cambridge: Cambridge University Press, 2004.

[13] M. Cao, A. S. Morse, and B. D. Anderson. Reaching a consensus in a dynamically changing environment: A graphical approach. *SIAM Journal on Control and Optimization*, 47(2):575–600, 2008.

[14] W. Cao, J. Zhang, and W. Ren. Leader-follower consensus of linear multiagent systems with unknown external disturbances. *Systems & Control Letters*, 82: 64–70, 2015.

[15] Y. Cao, W. Yu, W. Ren, and G. Chen. An overview of recent progress in the study of distributed multi-agent coordination. *IEEE Transactions on Industrial Informatics*, 9(1):427–438, 2013.

[16] W. K. V. Chan and C. P. Chen. Consensus control with failure-Wait or abandon? *IEEE Transactions on Cybernetics*, 46(1):75–84, 2015.

[17] G. Chen and T. Ueta. Yet another chaotic attractor. *International Journal of Bifurcation and Chaos*, 9(7):1465–1466, 1999.

[18] C. P. Chen, G. X. Wen, Y. J. Liu, and F. Y. Wang. Adaptive consensus control for a class of nonlinear multiagent time-delay systems using neural networks. *IEEE Transactions on Neural Networks and Nearning Systems*, 25(6):1217–1226, 2014.

[19] C.-T. Chen. *Linear System Theory and Design*. New York: Oxford University Press, 1998.

[20] T. Chen, X. Liu, and W. Lu. Pinning complex networks by a single controller. *IEEE Transactions on Circuits and Systems I: Regular Papers*, 54(6):1317–1326, 2007.

[21] W. H. Chen, J. Yang, L. Guo, and S. Li. Disturbance-observer-based control and related methods-An overview. *IEEE Transactions on Industrial Electronics*, 63(2): 1083–1095, 2016.

[22] Y. Chen, J. Lu, X. Yu, and D. J. Hill. Multi-agent systems with dynamical topologies: Consensus and applications. *IEEE Circuits and Systems Magazine*, 13(3):21–34, 2013.

[23] Y. Chen, J. Lü, X. Yu, and Z. Lin. Consensus of discrete-time second-order multiagent systems based on infinite products of general stochastic matrices. *SIAM Journal on Control and Optimization*, 51(4):3274–3301, 2013.

[24] L. Cheng, Z. G. Hou, M. Tan, Y. Lin, and W. Zhang. Neural-network-based adaptive leader-following control for multiagent systems with uncertainties. *IEEE Transactions on Neural Networks*, 21(8):1351–1358, 2010.

[25] L. Cheng, Y. Wang, W. Ren, Z. G. Hou, and M. Tan. Containment control of multiagent systems with dynamic leaders based on a PI^n-type approach. *IEEE Transactions on Cybernetics*, 46(12):3004–3017, 2015.

[26] Z. Cheng, D. Yue, S. Hu, H. Ge, and L. Chen. Distributed event-triggered consensus of multi-agent systems under periodic dos jamming attacks. *Neurocomputing*, 400:458–466, 2020.

[27] J. Cortés. Discontinuous dynamical systems. *IEEE Control Systems*, 28(3):36–73, 2008.

[28] L. Cui and Y. Yang. Disturbance rejection and robust least-squares control allocation in flight control system. *Journal of Guidance, Control, and Dynamics*, 34(6):1632–1643, 2011.

[29] L. Cui, S. Kumara, and R. Albert. Complex networks: An engineering view. *IEEE Circuits and Systems Magazine*, 10(3):10–25, 2010.

[30] L. Dal Col, I. Queinnec, S. Tarbouriech, and L. Zaccarian. Regional H_∞ synchronization of identical linear multi-agent systems under input saturation. *IEEE Transactions on Control of Network Systems*, 6(2):789–799, 2019.

[31] M. Darouach, M. Zasadzinski, and S. J. Xu. Full-order observers for linear systems with unknown inputs. *IEEE Transactions on Automatic Control*, 39(3):606–609, 1994.

[32] M. C. de Oliveira and R. E. Skelton. *Stability Tests for Constrained Linear Systems*. New York: Springer-Verlag, 2001.

[33] S. M. Dibaji and H. Ishii. Resilient consensus of second-order agent networks: Asynchronous update rules with delays. *Automatica*, 81:123–132, 2017.

[34] Z. Ding. Consensus disturbance rejection with disturbance observers. *IEEE Transactions on Industrial Electronics*, 62(9):5829–5837, 2015.

[35] T. T. Doan and C. L. Beck. Distributed resource allocation over dynamic networks with uncertainty. arXiv:1708.03543, 2018.

[36] L. Dong, C. Han, and S. Du. Adaptive sliding mode control for disturbed multirobot systems performing target tracking under continuously time-varying topologies. *International Journal of Advanced Robotic Systems*, 17(3):1729881420921018, 2020.

[37] X. Dong, B. Yu, Z. Shi, and Y. Zhong. Time-varying formation control for unmanned aerial vehicles: Theories and applications. *IEEE Transactions on Control Systems Technology*, 23(1):340–348, 2015.

[38] F. Dörfler, and F. Bullo. Synchronization in complex networks of phase oscillators: A survey. *Automatica*, 50(6):1539–1564, 2014.

[39] S. N. Dorogovtsev and J. F. F. Mendes. *Evolution of Networks: From Biological Networks to the Internet and WWW*. Oxford: Oxford University Press, 2003.

[40] H. Du, S. Li, and S. Ding. Bounded consensus algorithms for multi-agent systems in directed networks. *Asian Journal of Control*, 15(1):282–291, 2013.

[41] W. Du, L. Yao, D. Wu, X. Li, G. Liu, and T. Yang. Accelerated distributed energy management for microgrids. *In 2018 IEEE Power & Energy Society General Meeting*, 1–5, 2018.

[42] Z. Duan and G. Chen. Global robust stability and synchronization of networks with Lorenz-type nodes. *IEEE Transactions on Circuits and Systems II: Express Briefs*, 56(8):679–683, 2009.

[43] Y. K. Foo. $\mathcal{H}_\infty$ control with initial conditions. *IEEE Transactions on Circuits and Systems II: Express Briefs*, 53(9):867–871, 2006.

[44] J. Fu, Y. Lv, and T. Huang. Distributed anti-windup approach for consensus tracking of second-order multi-agent systems with input saturation. *Systems & Control Letters*, 130:1–6, 2019.

[45] J. Fu, Y. Wan, G. Wen, and T. Huang. Distributed robust global containment control of second-order multi-agent systems with input saturation. *IEEE Transactions on Control of Network Systems*, 6(4):1426–1437, 2019.

[46] J. Fu, G. Wen, T. Huang, and Z. Duan. Consensus of multi-agent systems with heterogeneous input saturation levels. *IEEE Transactions on Circuits and Systems II: Express Briefs*, 66(6):1053–1057, 2019.

[47] H. N. Gabow. Path-based depth first search for strong and biconnected components. *Information Processing Letters*, 74(3–4):107–114, 2000.

[48] V. Gazi and K. M. Passin. Stability analysis of social foraging swarms. *IEEE Transactions on Systems, Man, and Cybernetics, Part B (Cybernetics)*, 34(1):539–557, 2004.

[49] X. Ge and Q. L. Han. Consensus of multiagent systems subject to partially accessible and overlapping Markovian network topologies. *IEEE Transactions on Cybernetics*, 47(8):1807–1819, 2017.

[50] R. Guimerà and L. A. N. Amaral. Functional cartography of complex metabolic networks. *Nature*, 433:895–900, 2005.

[51] Y. Guo, W. Lin, and D. W. Ho. Discrete-time systems with random switches: From systems stability to networks synchronization. *Chaos*, 26(3):033113-1–033113-15, 2016.

[52] V. Gupta, B. Hassibi, and R. M. Murray. A sub-optimal algorithm to synthesize control laws for a network of dynamic agents. *International Journal of Control*, 78(16):1302–1313, 2005.

[53] J. K. Hale and S. M. Verduyn Lunel. *Introduction to Functional Differential Equations.* New York: Springer, 1993.

[54] Y. Hatano and M. Mesbahi. Agreement over random networks. *IEEE Transactions on Automatic Control*, 50(11):1867–1872, 2005.

[55] B. S. Heck and A. A. Ferri. Application of output feedback to variable structure systems. *Journal of Guidance, Control, and Dynamics*, 12(6):932–935, 1989.

[56] J. P. Hespanha and A. S. Morse. Stability of switched systems with average dwell-time. *In Proceedings of the 38th IEEE Conference on Decision and Control*, 3:2655–2660, 1999.

[57] W. He, C. Xu, Q. L. Han, F. Qian, and Z. Lang. Finite-time $\mathcal{L}_2$ leader-follower consensus of networked Euler-Lagrange systems with external disturbances. *IEEE Transactions on Systems, Man, and Cybernetics: Systems*, 48(110):1920–1928, 2018.

[58] W. He, Y. Dong, and C. Sun. Adaptive neural impedance control of a robotic manipulator with input saturation. *IEEE Transactions on Systems, Man, and Cybernetics: Systems*, 46(3):334–344, 2016.

[59] Y. Hong, L. Gao, D. Cheng, and J. Hu. Lyapunov-based approach to multiagent systems with switching jointly connected interconnection. *IEEE Transactions on Automatic Control*, 52(5):943–948, 2007.

[60] Y. Hong and X. Wang. Multi-agent tracking of a high-dimensional active leader with switching topology. *Journal of Systems Science and Complexity*, 22(4):722–731, 2009.

[61] Y. Hong, X. Wang, and Z. P. Jiang. Distributed output regulation of leadersCfollower multi-agent systems. *International Journal of Robust and Nonlinear Control*, 23(1):48–66, 2013.

[62] R. A. Horn and C. H. Johnson. *Matrix Analysis*. Cambridge: Cambridge University Press, 1985.

[63] Z. G. Hou, L. Cheng, and M. Tan. Decentralized robust adaptive control for the multiagent system consensus problem using neural networks. *IEEE Transactions on Systems, Man, and Cybernetics, Part B (Cybernetics)*, 39(3):636–647, 2009.

[64] J. Hu, J. Cao, J. Yu, and T. Hayat. Consensus of nonlinear multi-agent systems with observer-based protocols. *Systems & Control Letters*, 72:71–79, 2014.

[65] A. Jadbabaie, J. Lin, and A. S. Morse. Coordination of groups of mobile autonomous agents using nearest neighbor rules. *IEEE Transactions on Automatic Control*, 48(6):988–1001, 2003.

[66] J. Xiang and G. Chen. On the V-stability of complex dynamical networks. *Automatica*, 43(6):1049–1057, 2007.

[67] H. Kim, H. Shim, J. Back, and J. H. Seo. Consensus of output-coupled linear multi-agent systems under fast switching network: Averaging approach. *Automatica*, 49(1):267–272, 2013.

[68] P. P. Khargonekar, K. M. Nagpal, and K. R. Poolla. $\mathcal{H}_\infty$ control with transients. *SIAM Journal on Control and Optimization*, 29(6):1373–1393, 1991.

[69] J. Lai, X. Lu, X. Yu, A. Monti, and H. Zhou. Distributed voltage regulation for cyber-physical microgrids with coupling delays and slow switching topologies. *IEEE Transactions on Systems, Man, and Cybernetics: Systems*, 50(1):100–110, 2019.

[70] A. J. Laub. *Matrix Analysis for Scientists and Engineers*. Philadelphia: Siam, 2005.

[71] H. J. LeBlanc and X. Koutsoukos. Resilient first-order consensus and weakly stable, higher order synchronization of continuous-time networked multiagent systems. *IEEE Transactions on Control of Network Systems*, 5(3):1219–1231, 2018.

[72] H. J. LeBlanc, H. Zhang, X. Koutsoukos, and S. Sundaram. Resilient asymptotic consensus in robust networks. *IEEE Journal on Selected Areas in Communications*, 31(4):766–781, 2013.

[73] C. Li, W. Yu, and T. Huang. Impulsive synchronization schemes of stochastic complex networks with switching topology: Average time approach. *Neural Networks*, 54:85–94, 2014.

[74] X. Li, X. Wang, and G. Chen. Pinning a complex dynamical network to its equilibrium. *IEEE Transactions on Circuits and Systems I: Regular Papers*, 51(10):2074–2087, 2004.

[75] Y. Li, J. Xiang, and W. Wei. Consensus problems for linear time-invariant multi-agent systems with saturation constraints. *IET Control Theory & Applications*, 5(6):823–829, 2011.

[76] Z. Li, Z. Duan, G. Chen, and L. Huang. Consensus of multiagent systems and synchronization of complex networks: A unified viewpoint. *IEEE Transactions on Circuits and Systems I: Regular Papers*, 57(1):213–224, 2010.

[77] Z. Li, Z. Duan, and G. Chen. On $\mathcal{H}_\infty$ and $\mathcal{H}_2$ performance regions of multi-agent systems. *Automatica*, 47(4):797–803, 2011.

[78] Z. Li, Z. Duan, and G. Chen. Dynamic consensus of linear multi-agent systems. *IET Control Theory & Applications*, 5(1):19–28, 2011.

[79] Z. Li, X. Liu, M. Fu, and L. Xie. Global $\mathcal{H}^\infty$ consensus of multi-agent systems with Lipschitz non-linear dynamics. *IET Control Theory & Applications*, 6(13):2041–2048, 2012.

[80] Z. Li, W. Ren, X. Liu, and L. Xie. Distributed consensus of linear multi-agent systems with adaptive dynamic protocols. *Automatica*, 49(7):1986–1995, 2013.

[81] Z. Li, G. Wen, Z. Duan, and W. Ren. Designing fully distributed consensus protocols for linear multi-agent systems with directed graphs. *IEEE Transactions on Automatic Control*, 60(4):1152–1157, 2015.

[82] D. Liberzon. *Switching in Systems and Control*. Boston: Springer, 2003.

[83] P. Lin and Y. Jia. Average consensus in networks of multi-agents with both switching topology and coupling time-delay. *Physica A*, 387(1):303–313, 2008.

[84] P. Lin and Y. Jia. Consensus of second-order discrete-time multi-agent systems with nonuniform time-delays and dynamically changing topologies. *Automatica*, 45(9):2154–2158, 2009.

[85] P. Lin and Y. Jia. Consensus of a class of second-order multi-agent systems with time-delay and jointly-connected topologies. *IEEE Transactions on Automatic Control*, 55(3):778–784, 2010.

[86] P. Lin, W. Ren, and Y. Song. Distributed multi-agent optimization subject to nonidentical constraints and communication delays. *Automatica*, 65:120–131, 2016.

[87] P. Lin, W. Ren, C. Yang, and W. Gui. Distributed consensus of second-order multiagent systems with nonconvex velocity and control input constraints. *IEEE Transactions on Automatic Control*, 63(4):1171–1176, 2018.

[88] Z. Lin, M. Broucke, and B. Francis. Local control strategies for groups of mobile autonomous agents. *IEEE Transactions on Automatic Control*, 49(4):622–629, 2004.

[89] Z. Lin, T. Han, R. Zheng, and C. Yu. Distributed localization with mixed measurements under switching topologies. *Automatica*, 76:251–257, 2017.

[90] T. Liu and J. Zhao. Synchronization of complex switched delay dynamical networks with simultaneously diagonalizable coupling matrices. *Journal of Control Theory and Applications*, 6(4):351–356, 2008.

[91] Y. J. Liu, Y. Gao, S. Tong, and Y. Li. Fuzzy approximation-based adaptive backstepping optimal control for a class of nonlinear discrete-time systems with dead-zone. *IEEE Transactions on Fuzzy Systems*, 24(1):16–28, 2016.

[92] Y. Y. Liu, J. J. Slotine, and A. L. Barabási. Controllability of complex networks. *Nature*, 473:167–173, 2011.

[93] Z. W. Liu, X. Yu, Z. H. Guan, B. Hu, and C. Li. Pulse-modulated intermittent control in consensus of multiagent systems. *IEEE Transactions on Systems, Man, and Cybernetics: Systems*, 47(5):783–793, 2017.

[94] E. N. Lorenz. Deterministic nonperiodic flow. *Journal of the Atmospheric Sciences*, 20(2):130–141, 1963.

[95] J. Lu, X. Wu, and J. Lü. Synchronization of a unified chaotic system and the application in secure communication. *Physics Letters A*, 305(6):365–370, 2002.

[96] W. Lu and T. Chen. New approach to synchronization analysis of linearly coupled ordinary differential systems. *Physica D: Nonlinear Phenomena*, 213(2):214–230, 2006.

[97] W. Lu, F. M. Atay, and J. Jost. Synchronization of discrete-time dynamical networks with time-varying couplings. *SIAM Journal on Mathematical Analysis*, 39(4):1231–1259, 2007.

[98] W. Lu, F. M. Atay, and J. Jost. Chaos synchronization in networks of coupled maps with time-varying topologies. *The European Physical Journal B*, 63(3):399–406, 2008.

[99] W. Lu, X. Li, and Z. Rong. Global stabilization of complex networks with digraph topologies via a local pinning algorithm. *Automatica*, 46(1):116–121, 2010.

[100] J. Lü and G. Chen. A new chaotic attractor coined. *International Journal of Bifurcation and Chaos*, 12(3):659–661, 2002.

[101] J. Lü, X. Yu, and G. Chen. Chaos synchronization of general complex dynamical networks. *Physica A: Statistical Mechanics and its Applications*, 334(1–2):281–302, 2004.

[102] J. Lü, X. Yu, G. Chen, and D. Cheng. Characterizing the synchronizability of small-world dynamical networks. *IEEE Transactions on Circuits and Systems I: Regular Papers*, 51(4):787–796, 2004.

[103] J. Lü, and G. Chen. A time-varying complex dynamical network model and its controlled synchronization criteria. *IEEE Transactions on Automatic Control*, 50(6):841–846, 2005.

[104] Y. Mao, H. Jafarnejadsani, P. Zhao, E. Akyol, and N. Hovakimyan. Novel stealthy attack and defense strategies for networked control systems. *IEEE Transactions on Automatic Control*, 65(9):3847–3862, 2020.

[105] R. Marino and P. Tomei, *Nonlinear Control Design: Geometric, Adaptive and Robust*. Englewood Cliffs, NJ, USA: Prentice Hall, 1995.

[106] J. Mei, W. Ren, and G. Ma. Distributed containment control for Lagrangian networks with parametric uncertainties under a directed graph. *Automatica*, 48(4):653–659, 2012.

[107] J. Mei, W. Ren, and J. Chen. Distributed consensus of second-order multi-agent systems with heterogeneous unknown inertias and control gains under a directed graph. *IEEE Transactions on Automatic Control*, 61(8):2019–2034, 2016.

[108] A. S. Morse. Supervisory control of families of linear set-point controllers-Part I. Exact matching. *IEEE Transactions on Automatic Control*, 41(10):1413–1431, 1996.

[109] K. Narendra and S. Tripathi. Identification and optimization of aircraft dynamics. *Journal of aircraft*, 10(4):193–199, 1973.

[110] National Research Council. *Network Science*. Washington, DC: The National Academies Press, 2005.

[111] A. Nedić and A. Olshevsky. Distributed optimization over time-varying directed graphs. *IEEE Transactions on Automatic Control*, 60(3):601–615, 2015.

[112] A. Nedić, A. Olshevsky, and W. Shi. Achieving geometric convergence for distributed optimization over time-varying graphs. *SIAM Journal on Optimization*, 27(4):2597–2633, 2017.

[113] B. Ning, J. Jin, and J. Zheng. Fixed-time consensus for multi-agent systems with discontinuous inherent dynamics over switching topology. *International Journal of Systems Science*, 48(10):2023–2032, 2017.

[114] K. K. Oh, M. C. Park, and H. S. Ahn. A survey of multi-agent formation control. *Automatica*, 53:424–440, 2015.

[115] A. Okubo. Dynamical aspects of animal grouping: swarms, schools, flocks, and herds. *Advances in Biophysics*, 22:1–94, 1986.

[116] R. Olfati-Saber and R. M. Murray. Consensus problems in networks of agents with switching topology and time-delays. *IEEE Transactions on Automatic Control*, 49(9):1520–1533, 2004.

[117] R. Olfati-Saber. Flocking for multi-agent dynamic systems: Algorithms and theory. *IEEE Transactions on Automatic Control*, 51(3):401–420, 2006.

[118] R. Olfati-Saber, J. A. Fax, and R. M. Murray. Consensus and cooperation in networked multi-agent systems. *Proceedings of the IEEE*, 95(1):215–233, 2007.

[119] A. Olshevsky and J. N. Tsitsiklis. On the nonexistence of quadratic Lyapunov functions for consensus algorithms. *IEEE Transactions on Automatic Control*, 53(11):2642–2645, 2008.

[120] G. Palla, I. Derényi, I. Farkas, and T. Vicsek. Uncovering the overlapping community structure of complex networks in nature and society. *Nature*, 435:814–818, 2005.

[121] L. M. Pecora and T. L. Carroll. Master stability functions for synchronized coupled systems. *Physical Review Letters*, 80(10):2109–2112, 1998.

[122] Z. Peng, D. Wang, H. Zhang, and G. Sun. Distributed neural network control for adaptive synchronization of uncertain dynamical multiagent systems. *IEEE Transactions on Neural Networks and Learning Systems*, 25(8):1508–1519, 2014.

[123] F. D. Priscoli, A. Isidori, L. Marconi, and A. Pietrabissa. Leader-following coordination of nonlinear agents under time-varying communication topologies. *IEEE Transactions on Control of Network Systems*, 2(4):393–405, 2015.

[124] J. Qin, H. Gao, and W. X. Zheng. Second-order consensus for multi-agent systems with switching topology and communication delay. *Systems & Control Letters*, 60(6):390–397, 2011.

[125] J. Qin, W. Fu, W. X. Zheng, and H. Gao. On the bipartite consensus for generic linear multiagent systems with input saturation. *IEEE Transactions on Cybernetics*, 47(8):1948–1958, 2017.

[126] Z. Qu. *Cooperative Control of Dynamical Dystems: Applications to Autonomous Vehicles*. London: Springer, 2009.

[127] W. Ren and Y. Cao. *Distributed Coordination of Multi-agent Networks: Emergent Problems, Models, and Issues*. London: Springer, 2010.

[128] W. Ren and R. W. Beard. Consensus seeking in multiagent systems under dynamically changing interaction topologies. *IEEE Transactions on Automatic Control*, 50(5):655–661, 2005.

[129] W. Ren and E. Atkins. Distributed multi-vehicle coordinated control via local information exchange. *International Journal of Robust and Nonlinear Control*, 17(10–11):1002–1033, 2007.

[130] W. Ren and R. Beard. *Distributed Consensus in Multi-vehicle Cooperative Control*. London: Springer, 2008.

[131] D. Saldana, A. Prorok, S. Sundaram, M. F. M. Campos, and V. Kumar. Resilient consensus for time-varying networks of dynamic agents. Seattle: American Control Conference (ACC), 252–258, 2017.

[132] Q. Shen, P. Shi, Y. Shi, and J. Zhang. Adaptive output consensus with saturation and dead-zone and its application. *IEEE Transactions on Industrial Electronics*, 64(6):5025–5034, 2017.

[133] M. Siavash, V. J. Majd, and M. Tahmasebi. A practical finite-time backstepping sliding-mode formation controller design for stochastic nonlinear multi-agent systems with time-varying weighted topology. *International Journal of Systems Science*, 51(3):488–506, 2020.

[134] J. J. E. Slotine. Sliding controller design for non-linear systems. *International Journal of Control*, 40(2):421–434, 1984.

[135] B. Sinopoli, C. Sharp, L. Schenato, S. Schaffert, and S. S. Sastry. Distributed control applications within sensor networks. *Proceedings of the IEEE*, 91(8):1235–1246, 2003.

[136] Q. Song and J. Cao. On pinning synchronization of directed and undirected complex dynamical networks. *IEEE Transactions on Circuits and Systems I: Regular Papers*, 57(3):672–680, 2009.

[137] Q. Song, J. Cao, and W. Yu. Second-order leader-following consensus of nonlinear multi-agent systems via pinning control. *Systems & Control Letters*, 59(9):553–562, 2010.

[138] Q. Song, F. Liu, J. Cao, and J. Lu. Some simple criteria for pinning a Lur'e network with directed topology. *IET Control Theory & Applications*, 8(2):131–138, 2014.

[139] E. D. Sontag. Input to state stability: Basic concepts and results. In *Nonlinear and Optimal Control Theory*, Gianna & Stefani (Eds.), Heidelberg, Germany: Springer–Verlag, 2008.

[140] D. J. Stilwell, E. M. Bollt, and D. G. Roberson. Sufficient conditions for fast switching synchronization in time-varying network topologies. *SIAM Journal on Applied Dynamical systems*, 5(1):140–156, 2006.

[141] S. H. Strogatz. Exploring complex networks. *Nature*, 410:268–276, 2001.

[142] H. Su, M. Z. Q. Chen, J. Lam, and Z. Lin. Semi-global leader-following consensus of linear multi-agent systems with input saturation via low gain feedback. *IEEE Transactions on Circuits and Systems I: Regular Papers*, 60(7):1881–1889, 2013.

[143] H. Su, X. Wang, and G. Chen. A connectivity-preserving flocking algorithm for multi-agent systems based only on position measurements. *International Journal of Control*, 82(7):1334–1343, 2009.

[144] H. Su, Y. Ye, Y. Qiu, Y. Cao, and M. Z. Q. Chen. Semi-global output consensus for discrete-time switching networked systems subject to input saturation and external disturbances. *IEEE Transactions on Cybernetics*, 49(11):3934–3945, 2019.

[145] Y. Su and J. Huang. Two consensus problems for discrete-time multi-agent systems with switching network topology. *Automatica*, 48(9):1988–1997, 2012.

[146] Y. Su and J. Huang. Stability of a class of linear switching systems with applications to two consensus problems. *IEEE Transactions on Automatic Control*, 57(6):1420–1430, 2012.

[147] J. Sun and Z. Geng. Adaptive consensus tracking for linear multi-agent systems with heterogeneous unknown nonlinear dynamics. *International Journal of Robust and Nonlinear Control*, 26(1):154–173, 2016.

[148] J. Sun, Z. Geng, Y. Lv, Z. Li, and Z. Ding. Distributed adaptive consensus disturbance rejection for multiagent systems on directed graphs. *IEEE Transactions on Control of Network Systems*, 5(1):629–639, 2018.

[149] X. M. Sun, J. Zhao, and D. J. Hill. Stability and $\mathcal{L}_2$-gain analysis for switched delay systems: A delay-dependent method. *Automatica*, 42(10):1769–1774, 2006.

[150] X. M. Sun, G. P. Liu, D. Rees, and W. Wang. Stability of systems with controller failure and time-varying delay. *IEEE Transactions on Automatic Control*, 53(10):2391–2396, 2008.

[151] X. M. Sun, G. P. Liu, W. Wang, and D. Rees. $\mathcal{L}_2$-gain of systems with input delays and controller temporary failure: Zero-order hold model. *IEEE Transactions on Control Systems Technology*, 19(3):699–706, 2011.

[152] Y. G. Sun and L. Wang. Consensus of multi-agent systems in directed networks with nonuniform time-varying delays. *IEEE Transactions on Automatic Control*, 54(7):1607–1613, 2009.

[153] Z. Sun and S. S. Ge. *Stability Theory of Switched Dynamical Systems*. London: Springer, 2011.

[154] B. Sundararaman, U. Buy, and A. D. Kshemkalyani. Clock synchronization for wireless sensor networks: A survey. *Ad Hoc Networks*, 3(3):281–323, 2005.

[155] A. Tahbaz-Salehi and A. Jadbabaie. A necessary and sufficient condition for consensus over random networks. *IEEE Transactions on Automatic Control*, 53(3):791–795, 2008.

[156] A. Tahbaz-Salehi and A. Jadbabaie. Consensus over ergodic stationary graph processes. *IEEE Transactions on Automatic Control*, 55(1):225–230, 2010.

[157] R. Tarjan. Depth-first search and linear graph algorithms. *SIAM Journal on Computing*, 1(2):146–160, 1972.

[158] A. R. Teel. Semi-global stabilizability of linear null controllable systems with input nonlinearities. *IEEE Transactions on Automatic Control*, 40(1):96–100, 1995.

[159] A. R. Teel. On $\mathcal{L}_2$ performance induced by feedbacks with multiple saturations. *ESAIM: Control, Optimisation and Calculus of Variations*, 1:225–240, 1996.

[160] A. R. Teel and N. Kapoor. The $\mathcal{L}_2$ anti-winup problem: Its definition and solution. Brussels, Belgium, European Control Conference (ECC), 1897–1902, 1997.

[161] J. Toner and Y. Tu. Flocks, herds, and schools: A quantitative theory of flocking. *Physical Review E*, 58(4):4828–4858, 1998.

[162] H. L. Trentelman, K. Takaba, and N. Monshizadeh. Robust synchronization of uncertain linear multi-agent systems. *IEEE Transactions on Automatic Control*, 58(6):1511–1523, 2013.

[163] V. Ugrinovskii. Distributed robust filtering with $\mathcal{H}_\infty$ consensus of estimates. *Automatica*, 47(1):1–13, 2011.

[164] J. Usevitch and D. Panagou. Resilient leader-follower consensus to arbitrary reference values in time-varying graphs. *IEEE Transactions on Automatic Control*, 65(4):1755–1762, 2020.

[165] V. Utkin, J. Guldner, and J. Shi. *Sliding Mode Control in Electro-Mechanical Systems*. New York: CRC Press, 2009.

[166] M. E. Valcher and I. Zorzan. On the consensus of homogeneous multi-agent systems with arbitrarily switching topology. *Automatica*, 84:79–85, 2017.

[167] T. Vicsek, A. Czirók, E. Ben-Jacob, I. Cohen, and O. Shochet. Novel type of phase transition in a system of self-driven particles. *Physical Review Letters*, 75(6):1226–1229, 1995.

[168] Y. Wan, J. Cao, G. Chen, and W. Huang. Distributed observer-based cyber-security control of complex dynamical networks. *IEEE Transactions on Circuits and Systems I: Regular Papers*, 64(11):2966–2975, 2017.

[169] Y. Wan and J. Cao. Observer-based tracking control for heterogeneous dynamical systems under asynchronous attacks. *In 2017 International Workshop on Complex Systems and Networks*, 224–229, 2017.

[170] C. Wang, Z. Zuo, Z. Qi, and Z. Ding. Predictor-based extended-state-observer design for consensus of MASs with delays and disturbances. *IEEE Transactions on Cybernetics*, 49(4):1259–1269, 2019.

[171] D. Wang, J. Wang, and W. Wang. H^∞ controller design of networked control systems with Markov packet dropouts. *IEEE Transactions on Systems, Man, and Cybernetics: Systems*, 43(3):689–697, 2013.

[172] J. Wang, Y. Tan, and I. Mareels. Robustness analysis of leader-follower consensus. *Journal of Systems Science and Complexity*, 22(2):186–206, 2009.

[173] P. Wang, W. Yu, and X. Yu. Robust node-to-node consensus of linear multiagent systems with directed switching topologies subject to uncertain pinning communications. *International Journal of Robust and Nonlinear Control*, 28(5):1886–1900, 2018.

[174] W. Wang, D. Wang, Z. Peng, and T. Li. Prescribed performance consensus of uncertain nonlinear strict-feedback systems with unknown control directions. *IEEE Transactions on Systems, Man, and Cybernetics: Systems*, 46(9):1279–1286, 2016.

[175] X. F. Wang. Complex networks: Topology, dynamics and synchronization. *International Journal of Bifurcation and Chaos*, 12(5):885–916, 2002.

[176] X. F. Wang and G. Chen. Synchronization in small-world dynamical networks. *International Journal of Bifurcation and Chaos*, 12(1):187–192, 2002.

[177] X. F. Wang and G. Chen. Synchronization in scale-free dynamical networks: Robustness and fragility. *IEEE Transactions on Circuits and Systems I: Fundamental Theory and Applications*, 49(1):54–62, 2002.

[178] X. F. Wang and G. Chen, Pinning control of scale-free dynamical networks. *Physica A: Statistical Mechanics and its Applications*, 310(3–4):521–531, 2002.

[179] Y. W. Wang, J. W. Xiao, C. Wen, and Z. H. Guan. Synchronization of continuous dynamical networks with discrete-time communications. *IEEE Transactions on Neural Networks*, 22(12):1979–1986, 2011.

[180] Y. E. Wang, X. M. Sun, P. Shi, and J. Zhao. Input-to-state stability of switched nonlinear systems with time delays under asynchronous switching. *IEEE Transactions on Cybernetics*, 43(6):2261–2265, 2013.

[181] G. Wen, Z. Duan, W. Yu, and G. Chen. Consensus in multi-agent systems with communication constraints. *International Journal of Robust and Nonlinear Control*, 22(2):170–182, 2012.

[182] G. Wen, Z. Duan, H. Su, G. Chen, and W. Yu. A connectivity-preserving flocking algorithm for multi-agent dynamical systems with bounded potential function. *IET Control Theory & Applications*, 6(6):813–821, 2012.

[183] G. Wen, Z. Duan, Z. Li, and G. Chen. Stochastic consensus in directed networks of agents with non-linear dynamics and repairable actuator failures. *IET Control Theory & Applications*, 6(11):1583–1593, 2012.

[184] G. Wen, Z. Duan, Z. Li, and G. Chen. Consensus and its $\mathcal{L}_2$-gain performance of multi-agent systems with intermittent information transmissions. *International Journal of Control*, 85(4):384–396, 2012.

[185] G. Wen, G. Hu, W. Yu, J. Cao, and G. Chen. Consensus tracking for higher-order multi-agent systems with switching directed topologies and occasionally missing control inputs. *Systems & Control Letters*, 62(12):1151–1158, 2013.

[186] G. Wen, Z. Duan, W. Ren, and G. Chen. Distributed consensus of multi-agent systems with general linear node dynamics and intermittent communications. *International Journal of Robust and Nonlinear Control*, 24(16):2438–2457, 2014.

[187] G. Wen, W. Yu, M. Z. Chen, X. Yu, and G. Chen. $\mathcal{H}^{\infty}$ pinning synchronization of directed networks with aperiodic sampled-data communications. *IEEE Transactions on Circuits and Systems I: Regular Papers*, 61(11):3245–3255, 2014.

[188] G. Wen, Z. Duan, G. Chen, and W. Yu. Consensus tracking of multi-agent systems with Lipschitz-type node dynamics and switching topologies. *IEEE Transactions on Circuits and Systems I: Regular Papers*, 61(2):499–511, 2014.

[189] G. Wen, G. Hu, W. Yu, and G. Chen. Distributed $\mathcal{H}_{\infty}$ consensus of higher order multiagent systems with switching topologies. *IEEE Transactions on Circuits and Systems II: Express Briefs*, 61(5):359–363, 2014.

[190] G. Wen, W. Yu, G. Hu, J. Cao, and X. Yu. Pinning synchronization of directed networks with switching topologies: A multiple Lyapunov functions approach. *IEEE Transactions on Neural Networks and Learning Systems*, 26(12):3239–3250, 2015.

[191] G. Wen, G. Hu, J. Hu, X. Shi, and G. Chen. Frequency regulation of source-grid-load systems: A compound control strategy. *IEEE Transactions on Industrial Informatics*, 12(1):69–78, 2016.

[192] G. Wen, W. Yu, Z. Li, X. Yu, and J. Cao. Neuro-adaptive consensus tracking of multiagent systems with a high-dimensional leader. *IEEE Transactions on Cybernetics*, 47(7):1730–1742, 2017.

[193] G. Wen, W. Yu, Y. Xia, X. Yu, and J. Hu. Distributed tracking of nonlinear multiagent systems under directed switching topology: An observer-based protocol. *IEEE Transactions on Systems, Man, and Cybernetics: System*, 47(5):869–881, 2017.

[194] G. Wen, W. Yu, X. Yu, and J. Lü. Complex cyber-physical networks: From cybersecurity to security control. *Journal of Systems Science and Complexity*, 30(1):46–67, 2017.

[195] G. Wen and W. X. Zheng. On constructing multiple Lyapunov functions for tracking control of multiple agents with switching topologies. *IEEE Transactions on Automatic Control*, 64(9):3796–3803, 2018.

[196] G. Wen, P. Wang, T. Huang, W. Yu, and J. Sun. Robust neuro-adaptive containment of multileader multiagent systems with uncertain dynamics. *IEEE Transactions on Systems, Man, and Cybernetics: Systems*, 49(2):406–417, 2019.

[197] G. X. Wen, C. P. Chen, Y. J. Liu, and Z. Liu. Neural-network-based adaptive leader-following consensus control for second-order non-linear multi-agent systems. *IET Control Theory & Applications*, 9(13):1927–1934, 2015.

[198] C. W. Wu and L. O. Chua. Synchronization in an array of linearly coupled dynamical systems. *IEEE Transactions on circuits and systems I: Fundamental Theory and Applications*, 42(8):430–447, 1995.

[199] C. W. Wu. Synchronization in networks of nonlinear dynamical systems coupled via a directed graph. *Nonlinearity*, 18(3):1057–1064, 2005.

[200] C. W. Wu. Synchronization and convergence of linear dynamics in random directed networks. *IEEE Transactions on Automatic Control*, 51(7):1207–1210, 2006.

[201] C. W. Wu. *Synchronization in Complex Networks of Nonlinear Dynamical Systems*. Singapore: World Scientific, 2007.

[202] J. H. Wilkinson. *The Algebraic Eigenvalue Problem*. Oxford: Oxford University Press, 1965.

[203] W. M. Wonham and A. S. Morse. Decoupling and pole assignment in linear multivariable systems: A geometric approach. *SIAM Journal on Control*, 8(1):1–18, 1970.

[204] Z. G. Wu, P. Shi, H. Su, and J. Chu. Sampled-data exponential synchronization of complex dynamical networks with time-varying coupling delay. *IEEE Transactions on Neural Networks and Learning Systems*, 24(8):1177–1187, 2013.

[205] L. Y. Xiang, Z. X. Liu, Z. Q. Chen, F. Chen, and Z. Z. Yuan. Pinning control of complex dynamical networks with general topology. *Physica A: Statistical Mechanics and its Applications*, 379(1):298–306, 2007.

[206] X. Xu, L. Liu, and G. Feng. Consensus of discrete-time linear multiagent systems with communication, input and output delays. *IEEE Transactions on Automatic Control*, 63(2):492–497, 2018.

[207] T. Yang, Z. Meng, D. V. Dimarogonas, and K. H. Johansson. Periodic behaviors for discrete-time second-order multiagent systems with input saturation constraints. *IEEE Transactions on Circuits and Systems II: Express Briefs*, 63(7):663–667, 2016.

[208] T. Yang, J. Lu, D. Wu, J. Wu, G. Shi, Z. Meng, and K. H. Johansson. A distributed algorithm for economic dispatch over time-varying directed networks with delays. *IEEE Transactions on Industrial Electronics*, 64(6):5095–5106, 2017.

[209] X. Yang, J. Wang, and Y. Tan. Robustness analysis of leader–follower consensus for multi-agent systems characterized by double integrators. *Systems & Control Letters*, 61(11):1103–1115, 2012.

[210] V. K. Yanumula, I. Kar, and S. Majhi. Consensus of second-order multi-agents with actuator saturation and asynchronous time-delays. *IET Control Theory & Applications*, 11(17):3201–3210, 2017.

[211] D. Ye, M. Zhang, and A. V. Vasilakos. A survey of self-organization mechanisms in multiagent systems. *IEEE Transactions on Systems, Man, and Cybernetics: Systems*, 47(3):441–461, 2017.

[212] X. Yin, D. Yue, and S. Hu. Adaptive periodic event-triggered consensus for multi-agent systems subject to input saturation. *International Journal of Control*, 89(4):653–667, 2016.

[213] K. You, Z. Li, and L. Xie,. Consensus condition for linear multi-agent systems over randomly switching topologies. *Automatica*, 49(10):3125–3132, 2013.

[214] T. Yucelen and W. M. Haddad. Low-frequency learning and fast adaptation in model reference adaptive control. *IEEE Transactions on Automatic Control*, 58(4):1080–1085, 2013.

[215] W. Yu, J. Cao, and J. Lü. Global synchronization of linearly hybrid coupled networks with time-varying delay. *SIAM Journal on Applied Dynamical Systems*, 7(1):108–133, 2008.

[216] W. Yu, G. Chen, and J. Lü. On pinning synchronization of complex dynamical networks. *Automatica*, 45(2):429–435, 2009.

[217] W. Yu, G. Chen, M. Cao, and J. Kurths. Second-order consensus for multiagent systems with directed topologies and nonlinear dynamics. *IEEE Transactions on Systems, Man, and Cybernetics, Part B (Cybernetics)*, 40(3):881–891, 2010.

[218] W. Yu, G. Chen, and M. Cao. Consensus in directed networks of agents with nonlinear dynamics. *IEEE Transactions on Automatic Control*, 56(6):1436–1441, 2011.

[219] W. Yu, G. Wen, G. Chen, and J. Cao. *Distributed Cooperative Control of Multiagent Systems*. Singapore: John Wiley, 2016.

[220] W. Yu, H. Wang, H. Hong, and G. Wen. Distributed cooperative anti-disturbance control of multiagent systems: An overview. *Science China Information Sciences*, 60(11):110202, 2017.

[221] C. Zhao, E. Mallada, and F. Dörfler. Distributed frequency control for stability and economic dispatch in power networks. Chicago: American Control Conference (ACC), 2359–2364, 2015.

[222] J. Zhao, D. J. Hill, and T. Liu. Synchronization of complex dynamical networks with switching topology: A switched system point of view. *Automatica*, 45(11):2502–2511, 2009.

[223] H. Zhang, E. Fata, and S. Sundaram. A notion of robustness in complex networks. *IEEE Transactions on Control of Network Systems*, 2(3):310–320, 2015.

[224] H. Zhang, F. L. Lewis, and A. Das. Optimal design for synchronization of cooperative systems: State feedback, observer and output feedback. *IEEE Transactions on Automatic Control*, 56(8):1948–1952, 2011.

[225] H. Zhang, F. L. Lewis, and Z. Qu. Lyapunov, adaptive, and optimal design techniques for cooperative systems on directed communication graphs. *IEEE Transactions on Industrial Electronics*, 59(7):3026–3041, 2012.

[226] H. Zhang, Z. Li, Z. Qu, and F. L. Lewis. On constructing Lyapunov functions for multi-agent systems. *Automatica*, 58:39–42, 2015.

[227] J. Zhang, K. H. Johansson, J. Lygeros, and S. Sastry. Zeno hybrid systems. *International Journal of Robust and Nonlinear Control*, 11(5):435–451, 2001.

[228] W. A. Zhang and L. Yu. Stabilization of sampled-data control systems with control inputs missing. *IEEE Transactions on Automatic Control*, 55(2):447–452, 2010.

[229] Y. Zhao, Z. Duan, G. Wen, and G. Chen. Distributed $\mathcal{H}_\infty$ consensus of multi-agent systems: A performance region-based approach. *International Journal of Control*, 85(3):332–341, 2012.

[230] Y. Zheng and L. Wang. Distributed consensus of heterogeneous multi-agent systems with fixed and switching topologies. *International Journal of Control*, 85(12):1967–1976, 2012.

[231] K. Zhou and J. C. Doyle. *Essentials of Robust Control*. New Jersey: Prentice Hall, 1998.

[232] Z. Zhou, H. Fang, and Y. Hong. Distributed estimation for moving target based on state-consensus strategy. *IEEE Transactions on Automatic Control*, 58(8):2096–2101, 2013.

[233] J. Zhu, Y. Yang, W. A. Zhang, L. Yu, and X. Wang. Cooperative attack tolerant tracking control for multi-agent system with a resilient switching scheme. *Neurocomputing*, 409:372–380, 2020.

[234] W. Zhu. Consensus of multiagent systems with switching jointly reachable interconnection and time delays. *IEEE Transactions on Systems, Man and Cybernetics, Part A: Systems and Humans*, 42(2):348–358, 2012.

[235] Z. Zuo, W. Yang, L. Tie, and D. Meng. Fixed-time consensus for multi-agent systems under directed and switching interaction topology. *In 2014 American Control Conference*, 5133–5138, 2014.

Index